S0-GQF-062

// ANNALS OF
THE NEW YORK ACADEMY
OF SCIENCES

*Volume 1026*

EDITORIAL STAFF

*Director, Publishing and New Media*
SARAH GREENE

*Managing Editor*
JUSTINE CULLINAN

*Associate Editor*
JOYCE HITCHCOCK

*The New York Academy of Sciences*
*2 East 63rd Street*
*New York, New York 10021*

THE NEW YORK ACADEMY OF SCIENCES
(Founded in 1817)

BOARD OF GOVERNORS, September 2004 – September 2005

TORSTEN N. WIESEL, *Chairman of the Board*
GERALD D. FISCHBACH, *Vice Chairman*
MICHAEL SCHMERTZLER, *Treasurer*
ELLIS RUBINSTEIN, *Chief Executive Officer* [ex officio]

*Honorary Life Governors*
WILLIAM T. GOLDEN JOSHUA LEDERBERG

*Governors*

KAREN E. BURKE
VIRGINIA W. CORNISH
PETER B. CORR
R. BRIAN FERGUSON
RONALD L. GRAHAM
MARNIE IMHOFF
WENDY EVANS JOSEPH
JACQUELINE LEO
RODERT W. LUCKY
PAUL MARKS
BRUCE McEWEN
RONAY MENSCHEL
JOHN T. MORGAN
JOHN F. NIBLACK
SANDRA PANEM
PETER RINGROSE
DAVID D. SABATINI
JOHN SEXTON
DEBORAH WILEY

VICTORIA BJORKLUND, *Counsel* [ex officio]
LARRY R. SMITH, *Secretary* [ex officio]

# IMPACT OF ECOLOGICAL CHANGES ON TROPICAL ANIMAL HEALTH AND DISEASE CONTROL

56494018

ANNALS OF THE NEW YORK ACADEMY OF SCIENCES

*Volume 1026*

# IMPACT OF ECOLOGICAL CHANGES ON TROPICAL ANIMAL HEALTH AND DISEASE CONTROL

*Edited by Bob H. Bokma, Edmour F. Blouin, and Gervásio Henrique Bechara*

Q
11
N5
V. 1026

*The New York Academy of Sciences*
*New York, New York*
*2004*

SLC 1/5/05

OHIO UNIVERSITY
LIBRARY

*Copyright © 2004 by the New York Academy of Sciences. All rights reserved. Under the provisions of the United States Copyright Act of 1976, individual readers of the* Annals *are permitted to make fair use of the material in them for teaching or research. Permission is granted to quote from the* Annals *provided that the customary acknowledgment is made of the source. Material in the* Annals *may be republished only by permission of the Academy. Address inquiries to the Permissions Department (editorial@nyas.org) at the New York Academy of Sciences.*

Copying fees: *For each copy of an article made beyond the free copying permitted under Section 107 or 108 of the 1976 Copyright Act, a fee should be paid through the Copyright Clearance Center, Inc., 222 Rosewood Drive, Danvers, MA 01923 (www.copyright.com).*

♾ *The paper used in this publication meets the minimum requirements of the American National Standard for Information Sciences—Permanence of Paper for Printed Library Materials, ANSI Z39.48-1984.*

**Library of Congress Cataloging-in-Publication Data**

Impact of ecological changes on tropical animal health and disease control /
edited by Bob H. Bokma, Edmour Blouin, and Gervásio Henrique Bechara
p. cm. — (Annals of the New York Academy of Sciences : v. 1026)
"This volume is the result of a conference held by the Society for Tropical Veterinary Medicine in Iguazu Falls, Brazil on June 22 through 27, 2003."—Prelim.

Includes bibliographical references and index.
ISBN 1-57331-504-4 (cloth : alk. paper) — ISBN 1-57331-505-2 (pbk. : alk. paper)
1. Ecology—Tropics—Congresses. 2. Animal health—Tropics—Congresses.
3. Veterinary tropical medicine—Congresses. I. Bokma, Bob H. II. Blouin, Edmour.
III. Society for Tropical Veterinary Medicine. Meeting (2003 : Iguazu Falls, Brazil)
IV. Series.

Q11.N5 no. 1026
[QH84.5]
500 s—dc22
[636.089/698>

GYAT/PCP
*Printed in the United States of America*
**ISBN 1-57331-504-4** (cloth)
**ISBN 1-57331-505-2** (paper)
**ISSN 0077-8923**

ANNALS OF THE NEW YORK ACADEMY OF SCIENCES

*Volume 1026*
*October 2004*

# IMPACT OF ECOLOGICAL CHANGES ON TROPICAL ANIMAL HEALTH AND DISEASE CONTROL

*Editors*
BOB H. BOKMA, EDMOUR F. BLOUIN, AND GERVÁSIO HENRIQUE BECHARA

*Editorial assistance was received from:* CHRISTOPHER M. GROOCOCK, JOSE DE LA FUENTE, KATHERINE KOCAN, AND ROSANGELA ZACARIAS MACHADO

*Conference Organizing Committee*
EDMOUR F. BLOUIN, BOB H. BOKMA, GERVASIO HENRIQUE BECHARA, E. PAUL GIBBS, THOMAS WALTON AND VIVIANE ZUNCKELLER

This volume is the result of the 7th biennial conference of the Society for Tropical Veterinary Medicine, entitled **Impact of Ecological Changes on Tropical Animal Health and Disease Control**, held June 22–27, 2003, in Iguaçu Falls, Brazil.

CONTENTS

## Part II. Trends in the Study of Disease Agents

*Viruses*



*Bacteria*

*Bacteria*



*Fungi*



*Protozoa*



*Arthropods*



*Ticks and Tick-Borne Diseases*

## Part IV. Dedication



**Financial assistance was received from:**

- **UNITED STATES DEPARTMENT OF AGRICULTURE**
- **AGRICULTURAL RESEARCH SERVICE (ARS)**
- **ANIMAL AND PLANT HEALTH INSPECTION SERVICE (APHIS)**
- **COOPERATIVE STATE RESEARCH, EDUCATION, AND EXTENSION SERVICE (CSREES)**
- **CONSELHO NACIONAL DE DESENVOLVIMENTO CIENTIFICO E TECNOLOGICO (CNPq), BRAZIL**
- **COORDENAÇÃO DE APERFEIÇOAMENTO DE PESSOAL DE NÍVEL SUPERIOR (CAPES), BRAZIL**
- **FUNDAÇÃO DE AMPARO A PESQUISA DO ESTADO DE SÃO PAULO, BRAZIL**
- **VALEE, S.A.**
- **MERIAL**

The New York Academy of Sciences believes it has a responsibility to provide an open forum for discussion of scientific questions. The positions taken by the participants in the reported conferences are their own and not necessarily those of the Academy. The Academy has no intent to influence legislation by providing such forums.

# *In Memoriam*: Dr. Alain Provost

The 2001 biennial conference of the Society for Tropical Veterinary Medicine in South Africa was dedicated to Dr. Alain Provost, who, we are sad to report, died some months later in 2002. Dr. Provost was born in 1930 at Ezy-sur-Eure, a small village in Normandy, France, where he served as the deputy mayor for many years. He was a graduate of the National Veterinary School at Maisons-Alfort, near Paris. He went on to graduate in mycology, microbiology, and immunology at the Pasteur Institute in 1955.

From 1961 to 1962, he was a scientific advisor to Charles Mérieux, who created the famous Mérieux Institute. Following this, he went to Chad to work for IEMVT at the Farcha Research Laboratory in virology and, specifically, on rinderpest. From 1962 to 1969, he was the head of the virology unit at Farcha and developed the "Bisec" vaccine against both rinderpest and contagious bovine pleuropneumonia, largely used in Africa. From 1969 through 1976, he was Director of this laboratory and Regional Director of IEMVT for Central Africa.

From 1977 to 1988, Dr. Provost served as Director General of IEMVT and was largely responsible for the internationalization of the Institute, its development outside Africa, and the important evolution toward molecular biology.

Dr. Provost retired in 1988, but remained very active in many international organizations. In his career, he published more than 250 scientific papers and was a member or correspondent of many scientific societies, notably the French Veterinary Academy and the Overseas Scientific Academy, among others. Dr. Provost was a member of the editorial board of several journals, including *Veterinary Microbiology*, *Tropical Animal Health and Production*, *Revue Scientifique et Technique de l'Office International des Epizooties*, and *Revue d'Elevage et de Medecine Veternaire des Pays Tropicaux*. He was named "Extraordinary Professor" by the University of Pretoria.

Of the many awards and medals he received, we make particular mention of the "Chevalier de la Confrèrie du Taste-Fromage," linked to his origin in Normandy, and the Theiler Memorial Trust Award from South Africa. His scientific influence, his devotion to tropical veterinary medicine, and his deeply human character serve as an example for many people. This eminent scientist fought cancer for several years and finally succumbed just before Christmas 2002. He was married to Josette, had four children, and was also a grandfather.

To honor his memory, CIRAD has established support for the Alain Provost Award, which is targeted for students in developing countries to attend the STVM conference.

Jean-Charles Maillard
*CIRAD-EMVT*

**Ann. N.Y. Acad. Sci. 1026: xi (2004). © 2004 New York Academy of Sciences.**
**doi: 10.1196/annals.1307.052**

# Introduction

BOB H. BOKMA,[a] EDMOUR F. BLOUIN,[b] AND E. PAUL GIBBS[c]

[a]*National Center for Import and Export, Veterinary Services, Animal and Plant Health Inspection Service, United States Department of Agriculture, Riverdale, Maryland 20737, USA*

[b]*Department of Veterinary Pathobiology, Oklahoma State University, Stillwater, Oklahoma 74078, USA*

[c]*Department of Pathobiology, College of Veterinary Medicine, University of Florida, Gainesville, Florida 32611, USA*

This volume emanates from a conference of the Society for Tropical Veterinary Medicine (STVM) that was held in Brazil during June 2003 and entitled the "Impact of Ecological Changes on Tropical Animal Health and Disease Control". The conference follows the tradition of previous conferences of the STVM to highlight timely themes such as "Control and Prevention of Tropical Diseases in the Context of the New World Order" and "Wildlife and Livestock Disease and Sustainability: What Makes Sense?". The present conference continues this focus on the impact of sustained development and growth on animal health in the tropical regions.

It has become increasingly clear that continued development of resources and the ecological changes it brings truly have had a profound impact on tropical veterinary medicine, influencing the development of new and emerging diseases and the presentation of tropical animal and human diseases in the tropics.

In the keynote address at the 2003 conference, a connection was noted in the progression of STVM conference themes and linked events of animal viral diseases to human development and to human intrusion into otherwise relatively nonpeopled tropical ecosystems. The same can be said of disease caused by other types of agents, such as parasites and bacteria. With increased global trade, there is an increased risk of introducing foreign animal diseases. Society's rising welfare concern for the welfare of domestic animals and wildlife limits the speed and extent of "stamping out" (slaughter) policies. Without such policies, control is considerably more difficult for several diseases, particularly those considered "List A" by the Office International de Epizooties (O.I.E.). The threat of terrorism applied to animal agriculture was noted as an area of increasing concern.

Recently, as stated by Christopher Flavin, President of the Worldwatch Institute, "At their core, these disturbing events, such as 9/11, are powerful reminders that the ecological instability of today's world is matched by instability in human affairs that must be urgently addressed. Meeting basic human needs, slowing the unprecedented

Address for correspondence: Bob H. Bokma, National Center for Import and Export, Veterinary Services, Animal and Plant Health Inspection Service, United States Department of Agriculture, 4700 River Road, Unit 39, Riverdale, MD 20737. Voice: 301-734-8066; fax: 301-734-3222.

bob.h.bokma@aphis.usda.gov

**Ann. N.Y. Acad. Sci. 1026: xiii–xv (2004). © 2004 New York Academy of Sciences.**
**doi: 10.1196/annals.1307.051**

growth in human numbers, and protecting vital natural resources, such as fresh water, forests, and fisheries, are all prerequisites to healthy stable societies."

The STVM has recognized the wildlife/domestic animal interface as an important, but underappreciated and poorly funded area of study, particularly in view of emerging diseases. Relative to sustainability, tourism has been important, but the cultural value of wildlife to indigenous people is often ignored. A resolution by the participants of the 2001 biennial conference (joint STVM and Wildlife Disease Association meeting) called for an international funding agency to increase emphasis on and funding to study interrelationships that occur at the mentioned interface.

It has been stated that the environment has shaped the relationship between parasitic disease agents, notably viruses, and their hosts, through a long history of coevolution, such that equilibrium exists between the host and the parasite, whereby disease is of low incidence and new diseases appear only irregularly. While this may indeed have been the situation before medieval times, this can hardly be said to be true of the last 25 years. The diversity, geographical distribution, and clinical severity of many diseases are now in constant flux and several new diseases of epidemic potential affecting both animals and humans have been recognized within this time period. The term, "emerging diseases", is now commonly used to describe the situation (the term was first used by the Food and Agricultural Organization as early as 1966, but it is usually attributed to the 1992 report of the National Academy of Medicine in its landmark publication on microbial threats to human health in the United States).

The factors leading to the emergence of diseases are many and complex. While natural changes in climate have traditionally been viewed as the root cause of many diseases, we are now beginning to realize the extent and important influence of human activities. Intensive agriculture, deforestation, and irrigation, acting either independently or in concert, significantly contribute to the emergence of disease.

Carlos Morel (2001), UNDP/World Bank/WHO, writing about ecology and human disease, stated it thus: "Because the theme—the interface between ecosystem change and public health—is so extraordinarily complex, relevant literature and information sources are spread throughout a multitude of different disciplines, from biology, chemistry, and physics all the way to the social, economic, and behavioral sciences. As a consequence, students, faculty, and research workers in this area have lacked a primary source of inspiration."

We recognize that in today's world the movement of people and animals from one ecosystem to another probably contributes to the emergence of epidemics more than any other factor. As an illustration, consider that, in the early 1800s, it took nearly a year to circumnavigate the world; today, no two cities on the earth are more than a day apart by commercial air transport. Wilson (2001)[1] remarked that understanding how these and other factors influence the emergence of infectious diseases has taken on an urgency and importance rivaling the pressing issues of environmental conservation, natural resource utilization, population growth, and economic development. This is no less important in veterinary medicine.

Thus, a changing ecology results in changes in animal disease situations. Tropical animal diseases may become prevalent where there is climatic warming or where there has been an extension of a domestic animal population into tropical regions not previously inhabited by people with their livestock and poultry species. Many of these diseases are of zoonotic importance. The extension of humans into tropical regions or ecology changed by humans can bring with it poor sanitation and a very real risk for zoonotic diseases, although not all of these are tropical by any means.

We hope that you will enjoy reading the papers that follow as the authors address the theme of the conference.

## REFERENCE

1. Wilson, M.L. 2001. Ecology and infectious disease. *In* Ecosystem Change and Public Health: A Global Perspective, pp. 283–324. J.L. Aron & J.A. Patz, Eds. Johns Hopkins University Press. Baltimore.

# Conservation Medicine and a New Agenda for Emerging Diseases

PETER DASZAK,[a] GARY M. TABOR,[b] A. MARM KILPATRICK,[c]
JON EPSTEIN,[d] AND RAINA PLOWRIGHT[e]

[a]*Executive Director,* [c]*Senior Research Scientist,* [d]*Senior Program Officer, Consortium for Conservation Medicine, Wildlife Trust, Palisades, New York, USA*

[b]*Director, Yellowstone to Yukon Program, Wilburforce Foundation, Bozeman, Montana, USA*

[e]*Department of Veterinary Medicine and Epidemiology, University of California, Davis, California, USA*

**ABSTRACT: The last three decades have seen an alarming number of high-profile outbreaks of new viruses and other pathogens, many of them emerging from wildlife. Recent outbreaks of SARS, avian influenza, and others highlight emerging zoonotic diseases as one of the key threats to global health. Similar emerging diseases have been reported in wildlife populations, resulting in mass mortalities, population declines, and even extinctions. In this paper, we highlight three examples of emerging pathogens: Nipah and Hendra virus, which emerged in Malaysia and Australia in the 1990s respectively, with recent outbreaks caused by similar viruses in India in 2000 and Bangladesh in 2004; West Nile virus, which emerged in the New World in 1999; and amphibian chytridiomycosis, which has emerged globally as a threat to amphibian populations and a major cause of amphibian population declines. We discuss a new, conservation medicine approach to emerging diseases that integrates veterinary, medical, ecologic, and other sciences in interdisciplinary teams. These teams investigate the causes of emergence, analyze the underlying drivers, and attempt to define common rules governing emergence for human, wildlife, and plant EIDs. The ultimate goal is a risk analysis that allows us to predict future emergence of known and unknown pathogens.**

**KEYWORDS: conservation medicine; emerging diseases; zoonotic pathogens; Nipah virus; Hendra virus; West Nile virus; chytridiomycosis**

Address for correspondence: Peter Daszak, Executive Director, Consortium for Conservation Medicine, Wildlife Trust, 61 Route 9W, Palisades, NY, USA. Voice: 845-365-8595; fax: 845-365-8188.

Daszak@conservationmedicine.org; garyt@wilburforce.org;
kilpatrick@conservationmedicine.org; epstein@conservationmedicine.org;
rkplowright@ucdavis.edu

**Ann. N.Y. Acad. Sci. 1026: 1–11 (2004). © 2004 New York Academy of Sciences.**
**doi: 10.1196/annals.1307.001**

## INTRODUCTION

Emerging infectious diseases (EIDs) have become recognized as one of the most significant threats to public health over the last 30 years.[1,2] Emerging diseases are those that have recently: expanded in geographic range; moved from one host species to another; increased in impact or severity; undergone a change in pathogenesis; or are caused by recently evolved pathogens (see refs. 2–5 for more definitions). Some EIDs affect relatively few people, but represent a particular threat due to their high case fatality rates and lack of a vaccine or effective therapy (e.g., Ebola virus hemorrhagic fever, Nipah virus encephalitis, Lassa fever). Others (e.g., HIV/AIDS and pandemic influenza) have caused pandemics and are responsible for significant morbidity and mortality. These examples are all zoonotic and part of the 75% of human EIDs that are caused by zoonotic pathogens (those transmitted between animals and humans).[6] Combating these is a key goal of public health efforts nationally and globally which is hindered by the large pool of unknown agents that are yet to emerge.[7] Outbreaks of new zoonotic agents occur almost annually, with serious health and economic consequences. For example, SARS coronavirus caused over 700 deaths and $50 billion loss to the global economy in 2003 and appears to have wildlife origins.[8,9] Recently, a number of authors have started to widen the scope of EID research. Using the criteria that define EIDs affecting humans, they have identified emerging diseases of marine and terrestrial wildlife, domestic animals and plants.[3,10–14]

Emerging infectious diseases of humans, wildlife, and plants are linked by two common characteristics. First, by definition they are in a process of flux, either rising in incidence, expanding in host or geographic range, or changing in pathogenicity, virulence, or some other factor. Second, these changes are almost always driven by some type of large-scale anthropogenic environmental change (e.g., deforestation, agricultural encroachment, urban sprawl) or change in human population structure (e.g., increased density linked to urbanization) or behavior (e.g., increasing drug use, changes in medical practice, agricultural intensification, international trade).[2–5,7,15,16] These drivers often act via complex pathways that are poorly understood so that predicting the emergence of new pathogens or the spread of introduced pathogens is difficult. Furthermore, a series of anthropogenic changes that have only recently been linked to emerging diseases add to this complexity. For example, fragmentation generally leads to loss of biodiversity, and this has been linked to heightened Lyme disease risk in the northeastern US.[17–19]

Research in emerging diseases is beginning to address the fundamental rules that govern emergence. Predictive models based on climate analyses have been used for vector-borne diseases,[20,21] and papers have modeled host-pathogen dynamics with pathogen evolution to analyze the process of emergence.[22,23] However, theoretical approaches appear to be far ahead of experimental or field research. For example, few studies have analyzed the links between viral dynamics in wildlife and the environmental changes that have led to the emergence of new EIDs. Second, although the emergence of novel zoonotic agents is an important threat to public health, few studies are attempting to identify unknown agents that have the potential to emerge in the human population. In the following case studies, we demonstrate a novel approach to emerging diseases that consists of forming interdisciplinary teams to examine the underlying causes of emergence (for amphibian chytridiomycosis), develop risk analyses that enable prevention and control measures (for West Nile virus) and examine the likelihood that novel zoonotic pathogens from a newly discovered viral genus will emerge (Nipah virus). Each of these research approaches involves multidisciplinary teams of veterinarians, medical workers, public health researchers, ecologists, conservation biologists, and others.

## AMPHIBIAN CHYTRIDIOMYCOSIS

Amphibian population declines have occurred globally over the last two decades and have become a major conservation issue.[24] Although many of these are attributable to habitat loss, others have remained enigmatic until recently. In particular, amphibian declines in montane regions of the USA, Central and South America, and Australia were reported throughout the 1990s.[25–29] Hypotheses on the cause of these declines included pollution, increases in UV-B irradiation, unknown environmental "stressors," and climate change. However, these declines had occurred in areas outside the sphere of normal anthropogenic environmental changes: protected parks or remote montane forests with minimal human activity.[26,30,31] A breakthrough occurred in 1996, when amphibian carcasses were collected in the Tablelands National Park, Queensland, Australia. Just prior to this, a debate had begun in the literature over whether the pattern of amphibian declines in Australia resembled that which would be caused by a virulent pathogen.[32,33] Veterinary pathologists and parasitologists examined carcasses from Tablelands and other areas of Australia and determined that the cause of death was a previously unknown fungal pathogen that parasitized keratinaceous cells of the epidermis. Carcasses collected from Panama in 1997 were examined by the same group, and similar findings were reported. The new disease, amphibian chytridiomycosis, was proposed as the cause of mass mortalities related to population declines in tropical montane Australia and Central America.[34]

Since its description, chytridiomycosis has been reported as the cause of mass mortalities and population declines in North America,[35] Europe,[36] and New Zealand[37] and has been linked to at least one extinction.[38] The causative agent has now been described as *Batrachochytrium dendrobatidis* and Koch's postulates fulfilled.[39,40] Chytridiomycosis has been labeled as an emerging disease owing to its recent expansion in range and the likelihood that its impact has increased in recent years.[38,41] Because of the high profile of amphibian population declines, a number of research groups have begun to work towards understanding the life history, ecology, and impact of this pathogen. Notably, a group of over 20 researchers formed a collaborative group soon after the discovery of chytridiomycosis and have been working as a multidisciplinary team from 1999 onwards. This group has been funded by two National Science Foundation Integrated Research Challenges in Environmental Biology awards, a relatively new program (http://lsvl.la.asu.edu/irceb/amphibians/). Research has followed similar approaches to those used to study human EIDs.[42] The team has used a combination of molecular techniques, with experimental infections, experimental microbiology, outbreak investigations, and the formation of a global isolate collection to investigate the underlying environmental changes that are driving emergence and to map and understand its spread and impact. The collaborative group includes Australian and American ecologists, veterinarians, mycologists, parasitologists, pathologists, mathematical modelers, and conservation biologists. The results of their research include evidence that *B. dendrobatidis* is a recently emerged pathogen, with little variation in DNA sequence between isolates.[43] This and the finding of the pathogen in amphibians traded internationally for food, as pets, or for conservation purposes implicates anthropogenic introduction as a leading candidate for the cause of emergence.[38] Evidence from outbreak investigations, experimental infection studies, and ecologic studies suggests that the bullfrog (*Rana catesbeiana*) may be an efficient carrier of the pathogen and involved in its spread in some areas.[44–46]

## WEST NILE VIRUS

Since its first appearance in North America in 1999, West Nile virus (WNV) has spread across the continent and into Central America. It has infected more than 14,000 people and caused over 500 deaths, with the number of cases more than doubling in each of the last 3 years.[47] In addition, hundreds of thousands of birds of over 200 different species have died from WNV infection.[48] As a result, WNV has become a serious health and conservation concern both in places where it is established and in areas where it may soon spread such as Hawaii and South America.

Under the broad umbrella of research programs led by the New York State Department of Health, the Centers for Disease Control and Prevention, and others, the ecology of WNV in the new world is being studied through a combination of laboratory, field, and remote sensing approaches by teams of ecologists, climatologists, epidemiologists, and vector control personnel. Lab studies have provided data on the pathology and host competence of different bird species.[49] Similarly, new studies are underway to test mosquitoes and other vectors for their ability to transmit WNV after feeding on an infective host.[50] This information is combined with field data collected through arbovirus surveillance activities on patterns of infection of WNV in the field.[51–53] This enables the determination of which hosts and vectors are most important in amplifying the disease and transmitting it to accidental hosts, including humans. Finally, climatologists study the links between patterns of spatial variation in temperature, rainfall and vegetation and vector densities, dead birds infected with WNV, and human infections.[54,55] This multidisciplinary collaboration will lead to a broader understanding of the drivers of disease emergence than would be possible by any single group.

Understanding WNV emergence is extremely important to predict and prevent serious impacts on many threatened and endangered bird species. A recent study has shown that the impact of WNV on American crows in New York City was a 90%+ reduction of their population.[56] WNV has already led to significant declines in populations of some threatened species,[48] and its spread to Hawaii would almost certainly result in species extinctions of Hawaii's native avifauna that are naïve to vector-borne pathogens.[57] The most prudent approach to preventing extinction due to WNV is to prevent it from establishing wherever possible[58] and to minimize other threats to species that may be susceptible to this pathogen.

## NIPAH AND HENDRA VIRUS

In 1994, the first of a new genus of paramyxoviruses emerged in Australia.[59,60] Hendra virus, a zoonotic pathogen carried by Australian flying foxes (*Pteropus* spp.), was responsible for a fatal outbreak that killed 14 race horses and 2 humans. Five years later, a massive outbreak of a porcine respiratory disease in Malaysia caused the death of 105 pig farm or abbatoir workers and led to the discovery of a novel virus closely related to Hendra, called Nipah virus. Nipah virus, a febrile viral encephalitis in humans, had a 40% mortality rate in the Malaysian outbreak. Two species of pteropodid bat appear to act as the reservoir for this virus and one of these, *Pteropus hypomelanus*, has yielded a new Rubulavirus.[61,62]

A multinational collaborative group of scientists is currently studying the ecology of both Nipah and Hendra viruses to understand what factors caused their emergence (www.henipavirus.org). This group is using field studies of pteropodid bat serology and virus isolation, laboratory studies of virus transmission, and satellite telemetry of bat migration patterns to understand the dynamics of both viruses temporally and spatially in bat populations. In addition, the role of climate, deforestation, and other anthropogenic landscape changes in altering these dynamics is being investigated. For Nipah virus, mathematical models of viral dynamics are being used to predict the threshold density and management practices that would allow future emergence. These models will be parameterized with field and experimental data to further refine predictive capacity.

The experience with Hendra virus (HeV) emergence in Australia has shown that understanding the ecology of wildlife reservoirs can be integral to understanding the epidemiology of emerging infectious diseases. Field, experimental, and molecular investigations of HeV indicate that it is an endemic fruit bat virus that has probably co-evolved with its pteropid hosts.[63–65] Molecular epidemiology and sequencing have shown a rather conservative genetic past and, as such, the virus has not undergone major mutational changes prior to emergence.[66] Furthermore, the concurrent appearance of several other bat-associated viruses implies that changes in the ecology of fruit bats, as opposed to evolution of the pathogen itself, more than likely caused HeV to spill over into new hosts.

Bat biologists have noted changes in the ecology of pteropid bats in the regions where HeV outbreaks have occurred. Two of the three northeastern Australian flying fox species have experienced recent shifts in their ecologic ranges (P. Birt and L. Hall, personal communication). Extensive land clearing, which may have been exacerbated by climate change, has dramatically reduced fruit bat feeding resources,[67–69] bringing bats into closer association with human settlements. It is hypothesized that increased contact opportunities between fruit bats, domestic animals, and humans has led to the current HeV outbreaks. Understanding how land use change may be affecting the distributional ecology of fruit bats (using remote sensing and geographic information systems) is key to understanding the emergence of this disease. Finally, it is not clear which bat species are most important for determining disease risk to domestic animals and humans. Field data have shown us that there are different HeV dynamics in the various species of Australian pteropid bats (Field, personal communication). It is possible that some species act as maintenance hosts, whereas others act as temporary or "spillover" hosts. Mathematical modelers, bat ecologists, veterinary epidemiologists, and virologists are collaborating to determine which bat species maintain the virus in nature.

One hypothesis to emerge from recent work on henipaviruses is that henipaviruses have co-evolved with pteropodid hosts within which they naturally circulate. Pteropus species are common to all the outbreak sites of henipaviruses. To examine the broader risk of future heniparvirus emergence, we will be testing pteropodid bats for the presence or absence of henipaviruses and other novel, potentially zoonotic pathogens throughout major portions of their range. Pteropodids have a relatively ancient lineage (between 43 mya and 60 mya), a wide distribution, and a high degree of endemicity.[70–72] We predict a substantial diversity within the Henipavirus clade with a corresponding diversity of virulence and transmission potential within humans. The development of predictive models and assessment of viral biodiversity may therefore become a new predictive tool for the next unknown zoonotic pathogen of this group.

## CONSERVATION MEDICINE AND A NEW AGENDA FOR PUBLIC HEALTH AND CONSERVATION

These three examples demonstrate a new approach to investigating EIDs. All three involve multidisciplinary groups of scientists studying the ecology of an EID and testing hypotheses on the environmental changes that caused its emergence or on the primary factors that influence its transmission dynamics. They include a component of modeling, with the data from field studies and pathological and microbiological investigations providing data to parameterize these models. The underlying aim of these projects is to provide information that can be used to predict and control the emergence or spread of the disease or to predict future emergence of related pathogens. These studies enhance classic epidemiology by involving an array of medical, veterinary, health, and ecologic scientists and others in a dialog between model building, parameterization, and further refinement of models. The teams are brought together at the beginning of the study and actively collaborate throughout. Finally, the goals of improving public health and wildlife conservation are interchangeable and merge throughout all three studies.

These projects are examples of a newly evolving multidisciplinary approach, known as conservation medicine, that examines the ecologic determinants of disease. Better methods in data analysis, data synthesis, and field monitoring in the health and ecologic sciences are vividly demonstrating the connection between disease and environmental degradation. Many infectious and noninfectious diseases have ecologic drivers. From the climate change facilitated spread of dengue fever to the increased incidence of basal cell carcinoma due to ozone depletion, there is greater understanding of the ecologic aspects of the health and environment linkage. The field of conservation medicine has emerged as an integrative research and applied approach, bridging the health and ecologic sciences.[73–80]

The aim of conservation medicine is ultimately to develop a solution-oriented, practice-based approach in addressing health problems derived from environmental change. This builds upon existing knowledge frameworks in wildlife health, public health, epidemiology, ecology, conservation biology, and veterinary science. By working at a larger scale of perspective, conservation medicine provides context for more specialized disciplines to interact in a more effective manner. In this way, conservation medicine employs some of E.O. Wilson's concepts of "Consilience" by bringing together disciplines long separated by time and tradition.[81]

As with many emerging fields of multidisciplinary study, conservation medicine can be considered an evolving "work in progress." That much said, clear concepts are coming into focus that can provide a compass for others attempting to follow a similar path or pursue parallel methods. In the foregoing three cases, four "I"s can be identified that serve as guiding elements within conservation medicine: (1) Interdisciplinary interaction, the ability of individual researchers to understand how other disciplines work with their own to advance knowledge; (2) Individual collaboration, the formation of collaborative research teams of individuals from different disciplines; (3) Institutional cooperation, the building of institutional linkage and formal partnerships (consortia) to work in this collaborative way; and (4) Investigative innovation, the development by researchers of new approaches to doing integrated science. The acceptance of these concepts is challenged by well-known social barriers such as entrenched individual or institutional domains. Many times, just understanding the language of a discipline, whether existing or new, is a barrier unto itself.

For veterinarians as an example, this new way of investigating EIDs is an ideal opportunity to use the skills of their profession. Veterinarians work on both human and wildlife diseases and have a unique comparative perspective to bring to investigating zoonoses. Conservation medicine fieldwork is enhanced by an understanding of pathology. For example, the pathogen causing amphibian chytridiomycosis (*B. dendrobatidis)* is often highly prevalent in bullfrogs. This can be misinterpreted without pathological investigations that rapidly can indicate the very mild nature of lesions.[45] However, the dominance of domestic animal studies and the lack of population-scale focus of most veterinary curricula are a hindrance. Similarly, the lack of focus on disease ecology in most biology or ecology undergraduate programs or textbooks creates a complementary knowledge gap in these disciplines. With growing funding and interest in the study of human and wildlife EID ecology, there will likely be a dramatically increased demand for veterinarians, ecologists, modelers, and others who understand these integrated concepts.

Stories of the emergence of such viral diseases as Nipah, Hendra, and West Nile virus and the fungal disease, amphibian chytridiomycosis, demonstrate a new understanding of a pattern linking ecologic degradation and disease outcomes. At one level, the nature of this health concern is obvious but, unfortunately, only recently are these disease issues being recognized and addressed through more rigorous scientific examination. Conservation medicine builds upon the advances of knowledge in the health and ecologic sciences, so that future researchers do not remain oblivious to these obvious connections.

## REFERENCES

1. Binder, S., A.M. Levitt, J.J. Sacks & J.M. Hughes. 1999. Emerging infectious diseases: public health issues for the 21st century. Science **284:** 1311.
2. Lederberg, J., R.E. Shope & S.C.J. Oakes. 1992. Emerging Infections: Microbial Threats to Health in the United States. Institute of Medicine, National Academy Press. Washington, DC, USA.

3. DASZAK, P., A.A. CUNNINGHAM & A.D. HYATT. 2000. Emerging infectious diseases of wildlife: threats to biodiversity and human health. Science **287:** 443–449.
4. KRAUSE, R.M. 1992. The origins of plagues: old and new. Science **257:** 1073–1078.
5. KRAUSE, R.M. 1994. Dynamics of emergence. J. Infect. Dis. **170:** 265–271.
6. TAYLOR, L.H., S.M. LATHAM & M.E.J. WOOLHOUSE. 2001. Risk factors for human disease emergence. Phil. Trans. Roy. Soc. Lond. B **356:** 983–989.
7. MORSE, S.S. 1993. Examining the origins of emerging viruses. *In* Emerging Viruses. S.S. Morse, Ed.: 10–28. Oxford University Press. New York.
8. GUAN, Y., B.J. ZHENG, Y.Q. HE, *et al.* 2003. Isolation and characterization of viruses related to the SARS coronavirus from animals in Southern China. Science **302:** 276–278.
9. ROTA, P.A., M.S. OBERSTE, S.S. MONROE, *et al.* 2003. Characterization of a novel coronavirus associated with severe acute respiratory syndrome. Science **300:** 1394–1399.
10. ANON. 1998. Emerging diseases: preparing for the challenges ahead. Vet. Rec. **143:** 378.
11. DOBSON, A. & J. FOUFOPOULOS. 2001. Emerging infectious pathogens of wildlife. Philos. Trans. Roy. Soc. Lond. B **356:** 1001–1012.
12. HARVELL, C.D., K. KIM, J.M. BUKHOLDER, *et al.* 1999. Emerging marine diseases: climate links and anthropogenic factors. Science **285:** 1505–1510.
13. HAYES, R. & T. GOREAU. 1998. The significance of emerging diseases in the tropical coral reef ecosystem. Rev. Biol. Trop. **46:** 173–185.
14. NETTLES, V.F. 1996. Reemerging and emerging infectious diseases: economic and other impacts on wildlife. ASM News **62:** 589.
15. ANDERSON, P.K., A.A. CUNNIGHAM, N.G. PATEL, *et al.* 2004. Emerging infectious diseases of plants: crop homogeneity, pathogen pollution and climate change drivers. Trends Ecol. Evol. In press.
16. DASZAK, P., A.A. CUNNINGHAM & A.D. HYATT. 2001. Anthropogenic environmental change and the emergence of infectious diseases in wildlife. Acta Trop. **78:** 103–116.
17. ALLAN, B.F., F. KEESING & R.S. OSTFELD. 2003. Effect of forest fragmentation on Lyme disease risk. Cons. Biol. **17:** 267–272.
18. LOGIUDICE, K., R.S. OSTFELD, K.A. SCHMIDT & F. KEESING. 2003. The ecology of infectious disease: effects of host diversity and community composition on Lyme disease risk. Proc. Natl. Acad. Sci. USA **100:** 567–571.
19. OSTFELD, R.S. & F. KEESING. 2000. Biodiversity and disease risk: the case of Lyme disease. Conserv. Biol. **14:** 722–728.
20. LINTHICUM, K.J., A. ANYAMBA, C.J. TUCKER, *et al.* 1999. Climate and satellite indicators to forecast Rift Valley fever epidemics in Kenya. Science **285:** 397–400.
21. LINTHICUM, K.J., C.L. BAILEY, F.G. DAVIES & C.J. TUCKER. 1987. Detection of Rift-Valley fever viral activity in Kenya by satellite remote-sensing imagery. Science **235:** 1656–1659.
22. ANTIA, R., R.R. REGOES, J.C. KOELLA & C.T. BERGSTROM. 2003. The role of evolution in the emergence of infectious diseases. Nature **426:** 658–661.
23. BOOTS, M., P.J. HUDSON & A. SASAKI. 2004. Large shifts in pathogen virulence relate to host population structure. Science **303:** 842–844.
24. HOULAHAN, J.E., C.S. FINDLAY, B.R. SCHMIDT, *et al.* 2000. Quantitative evidence for global amphibian population declines. Nature **404:** 752–755.
25. CAREY, C. 1993. Hypothesis concerning the causes of the disappearance of boreal toads from the mountains of Colorado. Cons. Biol. **7:** 355–362.
26. LIPS, K.R. 1998. Decline of a tropical montane amphibian fauna. Conserv. Biol. **12:** 106–117.
27. LIPS, K.R. 1999. Mass mortality and population declines of anurans at an upland site in Western Panama. Conserv. Biol. **13:** 117–125.
28. RICHARDS, S.J., K.R. MCDONALD & R.A. ALFORD. 1993. Declines in populations of Australia's endemic tropical forest frogs. Pac. Conserv. Biol. **1:** 66–77.
29. TRENERRY, M.P., W.F. LAURANCE & K.R. MCDONALD. 1994. Further evidence for the precipitous decline of endemic rainforest frogs in tropical Australia. Pac. Conserv. Biol. **1:** 150–153.
30. MAHONY, M. 1996. The decline of the green and golden bell frog *Litoria aurea* viewed in the context of declines and disappearances of other Australian frogs. Aust. Zool. **30:** 237–247.

31. WILLIAMS, S.E. & J.-M. HERO. 1998. Rainforest frogs of the Australian wet tropics: guild classification and the ecological similarity of declining species. Proc. Roy. Soc. Lond. B **265:** 597–602.
32. ALFORD, R.A. & S.J. RICHARDS. 1997. Lack of evidence for epidemic disease as an agent in the catastrophic decline of Australian rain forest frogs. Cons. Biol. **11:** 1026–1029.
33. LAURANCE, W.F., K.R. MCDONALD & R. SPEARE. 1996. Epidemic disease and the catastrophic decline of Australian rain forest frogs. Conserv. Biol. **10:** 406–413.
34. BERGER, L., R. SPEARE, P. DASZAK, *et al.* 1998. Chytridiomycosis causes amphibian mortality associated with population declines in the rainforests of Australia and Central America. Proc. Natl. Acad. Sci. USA **95:** 9031–9036.
35. MUTHS, E., P.S. CORN, A.P. PESSIER & D.E. GREEN. 2003. Evidence for disease-related amphibian decline in Colorado. Biol. Conserv. **110:** 357–365.
36. BOSCH, J., I. MARTINEZ-SOLANO & M. GARCIA-PARIS. 2000. Evidence of a chytrid fungus infection involved in the decline of the common midwife toad in protected areas of Central Spain. *In* Conference Proceedings: Getting the Jump on Amphibian Disease. Cairns, Australia. August 26–30, 2000.
37. WALDMAN, B., V. ANDJIC, P. BISHOP, *et al.* 2000. Discovery of chytridiomycosis in New Zealand. *In* Conference Proceedings: Getting the Jump on Amphibian Disease, Cairns, Australia. August 26-30, 2000.
38. DASZAK, P., A.A. CUNNINGHAM & A.S. HYATT. 2003. Infectious disease and amphibian population declines. Divers. Distrib. **9:** 141–150.
39. LONGCORE, J.E., A.P. PESSIER & D.K. NICHOLS. 1999. *Batrachochytrium dendrobatidis* gen. et sp. nov., a chytrid pathogenic to amphibians. Mycologia **91:** 219–227.
40. PESSIER, A.P., D.K. NICHOLS, J.E. LONGCORE & M.S. FULLER. 1999. Cutaneous chytridiomycosis in poison dart frogs (*Dendrobates* spp.) and White's tree frogs (*Litoria caerulea*). J. Vet. Diagnost. Invest. **11:** 194–199.
41. DASZAK, P., L. BERGER, A.A. CUNNINGHAM, *et al.* 1999. Emerging infectious diseases and amphibian population declines. Emerg. Infect. Dis. **5:** 735–748.
42. COLLINS, J.P. & A. STORFER. 2003. Global amphibian declines: sorting the hypotheses. Divers. Distrib. **9:** 89–98.
43. MOREHOUSE, E.A., T.Y. JAMES, A.R.D. GANLEY, *et al.* 2003. Multilocus sequence typing suggests that the chytrid pathogen of amphibians is a recently emerged clone. Molec. Ecol. **12:** 395–403.
44. DASZAK, P., A. STRIEBY, A.A. CUNNINGHAM, *et al.* 2004. Experimental evidence that the bullfrog (*Rana catesbeiana)* is a potential carrier of chytridiomycosis, an emerging fungal disease of amphibians. Herpetol. J. In press.
45. HANSELMANN, R., A. RODRIGUEZ, M. LAMPO, *et al.* 2004. Presence of an emerging pathogen of amphibians in introduced bullfrogs (*Rana catesbeiana)* in Venezuela. Biol. Conserv. **120:** 115–119.
46. MAZZONI, R., A.A. CUNNINGHAM, P. DASZAK, *et al.* 2003. Emerging pathogen of wild amphibians in frogs (*Rana catesbeiana*) farmed for international trade. Emerg. Infect. Dis. **9:** 995–998.
47. CDC. 2003. West Nile virus http://www.cdc.gov/ncidod/dvbid/westnile/index.htm. Accessed on 12/8/2003.
48. MARRA, P. P., S. GRIFFING, C. CAFFREY, *et al.* 2004. West Nile virus and wildlife. Bioscience. **54:** 393–402.
49. KOMAR, N., S. LANGEVIN, S. HINTEN, *et al.* 2003. Experimental infection of north American birds with the New York 1999 strain of West Nile virus. Emerging Infect. Dis. **9:** 311–322.
50. TURELL, M.J., M.R. SARDELIS, M.L. O'GUINN & D.J. DOHM. 2002. Potential vectors of West Nile virus in North America. *In* Japanese Encephalitis and West Nile Viruses. Edited by J. Mackenzie, A. Barrett & V. Deubel.: 241–252. Springer-Verlag. Berlin.
51. BERNARD, K.A. & L.D. KRAMER. 2001. West Nile virus activity in the United States, 2001. Viral Immunol. **14:** 319–338.
52. KOMAR, N. 2000. West Nile viral encephalitis. Rev. Sci. Tech. Off. Int. Epizoot. **19:** 166–176.

53. USGS. 2004. Species found positive for WNV in surveillance efforts http://www.nwhc.usgs.gov/research/west_nile/wnvaffected.html Accessed on 1/15/2004.
54. ROGERS, D.J., M.F. MYERS, C.J. TUCKER, *et al.* 2002. Predicting the distribution of West Nile fever in North America using satellite sensor data. J. Am. Soc. Photogram. Remote Sensing 68.
55. SHAMAN, J., M. STEIGLITZ, C. STARK, *et al.* 2002. Using a dynamic hydrology model to predict mosquito abundances in flood and swamp water. Emerg. Infect. Dis. **8:** 6–13.
56. HOCHACHKA, W.M., A.A. DHONDT, K.J. MCGOWAN & L.D. KRAMER. 2004. Impact of West Nile virus on American crows in the Northeastern United States, and its relevance to existing monitoring programs. Ecohealth **1:** 60–68.
57. VAN RIPER, C., III, S.G. VAN RIPER, L.M. GOFF & M. LAIRD. 1986. The epizootiology and ecological significance of malaria in Hawaiian land birds. Ecol. Monogr. **56:** 327–344.
58. KILPATRICK, A.M., Y. GLUZBERG, J. BURGETT & P. DASZAK. 2004. A quantitative risk assessment of the pathways by which West Nile virus could reach Hawaii. Ecohealth **1:** 201–209.
59. MURRAY K., P. SELLECK, P. HOOPER, *et al.* 1995. A morbillivirus that caused fatal disease in horses and humans. Science **268:** 94–97.
60. SELVEY, L.A., R.M. WELLS, J.G. MCCORMACK, *et al.* 1995. Infection of humans and horses by a newly described morbillivirus. Med. J. Aust. **162:** 642–645.
61. CHUA, K.B., C.L. KOH, P.S. HOOI, *et al.* 2002. Isolation of Nipah virus from Malaysian Island flying-foxes. Microbes Infect. **4:** 145–151.
62. CHUA, K.B., L.F. WANG, S.K. LAM, *et al.* 2001. Tioman virus, a novel paramyxovirus isolated from fruit bats in Malaysia. Virology **283:** 215–229.
63. FIELD, H., P. YOUNG, J.M. YOB, *et al.* 2001. The natural history of Hendra and Nipah viruses. Microbes Infect. **3:** 307–314.
64. WANG, L.F., B.H. HARCOURT, M. YU, *et al.* 2001. Molecular biology of Hendra and Nipah viruses. Microbes Infect. **3:** 279–287.
65. WILLIAMSON, M.M., P.T. HOOPER, P.W. SELLECK, *et al.* 1998. Transmission studies of Hendra virus (equine morbillivirus) in fruit bats, horses and cats. Aust. Vet. J. **76:** 813–818.
66. HALPIN, K. 2000. Genetic Studies of Hendra Virus and Other Novel Paramyxoviruses. Ph.D. thesis, University of Queensland. Queensland, Australia.
67. BARSON M., C. RANDALL & V. BORDAS. 2000. Land cover change in Australia: results of the Collaborative Bureau of Rural Sciences project on remote sensing of land cover change. Bureau of Rural Sciences, Canberra, Australia.
68. COGGER, H., H. FORD, C. JOHNSON, *et al.* 2003. Impacts of Land Clearing on Australian Wildlife in Queensland. World Wildlife Fund, Australia.
69. TIDEMANN, C.R., M.J. VARDON, R.A. LOUGHLAND & P.J. BROCKLEHURST. 1999. Dry season camps of flying-foxes (*Pteropus* spp.) in Kakadu World Heritage Area, north Australia. J. Zool. **247:** 155–163.
70. COLGAN, D.J. & T.F. FLANNERY. 1995. A phylogeny of Indo-West Pacific Megachiroptera based on ribosomal DNA. Syst. Biol. **44:** 209–220.
71. MOHD-AZLAN, J., A. ZUBAID & T.H. KUNZ. 2001. Distribution, relative abundance, and conservation status of the large flying fox, *Pteropus vampyrus*, in peninsular Malaysia: a preliminary assessment. Acta Chiropterol. **3:** 149–162.
72. SPRINGER, M.S., W.J. MURPHY, E. EIZIRIK & S.J. O'BRIEN. 2003. Placental mammal diversification and the Cretaceous-Tertiary boundary. Proc. Natl. Acad. Sci. USA **100:** 1056–1061.
73. ALLCHURCH, A.F. 1999. Conservation medicine: an emerging science. Dodo **35:** 74.
74. ANON. 2000. Zoo and conservation medicine. Vet. Rec. **147:** 434–434.
75. DEEM, S.L., A.M. KILBOURN, N.D. WOLFE, *et al.* 2000. Conservation medicine. Trop. Vet. Dis. **916:** 370–377.
76. DIERAUF, L.A., G. GRIFFITH, V. BEASLEY & T.Y. MASHIMA. 2001. Conservation medicine: building bridges. J. Am. Vet. Med. Assoc. **219:** 596–597.
77. MCHENRY, T., W. BRADY, D. CANDLAND, *et al.* 1999. Wildlife Preservation Trust International's conservation perspective and strategic directions. Dodo **35:** 116–123.
78. MEFFE, G.K. 1999. Conservation medicine. Conserv. Biol. **13:** 953–954.

79. SPEAR, J.R. 2000. Conservation medicine: the changing view of biodiversity. Conserv. Biol. **14:** 1913–1917.
80. WEINHOLD, B. 2003. Conservation medicine: combining the best of all worlds. Envir. Health Persp. **111:** A524–A529.
81. WILSON, E.O. 1998. Consilience: The Unity of Knowledge. Random House. UK.

# Emerging Diseases and Their Impact on Animal Commerce

## The Argentine Lesson

B.G. CANÉ, L.F. LEANES, AND L.O. MASCITELLI

*Servicio Nacional de Sanidad y Calidad Agroalimentaria (SENASA), Buenos Aires, Argentina*

**ABSTRACT: As a result of the Argentine experience with foot-and-mouth disease (FMD) in 2001, a need was postulated for the establishment of efficient supranational schemes for continuous surveillance of the interrelations between tropical extractives livestock systems and the prairies that are optimal for the feeding of livestock in the southern region of South America. FMD in Argentina and in other countries, new or re-emerging risks from avian influenza with potential risks for public health, the spongiform encephalopathies, porcine reproductive and respiratory syndrome, and classical swine fever, among other animal diseases, have generated a strong reaction and evolution within the veterinary services of the country. These present lessons will influence decision-making within countries and should be accepted by the technical and scientific community. From the perspective of the official animal health sector and with the FMD eradication plan as a basis within the national territory, we have worked not only to achieve international recognition and credibility within animal health systems, but also to realize the formation of a regional block of countries that can be recognized internationally as an area with equivalent animal health status. We emphasize not only that this lesson is useful in FMD, but also that it is possible to apply the valuable conclusions reached for other emerging or re-emerging diseases.**

**KEYWORDS: emerging diseases; animal commerce; foot-and-mouth disease**

## INTRODUCTION

In 1990, Argentina had established animal disease control strategies based on the characterization of the production systems (including participation by all involved animal industry sectors) and through the use of oil-adjuvant–based vaccination. They allowed the elimination of clinical cases of foot-and-mouth disease (FMD) by 1994 and international access of bovine meat to most markets.

In 1990, Argentina started a two-phase eradication plan. The first phase lasted from 1990 until 1992 (control plan); the second phase ran from 1993 to 1997 (eradication plan), SENASA, 1993. On April 27, 1994, the last case of FMD was detected

Address for correspondence: Leonardo O. Mascitelli, Ramos Mejia 887, 6º "A," CP 1405, Ciudad de Buenos Aires, Argentina.
lmascitelli@ciudad.com.ar

**Ann. N.Y. Acad. Sci. 1026: 12–18 (2004). © 2004 New York Academy of Sciences.**
**doi: 10.1196/annals.1307.002**

in Rivadavia, in the province of Buenos Aires. In May 1997, in accordance with Chapter 2.1.1 of the International Animal Health Organization's sanitary code (Animal Health Code, OIE, 1997), Argentina achieved the classification of FMD free with vaccination. In May 2000, the status of FMD free without vaccination was achieved. This required the suspension of all vaccination during 1999 in order to complete this final phase of FMD eradication. These decisions were motivated by an erroneous perception of trade advantages and an underestimation of risks from the relevant ecosystems that influence the persistence of FMD virus. For reasons that have not been identified completely thus far, the disease reappeared in July 2000, perhaps due in part to the entry of the disease through the border regions with neighboring affected countries. During this epidemic, FMD was present and had a significant social and economic impact on the southern cone subregion of South America (FIGS. 1 and 2).

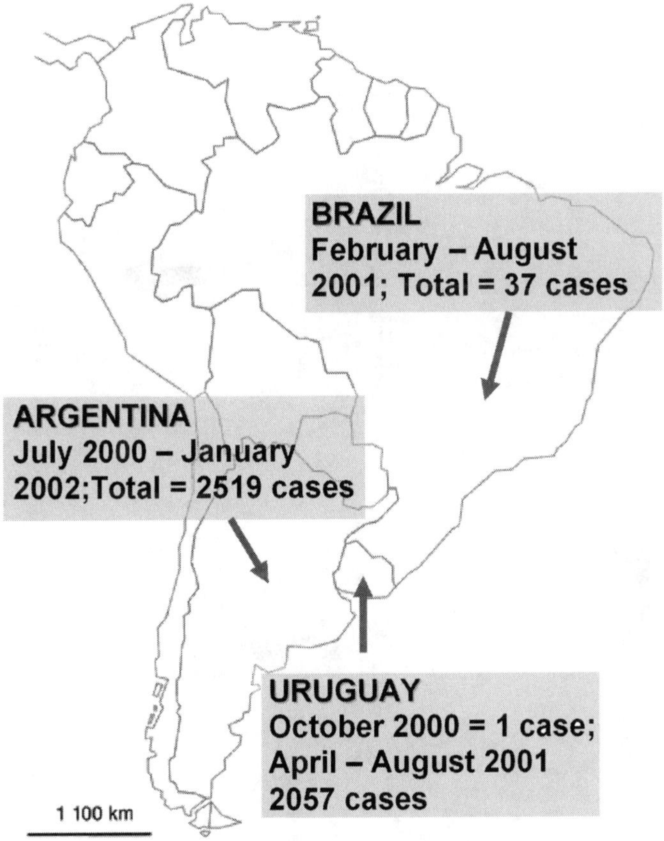


**FIGURE 1.** Foot-and-mouth disease epidemic in the southern cone region of South America, 2000 to 2001.

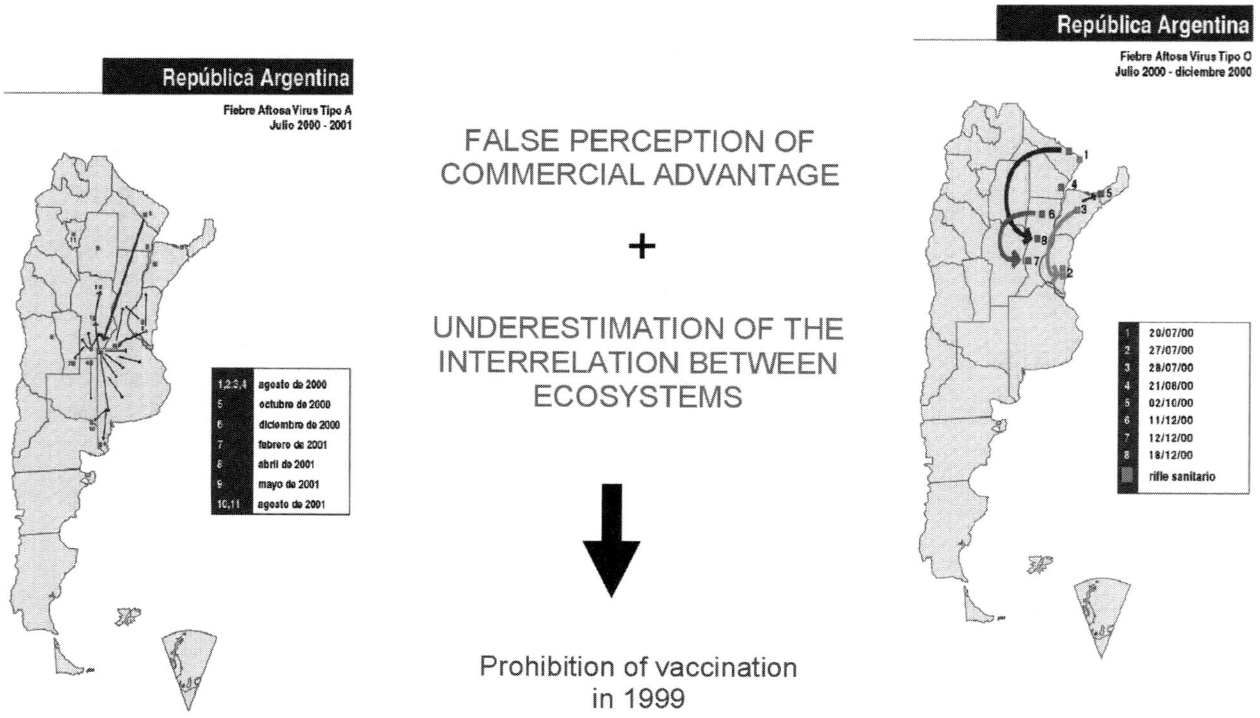


**FIGURE 2.** Hypothesis regarding the 2000–2001 epidemic of foot-and-mouth disease in Argentina.

On March 13, 2001, Argentina officially notified the OIE of cases of the disease in its national territory. Confronted with FMD and based on past experience, Argentina's Plan for FMD eradication was unveiled and implemented in April 2001 (SENASA Resolution No. 5, dated April 15, 2001). The goal of this plan was the clearly described eradication of FMD from the national territory, allowing for international recognition of the country as FMD-free. The goal would promote a program for the reinforcement of national and continental animal disease control and surveillance measures. In this way, we worked for the formation of regional areas of South America that could be internationally recognized as areas of equivalent animal health status (SENASA, 2001). Since January 2002, no cases of FMD have presented in Argentina.

It is also of interest to evaluate animal disease among different tropical ecosystems (or low-density extractive livestock) and those of high-density ecosystems (intensive fattening systems for livestock production). In this paper, we emphasize the importance of livestock movement and trade from extractive areas (including tropical systems) to fattening areas (high animal density). With this flow of animals highly transmissible disease agents meet native susceptible hosts such as newborn and still unvaccinated calves.

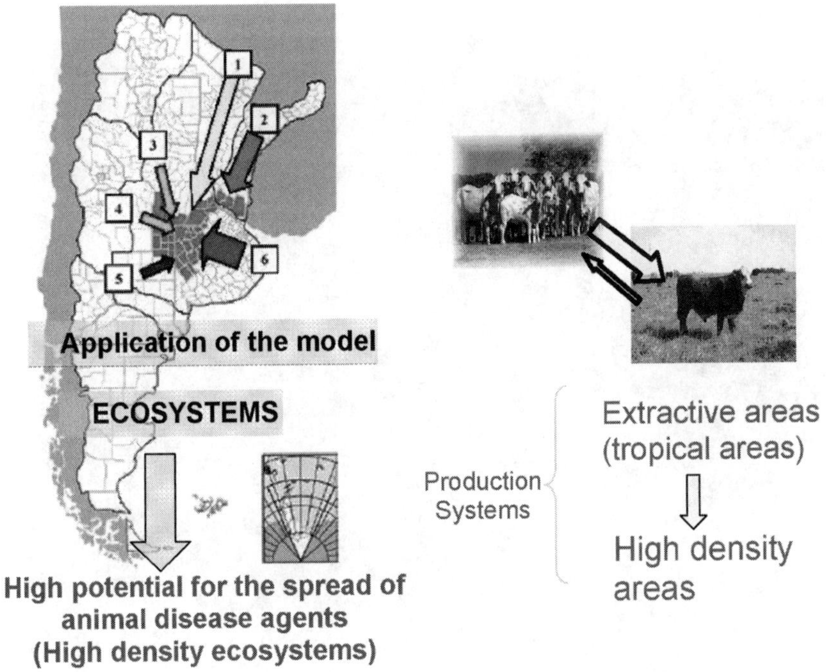


**FIGURE 3.** Model (bovine species) demonstrating the interrelation between low and high animal densities, which are established by the movement of animals.

## DISCUSSION

Within Argentina, the situation corresponding to FIGURE 3 was experienced. Cattle movement from marginal areas towards central areas of the country was observed. Interestingly, these marginal areas coincide with the extensive production systems of cattle that produce calves for fattening. Owing to thecharacteristics of these regions, which provide only the minimal nutritional needs for animal production, calves are moved away from these marginal areas to grow and fatten in the more productive, intensive cattle-rearing areas of the country. These areas have the higher value feeds to meet the nutritional needs of these animals. For this reason, the central area of the county receives most of the cattle from the marginal areas, making this a high animal density system.

The tropical ecosystems in which the extractive animal systems exist, with their high biodiversity, contain animal reservoir hosts for several pathogenic disease agents of humans and animals. These areas must be considered as favorable environments for the evolution of disease agents owing to the abundance of vectors and a lack of disease control mechanisms in these regions. Livestock trade in these areas, which may be formal or informal, continues to be one of the most frequent ways of dissemination of diseases within tropical areas. Also, trade has been one of the factors dictating the political realities for managing animal disease risk between countries.

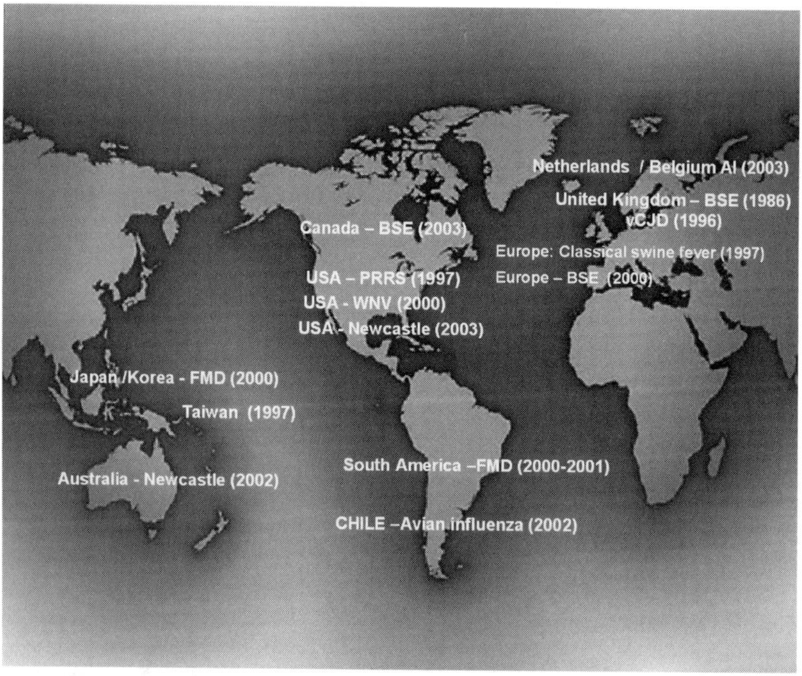


**FIGURE 4.** Evolution of veterinary services in different countries due to emerging and re-emerging animal diseases of high economic impact.

In ecosystems that have a high probability for the multiplication of disease agents, either due to high animal density or pertinent ecological conditions, the type of policy followed for animal disease control is a point of international interest.

Perhaps in the case of Argentina and as a consequence of experience with FMD, we observe that the application of internationally agreed upon animal health policies, as applied regionally between countries, is among the most efficient tools for disease control and success in sanitary plans. As a result, we propose the establishment of efficient supranational schemes for animal disease surveillance, with continuous monitoring of the disease interrelations between the tropical systems characteristic for livestock rearing and supply to the prairie areas used for intensive animal growing and feeding within the Southern Cone region of South America. Integration of systems and discussions of the polices and sanitary practices by regional members, including veterinary services, regional commissions, and institutions, will help in understanding the aspects and dynamics affecting the emergence of diseases and dissemination of pathogenic agents, particularly those with a high potential for spreading and those disease agents with great economic importance in tropical ecosystems.

The emergence or re-emergence of diverse diseases such as bovine spongiform encephalopathy or classical swine fever, which are among the diseases with economic significance as characterized by the OIE, have caused the generation or logical evolution of the veterinary services in the countries in which they appeared. This has resulted in lessons that not only should be learned by the technical and scientific community in order to make wise animal disease policies, but also that should be taken as experience from which to address animal disease emergencies.

## CONCLUSIONS

The epidemic of FMD in Argentina in 2001 left us with a positive attribute. We have a set of experiences emphasizing the impact that man can have in different but interrelated ecosystems—South America's Southern Cone. These include the tropical areas with poor and extractive systems for livestock production and the prairies that are optimal for feeding cattle under high animal densities.

We believe that this model applies to all conditions in ecosystems with a high potential for the recycling and buildup of pathogens. We disagree with the rules imposed on commerce that favor radical schemes for animal disease control without considering social and ecological interrelations within the production environment.

The current tendencies for international sanitary regulation, based on the principles of the World Trade Organization (WTO) and the Committee for the Application of Sanitary and Phytosanitary Measures of control (SPS), promote the use of and integration of rules for international commerce, which in the case of animal health is carried out by the International Animal Health Organization (OIE) through the International Zoo-sanitary Code. This guidance is so general and so global that it therefore becomes insufficient for the particular conditions of each country. If the particular case cannot be adapted to the general rules of international policies, then the countries should adapt the systems for animal disease control. This is not a problem for countries with significant resources, but it can be a big problem for developing countries that can find themselves faced with international trade discrimination due to the limitations that may affect official disease control.

For this to work, a serious and strong regional work commitment, based on the transparency, immediate notification, and knowledge of the sanitary situation, as well as the establishment of equivalent criteria for the prevention, control, and eradication of animal diseases of impact in animal health and international commerce is required.

## ACKNOWLEDGMENTS

The technicians in the Risk Analysis and Emerging Affaires area of SENASA are acknowledged for the review, contributions, and editorial assistance in the preparation of this manuscript.

## REFERENCES

1. BROWN, C. 2001. La importancia de las enfermedades emergentes para la sanidad animal, la salud pública y el comercio. 69 SG/9. Office International des Epizooties (OIE).
2. CANÉ, B.G. 2001. Desafios y oportunidades el mercado mundial de alimentos: su configuracion y requerimientos: la posición argentina frente al mercado mundial. Seminario: la organización sanitaria y agroalimentaria. Academia Nacional de Agronomia y Veterinaria. República Argentina. October 24, 2001.
3. Declaración de país libre con vacunación. Presentación ante la OIE. SENASA, 1996.
4. FARMER, P. 1996. Social inequalities and emerging infectious diseases. Emerg. Infect. Dis. **2:** 259–269.
5. Informe estadístico. 2001. SENASA.
6. Resolución SENASA N° 5/2001. Plan nacional de erradicación de la fiebre aftosa. SENASA. Buenos Aires, Argentina.

# Emergency Prevention System (EMPRES) for Transboundary Animal and Plant Pests and Diseases

## The EMPRES-Livestock: An FAO Initiative

VALDIR ROBERTO WELTE[a] AND MOISÉS VARGAS TERÁN[b]

[a]*Food and Agriculture Organization of the United Nations, Asuncion, Paraguay*

[b]*Food and Agriculture Organization of the United Nations, Santiago, Chile*

**ABSTRACT: The Food and Agriculture Organization of the United Nations (FAO) decided that the Organization should be focusing on the goal of enhancing world food security and the fight against transboundary animal diseases and plant pests. A mandate was obtained from the Governing Council and Conference to establish two new Special Programmes to address these fundamental issues. The first is the Special Programme on Food Security and the second is the Emergency Prevention System against transboundary animal and plant pests and diseases (EMPRES). EMPRES has two components, created after 1994 by a new policy of the Director-General of the FAO to better direct the FAO: the plant pest component focuses on the desert locust, whereas the animal diseases component focuses primarily on rinderpest but also on other epidemic diseases (e.g., contagious bovine pleuropneumonia, foot-and-mouth disease, peste de petit ruminants). For the program as a whole, a high-level EMPRES Steering Committee was established. This is chaired by the FAO Director-General and consists of the heads of key departments (Assistant Directors-General) and Divisional Directors. For the animal diseases component (hereafter referred to as EMPRES-Livestock Programme), FAO established a management unit within its Animal Health Service (AGAH), that is, the Infectious Diseases-EMPRES Group, to be responsible for implementation, including liaison with the Joint FAO–International Atomic Energy Agency (IAEA) Division in Vienna for some of the functions suballocated there. This paper briefly describes FAO EMPRES Livestock, its vision, its mission, and its activities to assist FAO developing member countries and regions in improving the ability of veterinary services to reduce the risks of introduction and/or dissemination of transboundary animal disease, by preventing, controlling, and eradicating those diseases, assisting countries in building their own surveillance/early warning systems, establishing contingency plans, and establishing a global information system for disease monitoring.**

**KEYWORDS: emergency prevention system (EMPRES); animal diseases; plant diseases; pests**

Address for correspondence: Valdir Roberto Welte, Oficina de Representación de la Organización de las Naciones Unidas para la Agricultura y la Alimentación Calle Ciencias Veterinarias y Ruta Mariscal. Estigarribia Km. 10,5, San Lorenzo, Paraguay. Fax: 595-21-574-342.
FAO-PY@fao.org; fao-paraguay@fao.org.py; registry@fao.org.py

**Ann. N.Y. Acad. Sci. 1026: 19–31 (2004). © 2004 New York Academy of Sciences.**
**doi: 10.1196/annals.1307.003**

## INTRODUCTION

Upon taking office in January 1994, the Director-General of the Food and Agriculture Organization of the United Nations (FAO) decided that the Organization should focus on enhancing world food security and fighting against transboundary animal diseases and plant pests, as outbreaks of such diseases or pests can result in food shortages, hunger, destabilization of markets, and the triggering of trade barriers. Therefore, a mandate was obtained from the Governing Council and Conference to establish two new Special Programmes to address these fundamental issues. The first is the Special Programme on Food Security and the second is the Emergency Prevention System against transboundary animal and plant pests and diseases (EMPRES).

EMPRES is a program with two components: the plant pest component focuses on the desert locust, whereas the animal diseases component focuses primarily on rinderpest but also on other epidemic diseases (contagious bovine pleuropneumonia [CBPP], foot-and-mouth disease, peste de petit ruminants, rift valley fever, Newcastle disease, lumpy skin disease, and African swine fever).

For the program as a whole, a high-level EMPRES steering committee was established that is chaired by the Director-General and consists of the heads of key departments (Assistant Directors-General) and Divisional Directors. For the animal diseases component (hereafter referred to as EMPRES-Livestock Programme), FAO has established a management unit within its Animal Health Service (AGAH), that is, the Infectious Diseases-EMPRES Group (IDG/EMPRES), to be responsible for implementation of the program including liaison with the Joint FAO–International Atomic Energy Agency (IAEA) division in Vienna for some of the functions suballocated there. IDG/EMPRES has embarked on a course that will assist countries in building their own surveillance/early warning systems and establishing contingency plans and will establish a global information system for disease monitoring.

### *EMPRES Vision and Mandate*

EMPRES is in a position to provide assistance in terms of training to national epidemiologists and will advise on the setting up of practical surveillance programs. The Group prefers to work through "clusters" of client countries to be as cost-effective as possible and also to promote international cooperation in disease management. Regional workshops on contingency planning and disease surveillance have already been held, and these will continue.

In cases of disease emergency, EMPRES will also intervene when requested by member countries and assist directly with disease combat through Technical Cooperation Programmes (TCP).[1] EMPRES is ideally placed to source donor funding and also to stimulate research. By means of consultative groups, technical consultations and expert consultations, EMPRES regularly assembles international experts in various areas of disease management and can pass on the best and latest information and advice to member countries.

EMPRES received its mandate from the World Food Summit, 1996 and its program strategy is guided by a panel of experts that meets annually as the EMPRES Expert Consultation group.

**TABLE 1. Documents supporting the EMPRES mandate**

| |
|---|
| CL 106/2 Director-General's Review of the Programmes, Structures and Policies of the Organisation |
| FAO–Committee on Agriculture[a] |
| Reports of Expert Consultations[b-c–d-e] |
| OIE Resolution XIII[f] |
| Emergency Prevention System (EMPRES) Livestock: Tracking trends and analyzing threats from epidemics of transboundary animal diseases |

[a]http://www.fao.org/unfao/bodies/COAG/COAG15/X0272E.htm
[b]http://www.fao.org/ag/AGA/AGAH/EMPRES/news/expertcons.htm
[c]http://www.fao.org/ag/AGA/AGAH/EMPRES/Info/other/EMP198.htm
[d]http://www.fao.org/docrep/W6319E/W6319E00.htm
[e]http://www.fao.org/DOCREP/004/W3737E/W3737E00.HTM
[f]http://www.fao.org/ag/AGA/AGAH/EMPRES/grep/oieres.htm/

**TABLE 2. Examples of global disease early warning message titles diffused by EMPRES-Livestock, 2002–2003**

| |
|---|
| Emergency Preparedness in Animal Health in Iraq and the Middle East (March 2003)[a] |
| Rift Valley fever in The Gambia; Tabaski is approaching (February 2003)[b] |
| Uganda and neighbouring countries under serious threat of ASF (November 2002)[c] |
| African Swine Fever in Ghana (October 2002)[d] |
| Suspicion of foot-and-mouth disease in Paraguay - The region under alert (October 2002) [e] |
| Monitoring of El Niño indicators in the Horn of Africa (September 2002) - (2)[f] |
| Foot-and-Mouth disease in the SADC region (June 2002)[g] |
| Monitoring of El Niño indicators in the Horn of Africa (May 2002) - (1)[h] |
| Crimean-Congo Haemorrhagic Fever (CCHF) warrants greater attention as an emerging disease zoonotic problem (February 2002)[i] |

[a] http://www.fao.org/ag/AGA/AGAH/EMPRES/warn_mes/warn15.htm
[b]http://www.fao.org/ag/AGA/AGAH/EMPRES/warn_mes/warn14.htm
[c]http://www.fao.org/ag/AGA/AGAH/EMPRES/warn_mes/warn13.htm
[d]http://www.fao.org/ag/AGA/AGAH/EMPRES/warn_mes/warn12.htm
[e]http://www.fao.org/ag/AGA/AGAH/EMPRES/warn_mes/warn11.htm
[f] http://www.fao.org/ag/AGA/AGAH/EMPRES/warn_mes/warn10.htm
[g]http://www.fao.org/ag/AGA/AGAH/EMPRES/warn_mes/warn9.htm
[h]http://www.fao.org/ag/AGA/AGAH/EMPRES/warn_mes/warn7.htm
[i]http://www.fao.org/ag/AGA/AGAH/EMPRES/crimea.htm

## *EMPRES Mission*

The mission of the EMPRES-Livestock program is to promote the effective containment and control of the most serious epidemic livestock diseases/transboundary animal diseases (TAD) as well as newly emerging diseases, by progressive elimination on a regional and global basis, through international cooperation involving early warning, early reaction, enabling research, and coordination.

**TABLE 3. Related Web sites**

| |
|---|
| FEWS (Famine Early Warning system)[a] |
| Eurosurveillance[b] |
| UNEP Project of risk evaluation, vulnerability, indexing and early warning (PREVIEW)[c] |
| The Disaster Relief Website[d] |
| FAO Emergency operations and Rehabilitation Division[e] |

[a]http://www.fews.net/
[b]http://www.eurosurveillance.org/index-02.asp
[c]http://www.grid.unep.ch/activities/earlywarning/preview/index.php
[d]http://www.disasterrelief.org/
[e]http://www.fao.org/reliefoperations/

## EARLY WARNING

Early warning is identified as all disease initiatives, based predominantly on epidemiological surveillance, that would lead to improved awareness and knowledge of the distribution of disease or infection and that might permit forecasting further evolution of an outbreak. The importance of this element has been re-enforced by OIE Resolution XIII[2] of the 66$^{th}$ General Session of the International Committee and in a document presented by Rweyemamu[3] at this same meeting.

As result of the initiative, EMPRES has created a Web page where animal disease warnings and special links to related web sites can be seen (TABLES 2 and 3).

### *EMPRES Early Warning Activities*

*The Global Rinderpest Eradication Programme (GREP)* is a time-bound program to eliminate rinderpest from the world by the year 2010. Strategies have been devised and programs implemented to reduce the clinical incidence of rinderpest to zero. Elimination of disease and infection will be confirmed by statistically valid active disease surveillance programs.

*The Regional Animal Disease Surveillance and Control Network for North Africa, the Middle East, and the Arab Peninsula RADISCON.* This has helped to develop an animal disease surveillance and control network through a process of strengthening of veterinary investigation, animal disease information, and laboratory services, continued professional development, and improved professional communication between neighboring countries. RADISCON is a joint FAO/IFAD effort targeted at 29 countries located in North Africa, the Sahel, the Horn of Africa, the Middle East, and the Persian Gulf, whose aim is to promote animal disease surveillance within and among countries. RADISCON implementation started in June 1996. The project has assisted each individual country to establish its National Animal Disease Surveillance System (NADSS) and communication between participating countries using Internet/electronic mail facilities where possible.

*Transboundary Animal Disease Information System*[4] (TADinfo ACCESS version). The software development for the Transboundary Animal Disease Information System is underway. TADinfo is structured in a three-tier system: national, regional, and global. TADinfo "national ACCESS version" with a user-friendly interface is available to countries. It provides capabilities for storage and management of animal disease data. The program can perform standard and custom analysis on data and depict information both in report format and geographically through an built-in map viewer. A new version of TADinfo based on JAVA language is being developed and will soon be available to countries willing to adopt the FAO developed information system.

*Information Dissemination: the EMPRES Quarterly Bulletins.* Quarterly reports and analyses of the risks to countries of spread from the EMPRES priority diseases and progress in the control of these diseases in the affected countries are available in the EMPRES Web pages through the Internet.[5]

*Technical Cooperation Programme (TCP).*[6] EMPRES also implements Early Warning activities through Technical Cooperation Programme projects.[7]

### *EMPRES Early Warning Training Material and Software*

*Manual on Livestock Disease Surveillance and Information Systems.* FAO/EMPRES has compiled a manual on surveillance and information management as a guide to assist veterinarians in member countries in running surveillance systems. A number of workshops on surveillance and early warning have also been held as part of FAO projects. As surveillance for diseases is the key to early detection and therefore early warning of a change in the health status of any livestock population, it is also essential to prove the absence of disease or to determine the extent of a disease that is known to be present. All of this information is essential for disease management. As a way to fill the gap in the literature on veterinary epidemiology and to provide access to the information available in this manual, this instructional material is also available from EMPRES-Livestock Web pages.[8]

*Report on New Technologies in the Fight against Transboundary Animal Diseases*–FAO, with the support of the Japan Racing Association (JRA) fund, to bring the attention of professional animal health workers from FAO Member Countries to the

**TABLE 4. EMPRES partners on early warning activities**

| |
|---|
| Joint FAO/IAEA Division[a] |
| Office International des Epizooties (OIE)[b] |
| International Livestock Research Institute (ILRI) |
| Veterinary Epidemiology and Economics Research Unit (VEERU) |
| Centre de coopération internationale en recherche agronomique pour le développement |
| Massey University[c] |
| OAU-IBAR |
| SADC[d] |
| Onderstepoort Veterinary Institute |
| University of Pretoria Faculty of Veterinary Science |

[a]http://www.iaea.or.at/programmes/nafa/d3/index.html
[b]http://www.oie.int
[c]http://epicentre.massey.ac.nz/
[d]http://www.sadc.int/

**TABLE 5. Publications on early reaction from animal diseases**

| |
|---|
| Recognizing Transboundary Animal Diseases |
| Manual on the Preparation of National Animal Disease Emergency Preparedness Plans |
| Manual on the Diagnostic of Rinderpest |
| Manual on the Diagnostic of Peste des Petits Ruminants |
| Preparation of Rift Valley Fever Contingency Plans |
| Rift Valley Fever Fact Sheet (Global Livestock Production and Health Atlas– Glipha) |
| Good Emergency Management Practices (GEMP) |

newer technology that may be available to them on disease surveillance and animal health information systems, on other methods for studying the epidemiology of disease outbreaks, on disease diagnosis and methods for the characterization of etiological agents, and on better vaccines for disease control and eradication programs, in their endeavor against transboundary animal diseases, has developed an FAO-Japan Cooperative Project report called "Collection of Information on Animal Production and Health."[9]

*The Advanced Veterinary Information System (AVIS) Consortium.* The AVIS consortium formed by FAO, OIE, Institute for Animal Health Pirbright and Compton, and Telos ALEFF Ltd. AVIS was founded in 1992 to create a unique multimedia approach to understanding and managing key challenges in animal health, especially those caused by the OIE list A diseases. It has already released major programs on foot-and-mouth disease, rinderpest, poultry diseases, and CBPP.[10]

EMPRES has also been associated and has cooperated with many partners that are involved in early warning activities (TABLE 4).

## EARLY REACTION

Early reaction is identified as all actions that would be targeted at rapid and effective containment leading to the elimination of a disease outbreak, thus preventing it from turning into a serious epidemic or becoming endemic in the country. This includes contingency planning and emergency preparedness. In this regard, EMPRES-Livestock has published a number of manuals on emergency preparedness and contingency planning (TABLE 5) and has also conducted several workshops on national animal disease emergency planning in Africa, Central Europe, and Asia (TABLE 6).

*Technical Cooperation Programme*[11] *(TCP) on Early Reaction.* Upon request of member countries and following a disease emergency, the EMPRES program also provides technical guidance on early reaction activities through Technical Cooperation Programme projects. As an example, a list of several projects assisted by EMPRES-Livestock in the period between 1992 and 1998 can be seen at EMPRES-Livestock Web pages.[12]

**TABLE 6. Workshop on Early Reaction on Animal Disease, Global, 1995–1999**

| Workshop (geographical focus) | Place | Date | Partners |
|---|---|---|---|
| Emergency Preparedness and Contingency Planning for Rinderpest and Other Epidemic Disease Emergencies (West Africa) | Bamako, Mali | 1995 | OAU IBAR |
| Emergency Preparedness and Contingency Planning for Foot-and-Mouth Disease and Other Epidemic Diseases of Livestock: Central Europe and the Near East) | Velingrad, Bulgaria | 1995 | European Commission for Control of Foot-and-Mouth Disease |
| Contagious Bovine Pleuropneumonia Prevention and Control Strategies in Eastern and Southern Africa (Eastern and Southern Africa) | Arusha, Tanzania | 1995 | OAU IBAR |
| Subregional Workshop on the Surveillance of Rinderpest and the OIE Pathway (West Africa) | Dakar, Senegal | 1997 | International Atomic Energy Agency, OAU IBAR |
| Foot-and-Mouth Disease and Contingency Planning (South, South-East, and East Asia) | Hanoi, Vietnam | 1997 | Animal Production and Health Commission for Asia and the Pacific |
| Subregional Workshop on Emergency Preparedness against Rinderpest and Other Transboundary Animal Diseases in Southern Africa (SADC Countries) | Harare, Zimbabwe | 1998 | International Atomic Energy Agency,OAU IBAR, SADC, OIE |
| Subregional Workshop on Emergency Preparedness and the Surveillance of Rinderpest for East Africa | Kampala, Uganda | 1998 | International Atomic Energy Agency, OAU IBAR |
| Expert Consultation on Rinderpest Eradication (South and Southeast Asia) | Kandy, Sri Lanka | 1999 | Animal Production and Health Commission for Asia and the Pacific |
| Animal Disease Contingency Planning | Kochi, India | 1999 | Animal Production and Health Commission for Asia and the Pacific |

**TABLE 7. FAO manuals for veterinarians and technicians**

| |
|---|
| Manual on Procedures for Disease Eradication by Stamping Out (2001)[h] |
| Manual on the Preparation of African Swine Fever Contingency Plans (2001)[a] |
| Manual on Participatory Epidemiology (Methods for the Collection of Action-Oriented Epidemiological Intelligence) (2000)[b] |
| Manual on the Recognition of African Swine Fever (October 2000) |
| Manual on the Preparation of Rinderpest Contingency Plans (October 1999)[c] |
| Recognizing Peste des Petits Ruminants–A Field Manual (October 1999)[d] |
| Reconnaître La Peste des Petits Ruminants–Un Manuel de Terrain (October 1999)[e] |
| Manual on Livestock Disease Surveillance and Information Systems (November 1999)[i] |
| Manual on the Preparation of National Animal Disease Emergency Preparedness Plans (October 1999)[f] |
| Manual on Bovine Spongiform Encephalopathy (1998)[g] |

[a]http://www.fao.org/DOCREP/004/Y0510E/Y0510E00.htm
[b]http://www.fao.org/DOCREP/003/X8833E/X8833E00.htm
[c]http://www.fao.org/DOCREP/004/X2720E/X2720E00.htm
[d]http://www.fao.org/DOCREP/003/X1703E/X1703E00.htm
[e]http://www.fao.org/DOCREP/003/X1703F/X1703F00.htm
[f]http://www.fao.org/DOCREP/004/X2096E/X2096E00.htm
[g]http://www.fao.org/DOCREP/003/W8656E/W8656E00.htm
[h]http://www.fao.org//DOCREP/004/Y0660E/Y0660E00.htm
[i]http://www.fao.org//DOCREP/004/X3331/X3331E00.htm

### *EMPRES Early Reaction Documentation*

*Training Manuals.* Recognizing animal diseases is essential to enable timely reaction and implement appropriate control measures. Therefore, EMPRES-Livestock has developed a series of manuals (TABLE 7) meant for veterinarians and technicians to improve their skill in the recognition of major transboundary diseases.

Many partners of EMPRES-Livestock have cooperated and also have been involved in early reaction activities (TABLE 8).

### *Enabling Research*

Enabling research is identified as a prime element of EMPRES; it emphasizes the collaboration between FAO and scientific centers of excellence in directing research efforts towards problem-solving. EMPRES maintains close contact with FAO reference and collaborating centres.

EMPRES has nominated a number of world class institutions as reference laboratories[13] and collaborating centers[14] for a number of diseases. These institutions serve as valuable research partners in the fight against transboundary animal diseases.

### *EMPRES Enabling Research Activities*

EMPRES convenes from time to time as technical consultations or consultative groups to identify research needs and priorities.

**TABLE 8. EMPRES partners involved in early reaction activities**

| |
|---|
| FAO Special Relief Operations Service[i] |
| Joint FAO/IAEA Division[a] |
| Institute for Animal Health[b] |
| Centre de coopération internationale en recherche agronomique pour le développement[c] |
| Office International des Epizooties (OIE)[d] |
| OAU-IBAR |
| SADC[e] |
| Onderstepoort Veterinary Institute[f] |
| University of Pretoria Faculty of Veterinary Science[g] |
| Massey University[h] |
| Bureau of Resource Sciences, Australia (also AUSVET plan)[j] |

[a]http://www.iaea.org/programmes/naal/agri/index.htm
[b]http://www.iah.bbsrc.ac.uk/
[c]http://www.cirad.fr/fr/index.php
[d]http://www.oie.int
[e]http://www.sadc.int/
[f]http://www.arc-ovi.agric.za/main/intro.htm#ovi
[g]http://www.up.ac.za/academic/veterinary/
[h]http://epicentre.massey.ac.nz/
[i]http://www.fao.org/tc/TCOR4.htm
[j]http://www.aahc.com.au/ausvetplan/

*CBPP Consultative Group (Rome, October 5–7, 1998).* The recent alarming spread of CBPP has highlighted the decreased control of the disease throughout Africa. The reasons for this include shortcomings in the basic understanding of the disease and the implementation of effective surveillance and control programs. This prompted FAO together with OIE, FAO/IAEA, and OAU–IBAR to convene a joint meeting of specialists to review the current situation and to suggest actions for the improvement of this situation. The meeting was held at FAO in Rome, Italy, from October 5–7, 1998. The main outcome of the meeting was the recognition of the need for a CBPP Consultative Group (CG) to continually update the current knowledge and advice on the progress of improved strategies for the control and eradication of the disease.

*GREP Technical Consultation (October 1998).* The Global Rinderpest Eradication Programme (GREP) Technical Consultation and EMPRES Expert Consultation, which were held in Rome, Italy, in September/October 1998, reviewed the progress made in rinderpest eradication and endorsed the view of the GREP secretariat that a more vigorous approach is required if global freedom is to be attained by the year 2010. Experts unanimously stressed the need for an intensified GREP to complement the existing activities and to focus on clarifying any remaining areas of uncertainty and to eliminate the last remaining foci of persisting infection in the shortest possible time.

*Technical Cooperation Programme (TCP) on Enabling Research.* TCP projects often gather information that is later useful for research purposes. A list of technical cooperation projects assisted by EMPRES-Livestock can be seen at EMPRES Web pages.[15]

**TABLE 9. EMPRES partners involved in research activities**

| |
|---|
| List of FAO Reference Laboratories and Collaborating Centres[a] |
| Joint FAO/IAEA Division[b] |
| Institute for Animal Health[c] |
| Centre de coopération internationale en recherche agronomique pour le développement[d] |
| CDC (Centers for Disease Control and Prevention)[e] |
| Onderstepoort Veterinary Institute[f] |
| University of Pretoria Faculty of Veterinary Science[g] |

[a]http://www.fao.org/ag/AGA/AGAH/EMPRES/live_vis/centres.htm
[b]http://www.iaea.org/programmes/naal/agri/index.htm
[c]http://www.iah.bbsrc.ac.uk/
[d]http://www.cirad.fr/fr/index.php
[e]http://www.cdc.gov/
[f]http://www.arc-ovi.agric.za/main/intro.htm#ovi
[g]http://www.up.ac.za/academic/veterinary

### *EMPRES Partners Involved in Research Activities*

As FAO does not do any research by itself, it has been necessary to involve as many and the best possible partners to develop this activity (TABLE 9).

## COORDINATION

EMPRES Coordination involves either coordination of global eradication or progressive control for an identified animal disease, such as rinderpest, through the Global Rinderpest Eradication Programme (GREP) or encouraging regional initiatives for eradication or progressive control of a given transboundary animal disease, such as the Hemispheric Plan for the Eradication of Classical Swine Fever from the Americas.

### *EMPRES Coordination Activities*

EMPRES actively promotes cooperation between regional clusters of countries facing similar disease problems. This is done by holding regional workshops where possible, facilitating e-mail contact between countries, and publishing regular bulletins. GREP is the flagship of EMPRES coordination activities.

*The Global Rinderpest Eradication Programme (GREP).* The GREP is a time-bound program to eliminate rinderpest from the world by the year 2010. Strategies have been devised and programs implemented to reduce the clinical incidence of rinderpest to zero. Elimination of disease and infection will be confirmed by statistically valid active disease surveillance programs.

*The Regional Animal Disease Surveillance and Control Network for North Africa, the Middle East, and the Arab Peninsula (RADISCON)* seeks to develop an animal disease surveillance and control network through a process of strengthening veterinary investigation, animal disease information and laboratory services, continued professional development, and improved professional communication between

**TABLE 10. Regional disease information systems and commissions**

| |
|---|
| The Pan-American Foot-and-Mouth Disease Center (PANAFTOSA)[f] |
| The Animal Production and Health Commission for Asia and the Pacific of the FAO (APHCA)[a] |
| The Pan-African Programme for the Control of Epizootics (PACE)[b] |
| The Regional Animal Disease Surveillance and Control Network (RADISCON)[c] |
| The European Commission for the Control of Foot-and-Mouth Disease (EUFMD)[d] |
| The Veterinary Biotechnology and Epidemiology Network for Central and Eastern Europe (CENTAUR)[e] |

[a]http://www.aphca.org/
[b]http://www.fao.org/ag/AGA/AGAH/EMPRES/grep/pace.htm
[c]http://www.fao.org/WAICENT/FaoInfo/Agricult/AGA/AGAH/ID/Radiscon/Default.htm
[d]http://www.fao.org/ag/againfo/commissions/en/eufmd/eufmd.html
[e]http://centaur.vri.cz/
[f]http://www.panaftosa.org.br/novo/index.htm

neighboring countries. RADISCON is a joint FAO–IFAD[16] endeavor targeted at 29 countries located in North Africa, the Sahel, the Horn of Africa, the Middle East, and the Persian Gulf. Its aim is to promote animal disease surveillance within and among countries. RADISCON implementation started in June 1996. It is hoped that the on-going activities will increase the efficiency of control programs. The project is assisting each individual country to establish its National Animal Disease Surveillance System (NADSS), and communication between participating countries will take place electronically, where possible, using Internet/electronic mail facilities.

*FAO Expert Consultation.* The EMPRES expert consultation is an international panel of livestock experts together with representatives from partner organizations, which meets annually to advise the Director-General on progress of EMPRES and possible future directions.

*European Commission for the Control of Foot and Mouth Disease (EUFMD)* was established to meet an urgent need to prevent the recurrence of heavy losses to European agriculture caused by repeated outbreaks of foot-and-mouth disease. The main objective of the EUFMD, a Commission within the framework of the Food and Agriculture Organization of the United Nations, is to promote national and international action with respect to preventive and control measures against foot-and-mouth disease in Europe. If needed, EMPRES may coordinate its FMD activities with the EUFMD as well as any regional commission or institution (such as the Animal Production and Health Commission for Asia and the Pacific [APHCA])[17] or the Pan-American Foot and Mouth Disease Center.[18]

*Regional Disease Information Systems and Commissions as a Regional Coordinated Effort.* The crucial need for member countries within a geopolitical or economic region to work closely in developing a regionally based, early warning system against epidemic diseases was reaffirmed during the 1997 EMPRES expert consultation. Regional blocks operating disease surveillance networks are already in existence (TABLE 10) and provide relevant information at a regional level. Nevertheless, the creation of new or improved networks should be stimulated.

**TABLE 11. EMPRES partners involved in coordination activities**

| |
|---|
| Joint FAO/IAEA Division[a] |
| Office International des Epizooties (OIE)[b] |
| International Fund for Agricultural Development (IFAD)[c] |
| OAU-IBAR |
| SADC[d] |

[a]http://www.iaea.org/programmes/naal/agri/index.htm
[b]http://www.oie.int/
[c]http://www.ifad.org/
[d]http://www.sadc.int/

### *EMPRES Partners Involved in Coordination Activities*

More than its previous activities, this is one in which EMPRES is always involved with other partners (TABLE 11).

## CONCLUSIONS

The creation of EMPRES-Livestock provided the conditions for FAO member countries to be directly or indirectly assisted by FAO in the domain of veterinary epidemiology, disease surveillance, prevention, control and eradication, as well as emergency preparedness and contingency planning. Several guidelines, manuals, and modern tools were already developed and others are in the process of development, to help veterinarians and veterinary epidemiologists to better perform their daily tasks.

Additionally, most of the material may be readily available through the Internet or, alternatively, by government request to FAO by other means. Last but not least, the availability of a network of independent international experts on the EMPRES-Livestock domain, added to the ones from FAO Collaborating Centers and FAO Reference Laboratories—internationally recognized centers of excellence—make available an unlimited number of capacities to help FAO member countries when technical assistance is needed.

## REFERENCES

1. http://www.fao.org/WAICENT/FAOINFO/TCD/tcd/tcdt/Default.htm
2. http://www.fao.org/ag/AGA/AGAH/EMPRES/grep/oieres.htm
3. RWEYEMAMY, M.M. 1988. Forecasting systems using the laboratory and epidemiology to prevent outbreaks of existing and emerging diseases. Presented as technical item I: 66th OIE General Session/May 1988; http://www.fao.org/ag/AGA/AGAH/EMPRES/live_vis/e_forec1.htm
4. http://www.fao.org/ag/AGA/AGAH/EMPRES/tadinfo2/e_tadinf.htm
5. http://www.fao.org/ag/AGA/AGAH/EMPRES/live_vis/bullet5.htm
6. http://www.fao.org/WAICENT/FAOINFO/TCD/tcd/tcdt/Default.htm
7. http://www.fao.org/WAICENT/FAOINFO/AGRICULT/AGA/AGAH/EMPRES/Info/other/tcp94-99.pdf
8. http://www.fao.org/WAICENT/FAOINFO/AGRICULT/AGA/AGAH/EMPRES/Info/other/surveill.htm

9. http://www.fao.org/WAICENT/FAOINFO/AGRICULT/AGA/AGAH/EMPRES/Info/other/newtech.pdf
10. http://www.fao.org/ag/AGA/AGAH/EMPRES/download/avisoft.htm
11. http://www.fao.org/WAICENT/FAOINFO/TCD/tcd/tcdt/Default.htm
12. http://www.fao.org/WAICENT/FAOINFO/AGRICULT/AGA/AGAH/EMPRES/Info/other/tcp94-99.pdf
13. www.fao.org/ag/AGA/AGAH/EMPRES/live_vis/centres.htm
14. www.fao.org/ag/AGA/AGAH/EMPRES/live_vis/centres2.htm
15. http://www.fao.org/WAICENT/FAOINFO/AGRICULT/AGA/AGAH/EMPRES/Info/other/tcp94-99.pdf
16. Internacional Fund for Agricultural Development; http://www.ifad.org/
17. http://www/aphca.org/activities/index_activities.html
18. http://www.panaftosa.org.br/novo/

# Animal Health Organizations

## Roles to Mitigate the Impact of Ecologic Change on Animal Health in the Tropics

BOBBY R. ACORD[a] AND THOMAS E. WALTON[b]

[a]*Animal & Plant Health Inspection Service, United States Department of Agriculture, Washington, DC 20250, USA*

[b]*Animal & Plant Health Inspection Service, United States Department of Agriculture, Centers for Epidemiology and Animal Health, Fort Collins, Colorado 80526-8117, USA*

**ABSTRACT: Production of livestock across North and South America is extensive. The opportunities for production, commerce, and thriving economies related to animal agriculture are balanced against the devastating threats of disease. Commitment by livestock and poultry producers in exporting countries to production methods, herd health management, and biosecurity in their operations must be coupled with an animal health and marketing infrastructure that allows the industries to thrive and offers assurances to trading partners that their livestock industries will not be jeopardized. National and international animal health organizations play a key role in providing this infrastructure to the industries that they serve. The incentive for the successful World agricultural production economies to provide direction and support for improving animal health and conveying principles for competitive and safe production to lesser developed nations is the assurance that the expanding economies of these nations offer an eager and hungry market for the products of the other industries of an export-dependent economy. The World Trade Organization (WTO) was established after the Uruguay Round of the General Agreement on Tariffs and Trade (GATT). The WTO provides the permanent international multilateral institutional framework for implementing dispute resolution agreements and the agreement on the application of sanitary and phytosanitary (SPS) measures. The SPS agreements allow for the protection of animal and plant health.**

**KEYWORDS: animal health; International Organization for Animal Health (OIE); World Trade Organization; Animal & Plant Health Inspection Service (APHIS)**

Address for correspondence: Thomas E. Walton, Animal & Plant Health Inspection Service, U.S. Department of Agriculture, Centers for Epidemiology and Animal Health, 2150 Centre Avenue, NRRC Building B, Fort Collins, CO 80526-8117. Voice: 970-494-7201; fax: 970-472-2668.

Thomas.E.Walton@aphis.usda.gov

**Ann. N.Y. Acad. Sci. 1026: 32–40 (2004). © 2004 New York Academy of Sciences.**
**doi: 10.1196/annals.1307.004**

## INTRODUCTION

From Denver, Colorado in the United States to Iguaçu Falls, Parana State, Brazil is a journey of some 10,500 kilometers by air. One passes through 4 time zones from West to East and traverses the Tropic of Cancer, the Equator, and the Tropic of Capricorn on the journey. While water certainly comprises a substantial part of the view from 10,900 meters in altitude, one is struck by the vast panorama of grasslands and rangelands, forests, and agricultural lands. From livestock production on rangeland, to aquaculture, confinement rearing, nomadic herding, as well as hunting and fishing, the utilization of animal species is a significant part of the world economy and important for human survival.

The production of cattle, swine, sheep, goats, and horses across North and South America is enormous. The opportunities for production, commerce, and thriving economies related to animal agriculture are balanced against the devastating threats of disease. In the last decade, control or elimination of dreaded livestock diseases such as rinderpest, hog cholera, and especially foot-and-mouth disease have opened the world marketplace and provided trade opportunities to producers in many countries. Nowhere has this been exemplified more dramatically that in the tropical, subtropical, and temperate cattle-producing areas of South America. Argentina, Brazil, Chile, and Uruguay have experienced profound growth in international markets for fresh, chilled, and frozen beef over the last 10 years. Freedom from foot-and-mouth disease has permitted the emergence of a sophisticated, highly competitive, and profitable industry that one day will challenge Australia, Canada, the European Union, New Zealand, and the United States for international exports of inexpensive, safe, and wholesome meat and poultry products. But, realization of these production and trade opportunities is predicated upon freedom from disease and maintaining that freedom over time. Commitment by livestock and poultry producers to production methods, herd health management, and biosecurity in their operations must be coupled with an animal health and marketing infrastructure that allows the industries to thrive and offers assurances to trading partners that their industries will not be jeopardized.

National and international animal health organizations play a key role in providing this infrastructure to the industries that they serve. The incentive for the successful world agricultural production economies to provide direction and support for improving animal health and conveying principles for competitive and safe production to lesser developed nations is the assurance that the expanding economies of these nations offer an eager and hungry market for the products of the other industries of an export-dependent economy.

## INTERNATIONAL ORGANIZATIONS

The World Trade Organization (WTO) was established after the Uruguay Round of the General Agreement on Tariffs and Trade (GATT). The WTO provides the permanent international multilateral institutional framework for implementing dispute resolution agreements and the agreement on the application of sanitary and phytosanitary (SPS) measures. Under the WTO, member nations accepted various obligations for their trade agreements. Almost all countries in the Western Hemisphere are signatories of the WTO. The SPS agreements allow for the protection of animal and plant health. The agreements assume that a member nation has a viable agricultural health infrastructure that is essential to facilitate trade. Under the WTO SPS agreement, member nations accept an obligation under the provisions of article 9 to assist other countries in reducing disease risk to a region, facilitating trade, and protecting and enhancing the competitiveness of their trading partners.

Numerous international animal health organizations play an important role in mitigating the impact of animal diseases. Moreover, the veterinary services and animal health infrastructures of most nations can provide valuable assistance in mitigating the adverse impact of ecologic change. While the following list is not all-inclusive, the United States Department of Agriculture (USDA), Animal & Plant Health Inspection Service (APHIS) collaborates with a number of international organizations in the western hemisphere to promote their roles in safeguarding animal health. The USDA and APHIS provide fiscal support and personnel resources to these organizations in support of their missions.

The WTO designated the International Organization of Animal Health (Office International des Epizooties, OIE) as the official organization to draft international SPS standards for animals and animal products. Member nations are required to base their sanitary import measures on science-based principles and guidelines. The WTO recognizes the OIE as the international forum for setting animal health standards, reporting global animal situations and disease status, and presenting guidelines and recommendations on sanitary measures relating to animal health. With the reduction in international tariffs, countries may try to apply import health requirements to protect their own industries and thus limit trade in animals and/or animal products. However, countries may not establish health standards that are more restrictive than those established for their domestic industries, although they may take measures necessary to protect human and animal health.

The OIE currently is composed of 162 member nations, each of which is represented by a delegate who, in most cases, is the chief veterinary officer of the country. Its mission is to prevent the spread of animal diseases. To facilitate prevention of the spread of animal diseases, OIE collects and disseminates information on the distribution and occurrence of animal diseases, coordinates research on contagious animal diseases, and develops international standards for the safe movement of animals and animal products in international trade.

The mission of the Food and Agriculture Organization (FAO) of the United Nations is to alleviate poverty and hunger by promoting agricultural development, improved nutrition, and the pursuit of food security.

The mission of the Pan American Health Organization (PAHO) is to lead strategic collaborative efforts among Member States and other partners to promote equity in health, to combat disease, and to improve the quality of, and lengthen, the lives of the peoples of the Americas.

The mission of the World Health Organization (WHO) is to attain the highest possible level of health for all people.

The mission of the Inter-American Institute for Cooperation on Agriculture (IICA) is to modernize agricultural health and food safety systems, update and harmonize sanitary and phytosanitary systems, encourage measures to protect consumers and facilitate trade, and provide early warning to identify emerging animal health and food safety issues. IICA issues alert bulletins as needed and strengthens the inter-American and regional focus for analyzing trends and opportunities in agricultural health and food safety.

The mission of the International Regional Organization for Plant and Animal Health (OIRSA) is to facilitate the economic and social development of the Central American region by coordinating and concentrating actions on the prevention, control, and eradication of pests and plagues of socioeconomic importance; modernizing the infrastructures of livestock and crop protection; assisting in the harmonization of laws, regulations, and policies; and supporting a global and open agricultural commerce.

The USDA partnership with animal health organizations from other countries, especially those organizations that have had experience in dealing with specific disease problems of concern to other countries, is critically important. Animal health organizations must develop contingency plans to assess and access vaccines or diagnostic reagents for pathogens that may become a problem, even though they may not have been a problem in the past. Multiple countries in a region may face common threats for which they can prepare together, whereas individually they would be unable to do so. The possibility of eradicating diseases makes more sense when entire regions band together in the effort. Rinderpest and foot-and-mouth disease are good examples. Regional and global organizations such as OIE, FAO, PAHO, WHO, IICA, OIRSA, and many others, including the livestock producers, have been potent allies in such efforts.

As the global dynamics change, that is, with greater and more rapid international movement of people and animals/products and expanded interactions among humans, animals, insects, and wildlife that may not have existed previously, animal health organizations must be more proactive in partnering with others to be prepared for and to address potential threats. This includes working closely with public health and wildlife organizations. Introductions of, or interactions with, new pests or wildlife species can have tremendous adverse impacts. We have seen this with West Nile virus introduced into the western hemisphere and, very recently, with monkeypox virus that very likely was introduced into the United States through the global movement of Gambian rats that transmitted the disease to prairie dogs and humans on another continent.

## NATIONAL ORGANIZATIONS

The Animal & Plant Health Inspection Service of the USDA is an example of a national animal health organization with international interactions. The primary role of national and international animal health organizations is to safeguard animal health and mitigate foreign animal disease threats. Within the United States, this role is carried out by the USDA-APHIS-Veterinary Services (VS) staff. VS is the principal national animal health organization whose mission is to protect and improve the health, quality, and marketability of our nation's animals, animal products, and veterinary biologics by preventing, controlling, or eliminating animal diseases and monitoring and promoting animal health and productivity. Not only does VS provide leadership, but also it is the authority that sets policy for animal health in the United States.

Within VS, there are three major program areas that interact directly with OIE. These consist of the National Center for Import and Export's (NCIE), Sanitary

International Standards Team in Riverdale, Maryland; the Center for the Diagnosis of Animal Diseases and Vaccine Evaluation for the Americas in Ames, Iowa; and the Centers for Epidemiology and Animal Health (CEAH) in Fort Collins, Colorado.

The Sanitary International Standards Team guides the development of sound international health standards for safe trade in animals and animal products by coordinating consensus-based comments on proposed modifications to chapters of the OIE International Animal Health Code. The Team maintains a database of disease and subject matter experts to review specific code chapters; monitors and evaluates reports and scientific data produced by the OIE; compiles the Annual Tabular and Narrative Disease Reports on World Animal Health Status of member countries; and submits monthly disease reports. They coordinate the import/export efforts to address the international movement of livestock, biological products, and animal products and to provide recommendations on the health conditions regarding the movement of animals and animal products.

Within VS, the Center for the Diagnosis of Animal Diseases and Vaccine Evaluation for the Americas has three components: the National Veterinary Services Laboratories (NVSL), the Center for Veterinary Biologics (CVB), and the Institute for International Cooperation in Animal Biologics (IICAB). As a component of the OIE Collaborating Center, the NVSL provides diagnostic assistance, supplies reference reagents to other laboratories, evaluates diagnostic reagents, conducts developmental projects to improve diagnostic techniques, and provides training on diagnostic tests.

CVB develops, distributes, and uses worldwide standard protocols for biologics evaluation; validates and provides standard reagents to biologics manufacturers and regulatory laboratories; reviews, develops, compares, and harmonizes testing protocols; and trains scientists throughout the world on these protocols.

IICAB is based at Iowa State University and concentrates its efforts on training, facilitatation of international communication, and harmonization related to the availability, safety, and efficacy of veterinary biologics.

As an OIE collaborating center, the CEAH shares its risk analysis and disease surveillance expertise with member countries. It provides them with technical assistance and expert advice on disease surveillance and control and risk analyses. As a collaborating center, CEAH fulfills the following objectives: reviews, evaluates, and adapts methods and approaches to enhance animal disease surveillance systems and the risk analysis process; promotes a harmonized approach to disease surveillance and risk analysis; provides technical assistance and training to OIE member countries as needed; improves the quality of animal disease surveillance and risk analysis by establishing a critical mass of trained individuals in OIE member countries; and networks with other OIE collaborating centers to coordinate activities.

The USDA views the U.S. export markets as critical to the economic health of our Nation. Our ability to export agricultural products overseas is a key component in the overall balance of trade and economic growth for the United States. However, there are multiple natural and intentional ways that an exotic pathogen could gain access to domestic livestock, making our agricultural products less acceptable for our trading partners and less competitive in the world market place.

Like the veterinary services of all nations, the USDA, APHIS, and VS have a national animal health emergency management system responsible for assisting the livestock and poultry industries to prevent, prepare for, respond to, and recover from introductions of foreign animal diseases and other animal health emergencies. Infections by pathogens such as foot-and-mouth disease recently have resulted in billions of dollars in losses to producers. Within VS, the Emergency Programs Staff coordinates efforts to prepare for and respond to such outbreaks. The EP staff monitors foreign animal health and, within the United States, provides leadership and support to all VS units and the States to eradicate OIE List A and some List B disease outbreaks. Its mission is to prevent destructive and harmful effects on the health of animal and human populations in the United States from epizootics of foreign and emerging animal diseases and from technologic disasters. They accomplish this task by developing and maintaining a high level of expertise and preparedness and leading and coordinating rapid response efforts.

To react to OIE List A animal disease outbreaks, animal health organizations need to have preplanned and practiced their responses to various types of events. Development and maintenance of response strategies are crucial. Experience and expertise can be gained by participating in animal health responses in other countries. The 2001 foot-and-mouth disease response in the United Kingdom was a good example in which members of animal health organizations from many countries worked together effectively and learned from each other.

Recent experiences in the United States with the emergency responses to exotic Newcastle disease and infectious salmon anemia, control of low pathogenic avian influenza and chronic wasting disease, expanded control programs for scrapie and Johne's disease, and the emergency control programs for brucellosis, tuberculosis, and pseudorabies have provided a far better response experience to U.S. animal health authorities than any test exercise. Similarly, assistance provided to the United Kingdom during the European outbreak of foot-and-mouth disease in 2001 and hog cholera in 1999 and to Canada during the recent discovery of a case of bovine spongiform encephalopathy in a bovine allowed U.S. veterinarians and animal health officials experiences in disease response, control, surveillance, and public relations. Likewise, similar experiences have been provided to Mexican animal health professionals who have worked in the United States during the exotic Newcastle disease outbreak in the Southwestern United States.

In recent years, the United States has invested more than $700 million of emergency funds into disease control activities, in addition to appropriated funds. There has been some confusion among U.S. trading partners about the terminology used in the United States for emergencies. The Secretary of Agriculture has the option to declare an Emergency and/or Extraordinary Emergency for animal and plant health events. These declarations should not be perceived as an indication of a disease situation or outbreak that is out of control. It is our official mechanism to facilitate the acquisition of emergency funds, to mobilize specialized support assistance from other Federal agencies, and to permit Federal authorities to provide direct assistance to the States.

In addition to being prepared for a response to an animal health event, another key element is the early detection of pathogens or pests prior to a large incident or outbreak, that is, having adequate diagnostic facilities and personnel. Not only is having diagnostic reagents critical, but having equipment and diagnosticians capable of doing the best tests and having contingencies for doing very large volumes of testing is also equally critical. An expansion of our laboratory infrastructure is currently underway with a $550 million construction and renovation project for NVSL, CVB, and the Agricultural Research Service, National Animal Disease Center at Ames, Iowa. At the Plum Island Animal Disease Center, major new construction is anticipated that will benefit the APHIS Foreign Animal Disease Diagnostic Laboratory and the ARS research unit. However, with the transition of the PIADC facility to the new U.S. Department of Homeland Security, no details are available about future plans. In addition to laboratories, animal health organizations need to develop efficient and defensible surveillance systems for animal diseases.

APHIS is expanding and strengthening the U.S. emergency response infrastructure that will help VS and Plant Protection & Quarantine, the plant world analogue to VS, to navigate these challenging times of increased national security, current disease events, and other issues yet to emerge. APHIS recently opened the APHIS Emergency Operations Center which serves as the national command and coordination center for APHIS emergency programs disaster management in Riverdale, Maryland; implemented the Incident Command System, a comprehensive incident management response system; and formalized Standard Operating Guidelines and Procedures for emergency preparedness and response activities. New Emergency Operations Centers are being designed and developed as well in the Eastern and Western Regional Hubs in Raleigh, North Carolina and Fort Collins, Colorado, respectively.

The role of the VS National Center for Import and Export (NCIE) is to support animal health and ensure that safe international agricultural trade opportunities are created. Their activities facilitate the domestic and international marketability of U.S. animals and animal products. The growing interest in agricultural trade in the global market has expanded the role of VS to ensure that safe agricultural trade opportunities are created while continuing to safeguard the nation's animal health.

The NCIE, Regionalization Evaluation Services Staff is responsible for evaluating information submitted by foreign governments in support of requests to be recognized as free of specific diseases or to export specific commodities to the United States. Such evaluations include risk assessments and site visits and frequently result in rulemaking. Import risk assessments are conducted using a team approach that includes: the NCIE Regionalization Evaluation Services Staff, which is responsible for the narrative aspects of the risk assessments and some qualitative assessments; the CEAH, which is responsible for the analytic processes of quantitative risk assessments and data acquisition; the Policy and Program Development Risk Analysis Systems group, which is responsible for quality control and consistency; and APHIS International Services (IS), which assists in gathering technical information needed for the risk assessments.

The number of requests to recognize disease freedom in foreign regions and/or define mitigations appropriate to specific commodities has increased dramatically over the years. In addition, it is becoming increasingly apparent that, for export purposes, VS needs to develop the capacity to define regions of different animal health status within the United States effectively. Consequently, VS is forming a new unit for Domestic Regionalization Services to manage the increasing number of import risk assessment requests from foreign governments and to implement domestic regionalization activities to facilitate U.S. exports of animals and animal products. Both of these units will interact closely with CEAH to conduct import risk assessments and to collect, evaluate, and manage domestic data to facilitate exports.

In 1997, APHIS published a final rule on Importation of Animals and Animal Products. This regulation has served as the template for information required by the United

States to assess risk and evaluate the health status of livestock and poultry of our trading partners. This rule has been used as a model by our trading partners as well. Eleven criteria are assessed by the United States when evaluating our trading partners:

- Authority, organization, and infrastructure of the veterinary services organization;
- Type, quality, and extent of disease surveillance;
- Capabilities of diagnostic laboratories;
- Disease status or presence of a disease pathogen;
- Extent of an active disease control program if a pathogen is present;
- Vaccination status;
- Disease status of adjacent regions;
- Separation from regions of higher risk through physical or other barriers;
- Extent and biosecurity of controls on the movement of animals and animal products from higher risk regions;
- Livestock demographics and marketing practices; and
- Policies and infrastructure for animal disease control, including emergency management and response capacity.

USDA-APHIS-International Services (IS) plays an important role with VS to ensure safe agricultural trade. Protecting agriculture today is a challenge that reaches beyond political and geophysical boundaries around the globe. Transporting plants, animals, and their agricultural byproducts in international commerce helps to meet the worldwide demand for food, but it increases the risks of introducing pests and diseases into new areas of the world. The IS team works outside of the United States, but compliments VS and plays a major role in dealing with many international and regional organizations concerned with animal health. IS shares responsibility with other international organizations and cooperates in a number of major surveillance and control programs in foreign countries, focusing on nations where economically significant pests or disease pathogens are found. Through foreign or international contacts, IS gathers and exchanges information on animal health. To facilitate U.S. agricultural exports through close contact with their counterparts, zoosanitary information is exchanged and risk mitigation measures are documented.

Other regulatory agencies, including the USDA, Food Safety and Inspection Service and the Department of Homeland Security, conduct food safety certification and make appropriate animal disease and public safeguarding decisions at U.S. airports and other ports of entry, respectively.

## CONCLUSION

In working together with other national and international animal health organizations, USDA is dedicated to an active role of mitigating the impact of foreign animal diseases and protecting animal health in the United States and around the World. While this presentation was not intended to convey the impression that the U.S. system is the best system or the only system useful in mitigating threats to animal health in the tropics, it is a system that serves as one possible model. Through our participation in the OIE Region of the Americas, our collaboration with our hemispheric trading partners under the North American Free Trade Agreement and bilateral agreements, participation in the Quadrilateral Animal Health Committee (Australia, Canada, New Zealand, and the United States), and the North American Animal Health Committee (Canada, Mexico, and the United States), the USDA provides assistance, leadership, training, and resources to protect and improve livestock and poultry health internationally.

# Tropical Diseases, Pathogens, and Vectors Biodiversity in Developing Countries

## Need for Development of Genomics and Bioinformatics Approaches

ALBERTO M.R. DÁVILA,[a,c] MÁRIO STEINDEL,[b,c] AND EDMUNDO C. GRISARD[b,c]

[a]*DBBM, Instituto Oswaldo Cruz, Rio de Janeiro, RJ, Brazil, CEP 21045-900*

[b]*MIP, Universidade Federal de Santa Catarina. Caixa postal 476, Trindade, Florianópolis, SC, Brazil. CEP 88040-900*

[c]*Laboratório de Bioinformática, Universidade Federal de Santa Catarina, SC, Brazil*

**ABSTRACT: The world's biodiversity, including many infectious, parasitic disease agents and their vectors whose impact on both human and animal health is significant, is largely retained in the developing countries of the tropics. Owing to the number of species involved and the relatively low-level exploration of pathogens and vectors biodiversity, several organisms are still waiting to be discovered and consequently explored in terms of genomics. Although some parasitic species of humans and animals have been studied through genomics and bioinformatics approaches, a significant number of relevant species are still to be addressed. Through the use of modern technologies, such as genomics and bioinformatics, for assessment of biodiversity and targeting tropical diseases, other relevant advantages of these initiatives for developing countries would be technology transfer and capacity building. Consequently, these initiatives could be critical to the development of the respective countries. Moreover, intra- and interhemispheric scientific collaboration should be encouraged and supported to increase the chances for success. In Brazil, the Ministry of Science and Technology has stepped forward to further such initiatives, co-supporting collaborative genomics and bioinformatics projects. The need for the establishment of working groups on genomics and bioinformatics in developing countries as well as the improvement and strengthening of collaborative research projects between developed and developing countries is discussed from our point of view. As these discussions remain open to debate, we encourage colleagues to promote further discussion on the subject.**

**KEYWORDS: genomics; bioinformatics; genomics divide; developing countries; biodiversity; tropical diseases**

Address for correspondence: Alberto M.R. Dávila, DBBM, Instituto Oswaldo Cruz, Fiocruz. Av. Brasil 4365, Rio de Janeiro, RJ, Brazil, CEP 21045-900. Voice: 55-21-3865-8229; fax: 55-21-2590-3495.

davila@fiocruz.br

<http://www.biowebdb.org>

**Ann. N.Y. Acad. Sci. 1026: 41–46 (2004). © 2004 New York Academy of Sciences.**
**doi: 10.1196/annals.1307.005**

## INTRODUCTION

In this essay, a general discussion of the current status of genome initiatives for parasites and vectors causing diseases in developing countries is presented. Because space and time do not allow us to deal exhaustively with this subject, readers may consider these comments as the basis for further discussions.

Biodiversity, which includes all plants, animals, and microorganisms, can be measured and expressed in different units such as genes, individuals, populations, species, ecosystems, communities, and landscapes (Convention on Biological Diversity, 2003).[1] The world's greatest biodiversity is mostly retained in developing countries located in the tropics and includes large numbers of etiological agents of infectious and parasitic diseases whose impact on both human and animal health has been significant. Because of the number of species involved and the relatively low level of exploration of biodiversity, several organisms are still waiting to be discovered and explored in terms of genetics and genomics, mainly for a better understanding of our own natural genetic resources. Several genome projects have been started and completed in recent years (<http://www.ncbi.nlm.nih.gov/genomes/>) and those that have targeted microbes are the most abundant, mainly because the relatively small genomes of these organisms facilitate sequencing of the whole genome in a short time at large sequencing centers. Consequently, eukaryotic genomes that are larger require significantly more time, manpower, and budgets. Specifically in the area of livestock, cattle, pigs, chicken, sheep, horses, rabbits, turkeys, and aquaculture, species have differing levels of ongoing genome initiative. While these represent some of the most important species with respect to impact on the livestock industry, many species remain to be studied in a genomic context.

Concerning human and animal pathogens, several efforts have been directed at the sequencing of the genomes of *Trypanosoma* spp., *Leishmania major*, *Plasmodium* spp., *Schistosoma* spp., *Theileria* spp., *Toxoplasma gondii*, *Eimeria tenella*, *Giardia lamblia*, *Entamoeba* spp., *Brugia malayi*, *Mycobacterium* spp., and *Candida albicans* (http://www.ncbi.nlm.nih.gov/entrez/query.fcgi?db = Genome). A few other examples of species with genome sequencing initiatives are *Theileria annulata*, *Cowdria (Ehrlichia) ruminantium,* and *Babesia bovis*. Regarding vectors of parasitic diseases, one of the most relevant species has been sequenced now—*Anopheles gambiae*. Meanwhile, *Aedes aegypti, Glossina morsitans morsitans, Culex pipiens,* and *Amblyomma americanum* genomes are in different sequencing stages. Initiatives for Triatominae species, vectors of human trypanosomiasis, are just starting, and a project on the transcriptome of some tissues of *Lutzomyia longipalpis* is being funded by the Brazilian Council for Research and Development (CNPq). Again, these represent very important species, but many others, affecting especially the developing countries, remain to be studied. They appear unlikely to enter the genomics area, because they affect typically poor countries where there is neither a market nor a profit for drug and vaccine development by private companies. These diseases have been called "neglected diseases,"[2] and an initiative to target those specifically affecting human health was recently begun (Drugs for Neglected Diseases initiative: <http://www.dndi.org/>). Considering this, one important question to ask is, "who is going to explore the genetics and genomics of relevant livestock species and pathogens that are restricted to developing countries, with no (apparent) relevance to developed countries?" The most probable answer is "nobody," unless scientists in developing countries receive training, financial support, and capacity building to do this for themselves. Otherwise, the problem will increasingly affect their livestock or humans.

## THE WORLD HEALTH ORGANIZATION (WHO) EXAMPLE

The United Nations Development Programme (UNDP)–World Bank–WHO Special Programme for Research and Training on Tropical Diseases (TDR) has moved to the forefront of the genomics age by supporting the sequencing of several human pathogens and the establishment of several regional centers for "Bioinformatics and Applied Genomics" in Africa, Asia, and South America (<http://www.who.int/tdr/grants/awards/bioinformatics-10-01.htm>). The main reason for these developments is that WHO experts have realized that developing countries will lose out on the health benefits of genomics research unless action is taken to strengthen their participation.[3] Knowledge from human and pathogen genomes can revolutionize health care and result in the development of new vaccines and drugs that could benefit millions, but without a critical mass of qualified people, proper regulation, and ethical guidelines in developing countries, the research could actually widen the gap between rich and poor. According to Pang,[4] infectious diseases represent a great threat to the economic survival of the poorest developing countries; however, the bulk of genomics research is not directed at these diseases. The reason behind this is that research is largely market driven by market forces and its products are probably unaffordable to the poor living in the developing countries. This discrepancy risks the creation of a "genomics divide" between rich and poor countries, such that 90% of those with health problems will not benefit from what is arguably the most golden era of biology and medicine.[5] In response to the WHO report, an international panel of experts, including many from developing countries, recently ranked the top 10 biotechnologies most likely to impact in the next 5 to 10 years.[6] Among such technologies, we find the development of affordable molecular diagnostic tests for infectious diseases, the genome sequencing of pathogens for the design of better and more effective antimicrobial drugs, and bioinformatics research. According to the same authors, some experts have suggested the formation of an international commission to govern the application of biotechnology. However, an initiative originating in the developed world may take the form of just another international commission, with token participation from only a few scientists from developing countries, and may not be the best way forward. A more acceptable approach may be to truly involve and empower scientists and institutions in developing countries, working together in equal partnership. An international initiative should also involve scientists and institutions from the more genomically advanced developing countries such as Brazil, China, Malaysia, and Thailand, which have made significant inroads and are shining examples of national commitment to the potential of genomics. It is difficult to imagine a genome-sequencing project without bioinformatics support; however, the contrary is possible and could qualify developing countries to participate in analyses of data coming from genome projects.

## UNDEREPRESENTATION OF DEVELOPING COUNTRIES

In a recent world survey on genomics research, 40 initiatives were mentioned (<http://www.stanford.edu/class/siw198q/websites/genomics/>) and demonstrated that funding in this area increased from about $720 million (US) in 1998 to $1.8 billion in 2000. Three initiatives from China are mentioned, with a total around $36 million (US) for the period 1998–2000. (This amount represents less than 1.5% of the world total investment in genomics research from 1998–2000.) Based on these data as well as with respect to bioinformatics initiatives, it is clear that developing countries are less represented in the genomics age than developed nations.

Although this limited representation is mainly explained by the lack of funding and less developed science in developing countries, it is not rare to see reports and comments on the lack of collaboration of scientists from a developed country with their colleagues in developing countries. In fact, Gerald Keusch, director of the National Institute of Health's Fogarty International Center, said that, "First-World scientists pursuing genomic research are generally not working in collaboration with endemic-area scientists and local institutions."[7] Moreover, a complaint from Winston Hide, Director of the South African National Bioinformatics Institute, pointed out that efforts to sequence the genomes of *Plasmodium falciparum* and *Anopheles gambiae* did not include developing-country groups. He also recognized that African groups are not equipped to undertake the sequencing themselves, and he stated that they could have been involved in the analysis, which would also have given them early access to unpublished information. Something important to keep in mind is that there are precedents for using genome and bioinformatics projects to build a scientific infrastructure in a developing country, as was done in Brazil with the sequencing of the first plant pathogen (*Xylella fastidiosa;* <http://watson.fapesp.br/onsa/Genoma3.htm>).[8] Additionally, completion of the genome of *Chromobacterium violaceum* and other ongoing prokaryote genome projects (<http://www.brgene.lncc.br>) has contributed to the building of a genomic network that places Brazil in the forefront of genomics and bioinformatics research in the South American region.[9]

The largest sequencing facilities, able to sequence bacterial genomes in a few days, are located in the United Kingdom and the United States, whereas other countries such as France, Japan, Australia, China, and Brazil also have large sequencing centers. Moreover, the largest bioinformatics centers are in the United States (National Center for Biotechnology Information, NCBI) and the United Kingdom (European Bioinformatics Institute, EBI). Most of the genome sequencing projects to date did not include the participation of developing country scientists. From reviewing a list of the gene mapping of livestock species, available on the Internet (<http://www.genome.iastate.edu/resources/other.html>), we realize that developed countries are doing all of those mapping projects. However, things appear to be changing slightly. The comparative genomic projects on *Plasmodium vivax* and trypanosomes of cattle at The Sanger Center lists as collaborators some scientists from developing countries, and other relevant projects such as the one dealing with the sequencing of the *Cowdria (Ehrlichia) ruminantium* genome (<http://www.sanger.ac.uk/Projects/C_ruminantium/>) do so as well. Moreover, during the last years, the Fogarty International Center (<http://www.fic.nih.gov/>) and the European Union (<http://europa.eu.int/comm/research/>) have been promoting collaborative research projects with developing countries. This shows some promise for a new model of collaborative genome initiatives.[10]

## ARE INTERNATIONAL AGENCIES PREPARED TO FACE THE "GENOMICS DIVIDE"?

To our knowledge, the only international agency of the United Nations having an active agenda for genomics and bioinformatics research seems to be WHO. In an exhaustive search of genomics and bioinformatics information in the websites of the Food and Agriculture Organization (FAO), Office International des Epizooties (OIE), and International Atomic Energy Agency (IAEA), we only found on the FAO website a report on "Forest genomics for conserving adaptive genetic diversity" published by Krutovskii and Neale (2001).[11] More recently, just before the publication of this essay, we received the good news that IAEA has strategically decided to create an "IAEA collaborating centre on Animal Genomics and Bioinformatics." While this initiative could certainly help to address some of the points discussed here, there is still a need for the urgent establishment of a working group on genomics and bioinformatics in those agencies. That working group would be composed mainly of scientists from developing countries and would be charged with promoting research, developing training courses, and carrying out capacity building in the field of genomics and bioinformatics in developing countries.

The most likely scenario to avoid or help bridge the genomics divide is the improved collaboration between scientists in developing and developed countries and the promotion of South–South collaboration, thus contributing towards the development of science in poor countries more in need of it. There is a major role for the UN and other international agencies to promote the establishment of the necessary working groups for this endeavor and to support their activities to ensure successful collaboration and transfer of technology.

## ACKNOWLEDGMENTS

We would like to thank the Conselho Nacional de Pesquisa e Desenvolvimento (CNPq) for financial support.

## REFERENCES

1. Convention on Biological Diversity. Report of the *Ad Hoc* Technical Expert Group on Biodiversity and Climate Change. UNEP/CBD/SBSTTA/9/INF/12, September 30, 2003. <http://www.biodiv.org/doc/meetings/sbstta/sbstta-09/information/sbstta-09-inf-12-en.pdf>
2. Morel, C.M. 2003. Neglected diseases: under-funded research and inadequate health interventions. Can we change this reality? EMBO Rep. **4:** S35–S38.
3. Brown, P. 2002. Poor countries could lose out on benefits of genomic research. Br. Med. J. **324:** 1053.
4. Pang, T. 2003. Equal partnership to ensure that developing countries benefit from genomics. Nat. Genet. **33:** 18.

5. Advisory Committee on Health Research. Genomics and World Health: A Report from the Advisory Committee on Health Research (World Health Organization, Geneva, 2002). <http://www3.who.int/whosis/genomics/genomics_report.cfm>
6. Daar, A.S., H. Thorsteinsdottir, D.K. Martin, *et al.* 2002. Top ten biotechnologies for improving health in developing countries. Nat. Genet. **32:** 229–232.
7. Butler, D. 2002. What difference does a genome make? Nature **419:** 426–428.
8. Simpson, A.J. *et al.* 2000. The genome sequence of the plant pathogen *Xylella fastidiosa*. The *Xylella fastidiosa* Consortium of the Organization for Nucleotide Sequencing and Analysis. Nature **406:** 151–157.
9. The Brazilian National Genome Project Consortium. A.T.R. Vasconcelos *et al.* 2003. The complete genome sequence of *Chromobacterium violaceum* reveals remarkable and exploitable bacterial adaptability. Proc. Natl. Acad. Sci. USA **100:** 11660–11665.
10. Dávila, A.M., P.A. Majiwa, E.C. Grisard, *et al.* 2003. Comparative genomics to uncover the secrets of tsetse and livestock-infective trypanosomes. Trends Parasitol. **19:** 436–439.
11. Krutovskii, K.V. & D.B. Neale. 2001. Forest genomics for conserving adaptive genetic diversity. Working Paper FGR/3 (E) FAO, Rome (Italy). <http://www.fao.org/DOCREP/003/X6884E/X6884E00.HTM>

# Impact of Avian Influenza on U.S. Poultry Trade Relations–2002

## H5 or H7 Low Pathogenic Avian Influenza

CHERYL HALL

*United States Department of Agriculture, Animal and Plant Health Inspection Service, Riverdale, Maryland 20737, USA*

**ABSTRACT: Avian influenza (AI) viruses are Type A influenza viruses of the *Orthomyxoviridae* family. There are 15 subtypes of the virus widespread in migratory waterfowl throughout the world. It has become increasingly evident that some low pathogenic avian influenza (LPAI) H5 or H7 viruses have the capacity to mutate into the more virulent strains that cause extensive economic losses and high mortality. Recent AI disease outbreaks in several countries have increased attention and concern over low pathogenic H5 and H7 AI viruses. This heightened international concern increases the risk of unnecessary trade bans. For the US poultry industry, avian influenza continues to be a challenge to the flow of trade. On one hand, there is the increased focus of world attention on the H5 and H7 low pathogenic AI virus and the possibility of mutation. On the other hand, there are the factors contributing to our finding of infected flocks. Among these, perhaps the most important is the ever-present reservoir of virus in the migratory waterfowl population. With the discovery of exposed flocks comes the threat of trade bans.**

**KEYWORDS: avian influenza; Type A influenza viruses; poultry disease; trade bans; mutation**

## INTRODUCTION

Type A influenza viruses of the *Orthomyxoviridae* family, widespread in migratory waterfowl throughout the world,[1,2] can be subtyped by the "H" or hemagglutinin on the surface. Of the 15 subtypes, H5 and H7 have shown an ability to mutate from low pathogenic viruses to highly pathogenic forms that cause high mortality and significant economic losses. Owing to increased international concerns over the possibility of mutation, U.S. poultry trade export restrictions have resulted from H5 or H7 low pathogenic avian influenza (AI) cases frequently in 2002. The U.S. has adopted several programs to reassure trading partners that the risk of transmission of low pathogenic AI through poultry products has been negated.

Address for correspondence: Cheryl Hall, D.V.M., M.A.M., A.C.P.V., United States Department of Agriculture, Animal and Plant Health Inspection Service, National Center for Import and Export, 4700 River Road, Unit 46, Riverdale, MD 20737. Voice: 301-734-8715; fax: 301-734-4982.

cheryl.i.hall@aphis.usda.gov

**Ann. N.Y. Acad. Sci. 1026: 47–53 (2004). © 2004 New York Academy of Sciences.**
**doi: 10.1196/annals.1307.006**

## PATHOGENICITY CHARACTERISTICS OF THE VIRUS

Avian influenza viruses are classified as one or the other of two distinct pathotypes: low pathogenic or highly pathogenic. The current Office International des Epizooties (OIE)[3] Avian Influenza chapter has specific criteria for designating an AI isolate as highly pathogenic strain and thus a List A virus. Requirements are:

Any influenza virus that is lethal for six, seven, or eight of eight 4- to 6-week-old susceptible chickens within 10 days following intravenous inoculation with 0.2 mL of a 1:10 dilution of bacteria-free, infectious allantoic fluid.

For all H5 and H7 viruses of low pathogenicity, if growth is observed in cell culture without trypsin, the amino acid sequence of the connecting peptide of the hemagglutinin must be determined. If the sequence is similar to that observed for other HPAI isolates, the isolate being tested will be considered to be highly pathogenic.

Subtype H5 or H7 AI isolates that do not meet these criteria are considered low pathogenic. Losses from highly pathogenic strains are much more devastating than those from low pathogenic viruses. It was not until the ability of LPAI to become highly pathogenic was recognized that LPAI infections became an issue.

The OIE is currently considering a proposal to change the definition of avian influenza viruses of concern. Unlike the current OIE definition of avian influenza, the proposed OIE revision does not differentiate between low and highly pathogenic H5 and H7 isolates with regards to reporting to OIE. In the future, all H5 and H7 AI isolates would be treated as notifiable disease viruses. If this revision is adopted, it is reasonable to expect that additional trade requirements and restrictions will be developed and imposed by U.S. poultry trading partners even though the virus will not have changed.

## MUTATIONS

With advances in diagnostics, recent outbreaks of low pathogenic avian influenza that mutated to the highly pathogenic form have been documented and studied extensively in several countries. Examples of mutation to a more pathogenic form were first found in the Pennsylvania outbreak of 1983–1984 and in other significant avian outbreaks of recent years: Mexico (1993), Australia (1994), Pakistan (1994), Italy (1999–2001), and Chile (2002). The actual molecular site and type of mutation can be identified, and shifts towards high pathogenicity can be followed through gene sequencing of the cleavage sites.

### *Mutation in Mexico*

After H5 avian influenza was recognized in Mexico in 1993, flocks in 11 states were found to be serologically positive for the H5 virus. In just over a year, chickens were dying from infection with the highly pathogenic H5 virus. Horimoto *et al.*[4] found the presence of multiple basic amino acids at the HA cleavage site, which confers greater cleavability. Greater cleavability allows for systemic replication beyond the respiratory and gastrointestinal tracts.

### *Experience in Italy*

From March until mid-December 1999, a low pathogenic H7 virus circulated in the poultry populations in Northern Italy. The virus was characterized and found to meet all the criteria for classification as an LPAI. By mid-December, the virus changed from the original PEIPKGR*GLF to PEIPKGSRVRR*GLF or multiple basic amino acids at the cleavage site known to be a feature of HPAI.[5] High mortality and great economic losses followed, particularly in the turkey industry.

### *AI Surveillance in the United States*

Over the years, it became increasingly evident that intense surveillance for AI was necessary to protect the nation's flocks. In response, production companies perform agar gel immunodiffusion (AGID) testing on blood taken at processing from most of the flocks slaughtered as an AI screening test. Also, State laboratories investigate any case presenting with signs that may be AI as well as other randomly selected cases. These work-ups may include AGID, enzyme-linked immunosorbent assay (ELISA), virus isolation, and hemagglutination inhibition (HI) testing. For standardization in testing, the National Veterinary Services Laboratories (NVSL) in Ames, Iowa produces the antigen used for almost all AGID testing done in the United States and performs most of the HI testing. If H5 or H7 positives are found at any stage, additional identification by NVSL is made through gene sequencing, intravenous pathogenicity index (IVPI), and growth in cell culture without trypsin. Just the AGID testing alone in the United States accounts for over 1 million tests annually.

Additional surveillance is also part of the National Poultry Improvement Plan (NPIP) which provides a means of certification for AI-clean flocks, which means the flock is free of avian influenza. This category includes breeders of commercial egg-layers and breeders of meat-type turkeys and meat-type chickens.[6]

Migratory waterfowl and other wild aquatic birds are known to be the natural reservoir for AI viruses.[1] Several studies in Texas and Louisiana along the Gulf Coast have shown that up to 10% of the migratory waterfowl returning from wintering grounds in warmer countries to the south are infected with AI.[7] The major flyways for migratory birds pass over some of the most concentrated areas of poultry-rearing in the United States, and most of these areas do not have a history of AI infections. One notable exception is AI-positive flocks in Minnesota coinciding with the appearance of juvenile migratory and recently infected ducks. When turkeys were raised on the range, these infections were a frequent event. Turkeys are reared inside now, and the incidence of AI has decreased dramatically.[1]

## DEALING WITH POSITIVE CASES OF LOW PATHOGENIC AVIAN INFLUENZA

Trading bans and concerns over the mutation of LPAI to HPAI have changed the way infected flocks are handled. Recently, there was an outbreak in Virginia. The US government, because of interest and concern among our industry, acknowledged the importance of LPAI and partnered with Virginia to eradicate this disease from birds in that state. The program took over 4 months to rid the commercial flocks of the Shenandoah Valley of the virus. This action demonstrated the level of importance attributed to low pathogenic AI control in the United States.

Essential parts of any LPAI control program are extensive surveillance and monitoring of all avian species around the infected premises and epidemiologically linked to the infected premises. It is necessary to determine the extent of the infection of this easily disseminated virus so quarantines can be enacted around the entire infected area. Good biosecurity is one of the best defenses in any control/eradication effort.

## TRADE RESTRICTIONS

Many countries that receive poultry and poultry products from the United States have placed bans that have significantly affected export markets. Although these low pathogenic AI serotypes have traditionally not had an impact on export trade, the increased awareness and numbers of outbreak events have raised concerns among the international animal health community. There has been and continues to be a significant amount of discussion within the international community on both the economic impact on trade and the regulatory implications.

It is difficult to predetermine which AI finding will generate a ban or what type of ban might follow. The trading partner notifies the United States Department of Agriculture that it is placing the ban and the extent of the area that is banned from exporting poultry products to their country. Bans were enacted on a nationwide basis for an LPAI-positive flock of 4,000 ducks supplying a local market. Bans were put in place on a state with a positive serological finding even though, despite multiple attempts, the virus was never isolated.

Equally as difficult is determining when a ban will end and how to best facilitate the resumption of trade. In 2002, new methods were used to satisfy various concerns raised by trading partners. Negotiations with a number of countries led to agreements to provide immediate notifications of AI disease findings, to provide certifications, to attend arbitration, and to enact regionalization. Other methods to satisfy concerns, such as compartmentalization, which means considering the different poultry industries as separate from each other, were discussed.

## COMMUNICATION PLAN

As part of the negotiations with trading partners for regionalization when an infected flock is found, APHIS has supplied current and timely notifications of all aspects of the disease outbreak. Notifications include the date of the initial finding, the subtype and pathogenicity of the virus, the species and type of poultry, the size of the flock, and the location. Immediately, quarantines are put in place and surveillance testing begins. Follow-up reports to the various countries include any additional positive findings and the depopulation date or the disease management plan.

### *Certification for Exports*

Additional information relative to the disease history of birds whose meat is to be exported is being required. Requirements vary from simple statements to very complicated regionalization schemes with extensive testing. All of the increased record-keeping for the system and increased testing to satisfy the requirements of the certificate have caused the human and economic resources to be redirected.

For some countries, an AI certification system has been devised. A monthly certificate must go from the State veterinarian and the Area Veterinarian in Charge to the Food Safety and Inspection Service (FSIS), Inspector in Charge (IIC) at the plant exporting poultry products. These certificates enable the IIC to sign the detailed veterinary certificate that goes with the exported product.

When any avian influenza is identified by NVSL located in Ames, Iowa, the Federal Area Veterinarian in Charge in the appropriate State and the Avian Influenza Coordinator (AIC) at the Animal and Plant Health Inspection Service headquarters are notified. The health certificate for avian influenza reflects those findings. The State veterinarian and the AVIC sign the certification statement each month. In addition to the "AI-Free" status, the statement has three main categories for positive findings and the resulting actions including extensive testing that must be taken for any positive results.

### *Testing for Export*

Trading partners have the right to protect the health of the poultry flocks in their country in addition to their allowing the flow of trade. Some countries have placed restrictions, and in other instances to maintain the market, we must meet other requirements. For example, the US poultry industry is faced with a requirement to test all poultry and flocks producing poultry products with an extensive 59 sample testing program by hemagglutination inhibition (HI) for H5 and any subtype of avian influenza that has been found in a state over an undefined period of time. This testing must be done in addition to the AGID test. It is difficult to understand the scientific rationale behind the requirement for the HI test on sera from AGID-negative testing. While it is true that the response to the test wanes in time, it is months before the AGID results become negative for a flock. Along with the duplication in testing is large sample size. Is there a need for 95% certainty of finding a 5% infection in the flock? These test results must accompany the product that is shipped.

### *What the Future Brings*

Discussions within the United States with industry, regulatory officials, and other interested parties have led to decisions to develop a national AI H5/H7 program. This program will include periodic testing of all commercial turkey, broiler and layer breeders, and egg-laying birds. Commercial meat birds, chickens, and turkeys will be tested also. In addition, backyard flocks and game fowl will be monitored.

A separate program is being developed to prevent, control, and eradicate H5 and H7 from live bird markets. The National Live Bird Market System Program for Low Pathogenic Avian Influenza has started with trials of bird identification systems to enable inspectors to check for premises identity and test certificates. All birds entering the markets are required to be from test-negative flocks.

## CONCLUSIONS

For the US poultry industry, avian influenza continues to be a challenge to the flow of trade. The circumstances around the issue have multiple contributors. On one hand, there is the increased focus of world attention on the H5 and H7 low pathogenic AI virus and the possibility of mutation. On the other hand, there are the factors contributing to our finding of infected flocks. Among these, perhaps the most important is the ever-present reservoir of virus in the migratory waterfowl population. With the discovery of exposed flocks comes the threat of trade bans. There does not appear to be any relief in sight.

## UPDATE

APHIS (Animal and Plant Health Inspection Service) has received funding for the live bird market avian influenza program. The program includes monitoring of source flocks, transporters and consolidators, and birds in the markets. Requirements of the program include market closures, downtime, sanitation, and registration. Haulers must have appropriate and working sanitation equipment. Only registered haulers may transport birds to the markets. Federal and state employees will perform monitoring activities to include sample collection and record reviews as well as visual inspections of the premises.

The expanded National poultry Improvement Plan includes "Avian Influenza Clean" commercial flocks through a nationwide monitoring program for layer, turkey, and broiler flocks produced in the United States.

## ACKNOWLEDGMENTS

The author wishes to thank the following who provided guidance throughout the construction of the presentation: Dr. Max Brugh, Dr. Glen Garris, and Dr. Charles Beard. Special thanks go to Linda Collier for her assistance.

## REFERENCES

1. Easterday, B.C., V.S. Hinshaw & D.A. Halvorson. 1997. Influenza in Diseases of Poultry, 10th edit.: 583–607. University Press. Ames, Iowa.
2. Stallkneckt, D. 1997. Ecology and epidemiology of avian influenza viruses in wild bird population: waterfowl, shorebirds, pelicans, cormorants, etc. Proceedings of the Fourth International Symposium on Avian Influenza. Athens, GA. D.E. Swayne & R.D. Slemons, Eds.: 61–70. Georgia Center for Continuing Education. The University of Georgia. Athens, GA.
3. Office International des Epizooties. 2000. Manual of Standards for Diagnostic Tests and Vaccines, 4th edit.: 212–220. OIE. Paris.
4. Horimoto, T., E. Rivera, J. Pearson, *et al.* 1995. Origin and molecular changes associated with emergence of a highly pathogenic H5N2 influenza virus in Mexico. Virology **213:** 223–230.
5. Capua, I. & S. Marangon. 2000. Avian influenza in Italy (1999–2000): a review. Avian Pathol. **29:** 289–294.

6. NATIONAL POULTRY IMPROVEMENT PLAN AND AUXILIARY PROVISIONS. 2000. U.S. Avian Influenza Clean Status. United States Department of Agriculture, Animal and Plant Health Inspection Service, Veterinary Services, Conyers, GA.: 26.
7. Personal communication. David Stallneckt. June, 2003.

# Situation of Classical Swine Fever and the Epidemiologic and Ecologic Aspects Affecting Its Distribution in the American Continent

MOISÉS VARGAS TERÁN,[a] NELSON CALCAGNO FERRAT,[b] AND JUAN LUBROTH[c]

[a]*Animal Health Officer, FAO–Regional Office for Latin America and the Caribbean, Santiago, Chile*

[b]*Veterinary Officer, Servicio Agrícola y Ganadero, Santiago, Chile*

[c]*Senior Officer, Animal Health Service, FAO - Headquarters, Rome, Italy*

**ABSTRACT: Classical swine fever (CSF) is a viral transboundary animal disease that is highly contagious among domestic and wild pigs, such as boars and peccaries. Today, far from being what was *classically* described historically, the disease is characterized as having a varied clinical picture, and its diagnosis depends on resorting to proper sample collection and prompt dispatch to a laboratory that can employ several techniques to obtain a definitive diagnosis. Laboratory findings should be complemented with a field analysis of the occurrence of disease to have a better understanding of its epidemiology. The disease is still present in various regions and countries of Latin America and the Caribbean, thus hindering production, trade, and the livestock economy in the region. Consequently, it is among the diseases included in List A of the Office International des Epizooties (OIE). Currently, there are epidemiologic and ecologic aspects that characterize its geographical distribution in the region such as: continued trends in the demand for pork and pork products; an increase in swine investment with low production costs which are able to compete advantageously in international markets; the convention of associating CSF in the syndrome of "swine hemorrhagic diseases" owing to the historical description of its acute presentation and not to the new and more frequent subacute presentations or the diseases with which it may be confused (notably, porcine reproductive and respiratory syndrome and porcine dermopathic nephropathy syndrome, among others); dissemination of the virus through asymptomatic hosts such as piglets infected *in utero;* frequent lack of quality control and registration of vaccines and vaccinations; feeding of swine with contaminated food waste (swill); the common practice of smuggling animals and by-products across borders; the backyard family production system or extensive open field methods of swine rearing with minimal input in care and feeding; poor understanding of the epidemiologic role that boars and peccaries could have in the transmission and maintenance of the disease in the Americas; and new procedures in animal welfare that some countries are adopting for the production, transport, and slaughter of domestic animals. Consequently, many**

Address for correspondence: Moisés Vargas Terán, FAO–Regional Office for Latin America and the Caribbean, Chile. Fax: (562) 337-2101.
moises.vargasteran@fao.org

**Ann. N.Y. Acad. Sci. 1026: 54–64 (2004). © 2004 New York Academy of Sciences.
doi: 10.1196/annals.1307.007**

**countries (i.e., Canada, USA, Chile, Belize, Costa Rica, Panama, and Mexico, where 13 of 32 States are disease free) have given priority to the control and progressive eradication of CSF. In other parts of the Americas, the disease appears under control, as is the case of the five countries of the Andean Region and the 12 northern States of Brazil. In South America, Chile, Uruguay and 13 States in Brazil are disease free. Argentina has mounted a national campaign and is in the process of eradicating the disease. No recent information on its presence or distribution in Paraguay is available. With no master strategy to harmoniously progress in the control and eradication of the disease, 17 countries of the region, jointly led by the Food and Agriculture Organization of the United Nations, developed the Continental Plan for the Eradication of CSF whose objective is expected to be reached by 2020.**

KEYWORDS: **classical swine fever; hog cholera; swine rearing; transboundary disease transmission**

## INTRODUCTION

Classical swine fever (CSF) is also known as hog cholera. It is a highly contagious viral disease affecting domestic and wild pigs, boars (*Sus scrofa ferus)* and peccaries *(Tayassu tajacu* and *T. pecari*).[1] It is most commonly transmitted directly by contact with infected and susceptible animals and persons visiting farms, ingestion of contaminated foods (swill), and transplacentally, and indirectly through contaminated objects or veterinary instruments present or used in the rearing facilities. Postinfection, the virus may be found in virtually every organ, secretion, and excretion of live animals and corpses, although there is a tropism for lymphoid or epithelial tissues. Infected piglets that do not die of the disease can excrete the virus continuously for months. The infection gains entry through the conjunctival mucosa, skin or oral abrasions, natural or artificial insemination, transplacentally, and orally. The incubation period is generally 2–14 days. Clinical and pathological manifestations can be classified as acute, chronic, or congenital. Definitive diagnosis should be made with the assistance of a laboratory capable of differentiating this disease from the viral infections caused by African swine fever, bovine viral diarrhea, border disease, porcine reproductive and respiratory syndrome, porcine dermopathic nephropathy syndrome, and the bacteria that cause salmonellosis or erisipelas. Some of the laboratory methods recommended to detect the causative agents are: direct immunofluorescence, viral or bacterial isolation, and monoclonal antibody-based ELISAs, all complemented by molecular techniques such as the polymerase chain reaction (PCR). There is no treatment for this disease, and infected pigs should be sacrificed and destroyed. Although CSF is not a human pathogen, animals that die or are destroyed because of CSF infection should not enter into the human food chain, because it is not considered a wholesome pork product. This disease limits animal production and international trade and is included among the diseases on List A of the OIE .[2]

## SITUATION WITH CLASSICAL SWINE FEVER IN 2003

The geographic distribution of CSF in the Americas through May 2003, based on the official information sent by member countries to the OIE,[2] is as follows.

### *North America*

Canada continues to be disease-free, the last case having been reported in 1963. Similarly, the United States is disease-free, the last case being reported in 1976. In Mexico, 13 of the 32 states continue to be disease-free and have a concerted epidemiologic surveillance program. Whereas 150 outbreaks were documented in 1997, only 6 outbreaks were reported in 2002 within its controlled area and none in 2003.

### *Central America*

Belize, Costa Rica, and Panama continue to be disease-free and have continuous epidemiologic surveillance programs. The CSF situation in El Salvador is that of progressive control, and the country is expected to extend its national program activities to the western and central parts of the country, from the western border of the Lempa River up to its border with Guatemala. In 2001 there was an outbreak in the Department of San Salvador, with no outbreaks registered in 2002 or 2003. In Guatemala, progressive control activities continue in El Petén, where no official vaccination has taken place over the last 2 years. Concurrent with a no-vaccination policy, serological surveys are undertaken to identify possible circulating virus activity as well as to implement two animal mobilization control stations in the area. In the rest of Guatemala, 11 outbreaks were reported in 2002 and only two in 2003. Honduras is planning to place the entire Atlantic coast under a national control scheme; no outbreaks or cases have been reported in 2003. Likewise, Nicaragua has embarked on a national control plan; in 2002, the Department of Rivas in the southern part of the country was severely affected by the disease with 217 reported outbreaks throughout the country. An eradication program was elaborated through vaccinations, and major coverage was given to the free northern zone along the Atlantic departments, covering a surface of approximately 8,000 $km^2$. No outbreaks were reported through May 2003.

### *The Caribbean*

Cuba continues with intensive local control schemes and reported 113 outbreaks in 2002 and only 6 in 2003. The Dominican Republic reported 87 outbreaks in 2001 and 72 in 2002. Although no official reports were made in 2003, the disease continues to smolder. Haiti recognizes that it has the disease, but it has limited ability to document its occurrence and distribution. However, local control activities are being implemented, but infrastructure and veterinary outreach are weak. In the rest of the Caribbean, countries have not reported the disease and may be considered disease-free if disease surveillance programs are in place.

### *Andean Region*

Colombia currently has instituted local massive vaccination campaigns with the intention of covering not only intensive but also extensive production systems through deployment of vaccinating brigades, including the identification of vaccinated pigs. The campaign is being carried out through a strategic alliance between the government's Instituto Colombiano Agropecuario and the private producer syndicate Asociación Colombiana de Porcicultores. In 2002, 3,000 cases of CSF were reported. In 2003, up to the preparation of this article, only 12 cases were reported. Peru counts with a zoo-sanitary CSF control program on a local scale. Thirty-seven cases were reported in 2002 and 3 in 2003. The latest cases in Bolivia were reported in August 1999. In 2002 and to date, no official reports have been made regarding the disease. However, the Departments of Chuquisaca and Cochabamba have been affected by CSF. There have been no reports of outbreaks in 2002 and 2003 in Ecuador. Few outbreaks of CSF have been reported from Venezuela for several years—37 in 2002 and none through 2003 when this report was prepared.

### *Amazon Region*

Guyana, French Guyana, and Suriname have not reported any cases and continue to be disease-free. Reports from Brazil in 2001 identified 12 outbreaks (Pernambuco, 6; Ceará, 3; Paraíba, 2; and Rio Grande do Norte, 1) in 4 of the 12 states under a national control program with the last outbreak reported in August 2001. No cases were reported in 2002 and 2003. In other parts of Brazil (geographically incorporated into the Southern Cone [see below]), 14 of its 26 states continue disease-free.

### *Southern Cone*

Argentina's concerted national program has limited the occurrence of CSF, with its last case in May 1999. No cases have been reported in 2002 and 2003. Chile is considered disease-free with the last case reported in August 1996. As with Chile, Uruguay is considered disease-free, the last case being reported in November 1991. In Paraguay, no cases were reported in 2001, 2002, and 2003. The last case was reported in July 1995.

## EPIDEMIOLOGY

Classical swine fever is still present in some regions and territories of various countries of the Americas. National and regional concerted efforts have made outstanding progress in the last 5 years, but because of the transboundary nature of the disease, any negligence or laxness in control or surveillance activities should be of regional concern. This report provides information on the actual epidemiologic and ecologic factors that have prevented complete progress in the control and eradication of the disease in the Americas. At the same time, it provides information on the new regional strategy to progressively eliminate CSF from the region.

Pork produced under modern industrial conditions is one of the safest meats for human consumption. This results from the hygienic and sanitary conditions in which the animals are raised and slaughtered; by-products are carefully prepared and sold through supermarket chains or exported to other countries in the form of special cuts previously requested by the importer. The nutritional characteristics of pork meat derived from current breeds are considerably better than from that produced only 10 years ago. Consequently, pork has been converted into a preferred consumer request.[3]

Pig meat and pork products are achieving record gains in worldwide consumption, and in many regions they are in first place. Pork products occupy third place on the American continent with a positive trend.[4] Because of the increasing demand in local and international markets, enormous agricultural capital investments have been made to produce swine in the American region to satisfy consumers and obtain economic yields. This environment has led new entrepreneurs to increase industrial output and improve sanitary conditions in their swine herds. In 2002, the production cost per kilogram of fresh meat was US 0.62 cents in Brazil, 0.70 in Chile, 0.77 in the US, 0.95 in Mexico, and more than 1 US dollar in the Dominican Republic. The highly technical industrial system has allowed significant progress in fighting CSF, facilitating its eradication in Chile, where 75% of the pork production is under intensive systems. Forty percent of Brazil's production is considered to be under highly technical intensive systems and Mexico's inductrial swine production is at 50%. In the latter two countries, there are still vast areas of family backyard production, using semi-confinement systems with open range practices.[5] This constitutes a favorable environment for disease transmission and maintenance due to the difficulty in effectively implementing CSF control and eradication measures over a large swine population distributed over an extensive area.

The signs of clinical CSF have been known for decades. Nonetheless, currently it is difficult to define them clearly, because the now more common circulating virus strains in the American region appear to be moderate in virulence and produce subacute or chronic infection,[6,7] which is radically different from acute infection, commonly characterized by cyanosis of the skin, pantropic visceral hemorrhage, diarrhea, and a high mortality. In addition, the appearance of newly described and emerging swine diseases such as the respiratory and reproductive syndromes (PRRS) and the skin nephropathic syndromes (PDNS)[8] enormously complicate accurate clinical diagnosis. Consequently, it has been indispensable to resort to laboratory confirmation for a clearer picture of the epidemiologic situation in a determined area, something that is not conducted within the region with the required stringency.

Another reason that the CSF virus is still not under better control in the swine population in Latin American and Caribbean countries is the sale or transportation of piglets who have survived *in utero* infection, who are immunotolerant to the CSF virus, and who do not respond effectively to vaccination. The importance of transmission mechanisms cannot be underestimated, as surviving piglets shed the CSF virus through excretion, thus contaminating the environment and transmitting the virus to susceptible animals, while they themselves are serologically negative. Vaccination of pigs under these epidemiologic conditions, without subsequent reiterative monitoring with proper identification methods will maintain the virus in a locality because it will not respond to preventive vaccines. A serious epidemiologic issue in this situation is that the only trustworthy diagnosis is the identification of the antigen, because serological tests can incorrectly catalogue the animals as *negative*.[9] In addition, clinically ill animals may be partially immune, yet they continue to eliminate the virus, thus constituting a source of infection for healthy animals.[10] In the context of varied production systems and slaughter in many parts of Latin America, weak or sick pigs are often sold to markets or middlemen, thereby exacerbating disease diffusion.

Vaccines used in several areas and countries of the region where CSF is enzootic are likely produced with an attenuated virus strain in order to counteract losses caused by the disease. Nevertheless, vaccine production and use are not highly controlled in most cases, thus making it difficult to have a clear indication of the distribution of CSF in a determined area and to count on efficient and strategic measures for its control and eradication. The administration of vaccines in areas of virus persistence or during outbreaks is commonplace, but it does little to avert infection when the virus is circulating. Elimination of carrier animals must be emphasized during any intervention measure, and vaccine should only be applied to infection-free herds.

In various regions of Latin America and the Caribbean, countries still commonly use human food waste to feed swine. This practice is especially pronounced, but not limited to, production systems that are semi-confined and backyard pigs. Many countries have banned the practice of feeding swill to pigs (or other animals), but this is difficult to implement, even though cooking the foodstuffs would inactivate the CSF virus. Ensuring that cooking or boiling the swill has been done in a correct manner is virtually impossible for regulatory officials to enforce or audit. If this food has not been sufficiently cooked and contains the remains of meat or meat products from infected pigs, the herd can be affected by the disease and it can eventually lead to an outbreak of CSF with subsequent spread to neighboring areas or markets. This form of transmission constituted 22% of the outbreaks in 1973 during the CSF eradication program in the United States[11] and is also the most likely way for it to spread to those territories and countries free of the disease.

Over the last several years, there has been an increased trend in business associations within the trade blocks inside the American continent, such as the North America Free Trade Agreement (NAFTA), Andean Nation Community (ANC), Mercado Común del Sur (MERCOSUR), and Caribbean Community (CARICOM). These trade blocks individually have established animal and animal product related sanitary regulations in order to prevent the spread of the CSF virus through the marketing of swine and pork products. Even though control is exercised, informal trade and smuggling still continue along the extensive international borders of the countries or travelers carry food products with potentially CSF virus-contaminated pig meat, which constitutes a permanent risk for the spread of CSF to those areas or countries free of the disease.

As a domesticated species, swine significantly contribute towards food security in many rural areas of the region. Not only is pig meat an important source of protein in rural family diets and central to the role of women and youth in animal husbandry throughout the vast agricultural landscape, but pigs are also frequently a means for low-income families to invest in the future or to meet immediate needs. Often, persons engaged in family-raising pigs lack the technical awareness and knowledge of good farming practices to prevent or limit disease incursions, including CSF. For instance, preventable public health problems may arise, such as trichinosis, the parasitic disease caused by *Trichinella spiralis*. This parasite causes persistent outbreaks of trichinosis among the human population and poses a negative risk for large and modern industrial investments to work with local communities. In most cases, the result is the persistence of CSF and other swine diseases in extensive areas of the region.

Cultural factors have also contributed towards the persistence of CSF. Traditionally, and based on the severe outbreaks that occurred during the last half of the last century in Latin America and the Caribbean, veterinarians, livestock technicians, and producers include CSF among the hemorrhagic diseases affecting swine. When the chronic subacute signs of CSF appear, which is currently increasingly more common in the region, complacency or ineffective veterinary systems are not integrated so that rapid diagnosis cannot be made or counter epizootic measures taken, leading to further spread and costly outbreaks of the disease. As such, official reporting to the OIE from member countries should be interpreted as the "tip of the iceberg" regarding the true incidence and distribution of disease. Strengthening of veterinary health systems and reporting are paramount to defining strategies to combat CSF and other diseases.

## ECOLOGY

The present tendency in most countries of the region is to promote swine production without contaminating the soil, air, and water. Consequently, swine exploitation ventures not abiding by these policies will be excluded from promising livestock rearing opportunities and activities. Good farming practices, as described in numerous publications, including those of FAO and the FAO/WHO *Codex Alimentarius*, will help diminish the risk of easily spreading the CSF virus and infecting other pig farms.

Wildlife are also hosts of CSF virus. The virus has been found in wild boars, and they may be an important mechanism for maintenance as is the case in central Europe,[12] and is likely to have pertinent implications in the United States if the disease were to enter. Peccaries are also susceptible to infection, yet they are not likely to develop the disease or maintain it. The natural habitats of peccaries in the region are as follows: *Tayassu tajacu* (collared peccary) is found from Arizona to Texas, throughout Mexico, Central and South America, down to the northern parts of Argentina. *T. pecari* (white-lipped peccary) is found from the southeastern part of Mexico, Central and South America, as far south as the northeastern part of Argentina.[13] Other animals such as rodents and sucking and biting insects (flies, hornets) can transmit the CSF virus under experimental conditions and as noted in some anecdotal reports. Consequently, these species are a potential risk to consider for domestic pigs because they may become infected either directly or through contaminated food. CSF control and eradication measures affecting boars and peccaries are complex. In Europe, oral immunization and hunting strategies have been applied with limited success.[14] With regard to the prevention of animal diseases, with the application of adequate biosafety measures and good farming practices in swine production systems, efficient prevention and control methods can easily be established.

Furthermore, when considering animal welfare, the violent stress during production, mixing of groups, long distance transport, and slaughter procedures for pigs destined to enter the human food chain is a matter of concern to consumers and the governments of developed countries, such as the European Union (which recently introduced legislation that the uninterrupted transportation of pigs should not exceed 8 hours). Consequently, if adapted to other regions of the world, resting of transported animals in specially assigned areas may contribute to infected pigs, or those incubating the disease, to come into contact with susceptible swine and thereby transmit infection, as aerosol transmission may occur at close distances. Ensuring that transport vehicles from one region are adequately separated from others under the same circumstances should be envisioned in applying similar policies. In Latin

America and the Caribbean, some countries have already started discussions between the producers and the consumers in order to develop and design pig welfare polices.

## ERADICATING CLASSICAL SWINE FEVER

In Latin America and the Caribbean, progress has been achieved in the control of CSF, but most countries that share borders continue to be affected by trickling outbreaks of disease due to the introduction of animals or infected pork products, thus limiting progress of local or national control programs. In addition, there is a lack of international coordination between national and subregional programs for concerted action against this limitation to healthy swine production.

In view of the foregoing, in March 2000, during the XVth Conference of the Regional Commission of the OIE, with full consensus from its delegates, the FAO-led Continental Plan for the Eradication of CSF in the Americas was created, with FAO being responsible for its Technical Secretariat.[15,16] At present, this Plan counts on the official support of the following 17 countries: Bolivia, Brazil, Chile, Colombia, Costa Rica, Cuba, Dominican Republic, Ecuador, El Salvador, Guatemala, The Bahamas, Mexico, Nicaragua, Peru, Paraguay, Uruguay, and Venezuela.

The Continental Plan is a common regional strategy used to control and eradicate the disease by facilitating the harmonization of technical, financial, and human efforts in the countries of the region. It helps to coordinate its control and eventual eradication from the endemic countries and to progressively increase the number of countries that are free of the disease.

The Plan contemplates the following three zoo-sanitary areas: (a) *Control*, covering the territory where the disease is endemic and where vaccinations cannot be suspended; (b) *Eradication*, covering territories where outbreaks no longer exist and where measures must be undertaken to accelerate the eradication made possible by the suspension of vaccination; and (c) *Free*, in places where the disease has not been present for at least 2 years and where no indication of the CSF virus has been demonstrated through serological surveillance and syndrome/disease reporting and diagnosis.

The design of the Plan was based on regional experience obtained in the control and progressive eradication of foot-and-mouth disease (an aphthovirus of the Picornaviridae family) and New World screwworm (caused by myasis of the dipteran, *Cochliomyia hominivorax*) which is also coordinated by FAO in close collaboration with national institutions and international organizations, such as International Regional Organization for Animal and Plant Health (OIRSA), OIE, Pan-American Association of Veterinary Sciences (PANVET), Pan-American Health Organization (PAHO) with its Pan-American Centre for the Control of Foot-and-Mouth Disease (PANAFTOSA) and the Pan-American Institute for the Protection of Food and Zoonoses (INPAZ), the Joint Division of FAO and the International Agency of Atomic Energy (IAEA), the Inter-American Agricultural Co-operation Institute (IICA) and the US Department of Agriculture (USDA).

The objectives of the Plan are to: (a) eradicate CSF from the American Continent by 2020; (b) reinforce national programs to eradicate CSF; (c) create an epidemiologic network of transboundary swine diseases; (d) reinforce security in international marketing of pigs and by-products; and (e) increase healthy animal production.

The main activities of the Plan in the Region are to determine the epidemiology of CSF; increase those areas officially declared disease-free; coordinate activities between national and subregional programs; promote the financial participation of the private sector through "Strategic Alliances" within the implementation of the Plan; design specific procedures to control subregional conflictive situations; and sensitize public opinion regarding the importance of eradicating CSF.

To meet its objectives, the Plan's scheme is to facilitate and coordinate the activities of the national and subregional programs. OIRSA has accepted responsibility for coordination of the Plan in Central America, and the Institute of Veterinary Medicine of Cuba is responsible for the Caribbean. Coordination in the Andean, Amazonian, and Southern Cone subregions is carried out directly by the countries and FAO.

## RECOMMENDATIONS

In view of the foregoing and to improve CSF control and eradication in Latin America and the Caribbean, the following recommendations are made:

- The role of pigs as an important part of food security in rural areas should be supported through strategic alliances with industrial producers to improve CSF control and public health for the families involved in the production of this domestic animal.
- To design successful national CSF control and eradication programs, it is indispensable, among other issues, to have an accurate census of swine populations and production practices and their distribution. It is indispensable to have accurate accounts using verified epidemiologic information systems including timely outbreak information, population affected and at risk, vaccination registries, and personnel.
- Procedures should be established to discourage feeding pigs with human food waste, yet to promote proper cooking to destroy the CSF virus.
- Multilateral national border cooperation should be increased to improve control of transportation and quarantine control of animals and by-products.
- Laboratory diagnosis for conditions that may be confused with CSF should be improved.
- Swine production with governments and organized regional swine producers should be promoted with strict respect for the environment.
- It should be emphasized that not only is eradication of CSF required to increase the export of swine products, but also that other issues such as competitive production costs, efficient zoo-technical management practices, animal welfare, and consumer safety during raising and production should be included in strategic national schemes.
- To clearly establish the epidemiologic role of wild boars and peccaries in the transmission of CSF on the American continent, it is convenient to promote

the study of wildlife in the etiology, transmission, and shedding potential of these species as well as to validate diagnostic tests in these species.

- The need to implement alternatives in the control and eradication of CSF must be recognized, considering the local conditions prevailing in the different countries.
- To reduce the risk of spreading CSF virus and to improve animal welfare conditions during transportation, it would be advantageous to promote the establishment of slaughterhouses closer to production areas, with the meat being transported in refrigerated vehicles to large consumption centers or to better improve certified export markets and promote notions of production system compartmentalization.
- Countries must give top priority to the strengthening of national programs to prevent disease introduction and to include early reaction components if a disease were to be detected
- Political commitment and financial support must be optained from the governments and swine-producing organizations of the countries of the region to participate and support the technical and logistic procedures of the Continental Plan, to have CFS eradicated in the Americas by 2020.

## REFERENCES


1. KARSTAD, L.H. 1970. Miscellaneous viral diseases: hog cholera. *In* Infectious Diseases of Wild Animals. J.W. Davis *et al.,* Eds.: 168–171. Iowa State University Press, Ames.
2. OFFICE INTERNATIONAL DES EPIZOOTIES. 2003. Handistatus II, monthly animal diseases status. OIE database <http://www.oie.int/hs2/report.asp>, May 27, 2003. Paris, France.
3. ROPPA, L. 2003. Mitos y verdades sobre la carne del cerdo. Memoirs, VI Central American and Caribbean Congress on Swine, May 2003. Guatemala, Guatemala.
4. DELGADO, C., M. ROSEGRANT, H. STEINFELD, *et al.* 1999. Livestock to 2020: the next food revolution. International Food Policy Research Institute (IFPPI), Food and Agriculture Organization of the United Nations and International Livestock Research Institute. IFPPI Food, Agriculture and Environment Discussion, paper 28. IFPPI. Washington, DC.
5. PINTO CORTES, J. 2003. Estimación del Impacto de la Peste Porcina Clásica en Sistemas Productivos Porcinos en América Latina: Estudios de Casos en tres Países Latinoamericanos. (Manuscript) FAO Regional Office for Latin America and the Caribbean, February 2003. Santiago, Chile.
6. JONES, T.C. & R.D. HUNT. 1983. Hog cholera (swine fever, swine plague). *In* Veterinary Pathology, 5th edit.: 412-420. Lea &Febiger. Philadelphia.
7. VAN OIRSCHOL, J.T. 1992. Hog cholera. *In* Diseases of Swine, 7th edit. A.D. Leman, B.E. Straw, W.L. Mengeling, *et al.*, Eds. **20:** 274–285. Wolfe Publishing, Ltd. UK.
8. MOENNIG, V., G. FLOEGEL-NIESMANN & I. GREISNER–WILKE. 2003. Clinical signs and epidemiology of classical swine fever: a review of new knowledge. Vet. J. **165:** 11–20. Elsevier Editions. UK.
9. LUBROTH, J. 1999. Epidemiología, Virulencia y Peste Porcina Clásica en las Américas. Proceedings, FAO Consultation of Classical Swine Fever Experts, 1999. FAO Regional Office for Latin America and the Caribbean. Santiago, Chile.
10. VAN OIRSCHOL, J.T. & C. TEPSTRA. 1997. Hog cholera virus. *In* Virus Infections of Porcines. M.B. Pensaert, Ed.: 113–130. Elsevier. Amsterdam.
11. USDA. 1981. Hog cholera and its eradication: a review of U.S. Experience. United States Department of Agriculture, Animal Health Inspection Service, APHIS 91-55, Washington, DC, USA.

12. FRITZEMEIER J., I. GREISER-WILKER, K. DEPNER & V. MOENNING. 1998. Characterization of CSF virus isolates originating from German wild boar. *In* Report on Measures to Control Classical Swine Fever in European Wild Boar, Perugia, Italy. Commission of the European Communities, Document VI/7196/98-AL.: 107–109.
13. FOWLER, M.E. 1996. Husbandry and diseases of captive wild swine and peccaries. Scientific and Technical Review, OIE, vol.15 , No. 1. Paris, France.
14. RÜMENAPF, T., R. STARK, G. MEYERS & H.J. THIEL. 1991. Structural proteins of hog cholera virus expressed by vaccinia virus: further characterization and introduction of protective immunity. J. Virol. **65:** 589–597.
15. OFICINA INTERNACIONAL DE EPIZOOTIAS. 2000. XV Conference of the Regional Commission of the OIE for the Americas. Final Report, 73 pp. Paris, France.
16. FAO. 2000. Plan Continental para la Erradicación de la Peste Porcina Clásica de las Américas. FAO Regional Office for Latin America and the Caribbean. Document 23 pp. <http://www.rlc.fao.org/prior/segalim/animal/ppc/default.htm> Santiago, Chile.

# Foot-and-Mouth Disease in Tropical Wildlife

ARAMIS AUGUSTO PINTO

*Faculdade de Ciências Agrárias e Veterinárias, Departamento de Patologia Veterinária, Universidade Estadual Paulista–UNESP, 14884-900 Jaboticabal, SP, Brazil*

**ABSTRACT: This review of foot-and-mouth disease in cloven-hoofed, free-living animals, describes the disease, the wide range of the hosts, the carrier state, and the interrelationship between disease in domestic livestock and wildlife. This information becomes even more crucial to the development of control strategies when linked to the process of pathogenesis and the epidemiology of the disease.**

**KEYWORDS: foot-and-mouth disease; tropical wildlife; epidemiology; trade in animals**

## INTRODUCTION

Worldwide, foot-and-mouth disease (FMD) is considered one of the most serious viral diseases of cloven-hoofed (Artiodactyla) animals still prevalent in the world. It is extremely contagious and spreads rapidly. Although eradicated and controlled by several countries, FMD can cause enormous economic losses owing to its effect on the livestock industry and wildlife community. Trade of livestock and animal products is blocked, and a wide range of agricultural products is banned from export to other countries. North and Central America, Chile, Australia, New Zealand, and some Asian countries are free of the disease. FMD is present in two thirds of member countries of the Office International of Epizooties (OIE). It is still a major problem in the Middle East, part of Southeast Asia, Africa, and South America, having FMD-free countries where vaccination is not practiced and free countries with and without vaccination. In many of these countries, outbreaks in free areas occur sporadically. Consequently, FMD global distribution is constantly changing. It can occur anywhere, so that countries where the disease does not occur are not completely free and safe from an FMD epidemic at any time, similar to the ones in Uruguay, South Korea, Japan, the Netherlands, and the United Kingdom in 2001. Therefore, its appearance is completely unpredictable. In countries free from FMD normally, vaccination is forbidden, and in the event of an outbreak, all domestic and nondomestic animals, diseased and in-contact susceptible stock on the affected area, and also dangerous contacts must be destroyed immediately ("stamping-out" policy) to eliminate the principal source of the virus and reduce the risk of contagion. This practice, to-

Address for correspondence: Aramis Augusto Pinto, Faculdade de Ciências Agrárias e Veterinárias, Departamento de Patologia Veterinária, Universidade Estadual Paulista–UNESP, 14884-900 Jaboticabal, SP, Brazil. Voice: +55-16-32092662; fax: +55-16-32024275.
aramisap@fcav.unesp.br

**Ann. N.Y. Acad. Sci. 1026: 65–72 (2004). © 2004 New York Academy of Sciences.
doi: 10.1196/annals.1307.008**

gether with other essential measures, such as the restriction of transport, are taken to avoid the spread of the virus from one place to another mechanically or by wind, as soon as the presence of disease is confirmed. The major disadvantages of the stamping-out policy are the cost, the logistics, and the permanent loss of genetic potential, mainly of wild populations of rare species in their own ecological environment. Another negative feature is the dislike by conservationists and local communities of the brutal massive animal sacrifices and the possibility of danger from airborne emissions of the carcinogenic dioxin from burning dead animals covered with chlorine-bearing disinfectants or other chemicals.

## FOOT-AND-MOUTH DISEASE

Foot-and-mouth disease is a viral disease of domestic livestock—cattle, sheep, pigs, and goats, but it can eventually affect a wide range of cloven-hoofed wildlife species, causing severe disease. The clinical signs and lesions of FMD in nondomestic animals are very similar to those commonly observed in domestic livestock. The disease, which is extremely contagious, is characterized by fever, intense salivation, formation of vesicles with subsequent erosion on the mucous membrane of the mouth, especially the tongue, occasionally the nose, in suids, and on the interdigital space and surrounding the hooves or teats. The severity of the disease varies between different animal species. In some cases, the disease runs a severe course and the animal can die. In sheep, goats, buffalo as well as some wild animal species, the disease is relatively mild, and trivial interdigital and dental pad lesions have been described. In *Cervidae*, the severity of the disease can vary from mild or inapparent in some species to more severe in others. For example, the disease is mild in red and fallow deer but more severe and sometimes fatal in western roe and muntjac deer. In the kudu and impala, the disease is clinically severe and spreads rapidly. The severity of the disease also depends on the dose and pathogenicity of the virus strain and the suitability and health of the host. Some strains of FMD virus (FMDV) have shown preferential infectivity for a particular species such as cattle or pigs but not for water buffalo (*Bubalus bubalis*), African buffalo, and deer, and vice versa. Strains that affect water buffalo may not affect cattle with regularity, and the reverse is also true. A mild strain in cattle or sheep can infect pigs severely.

## WILDLIFE AS CARRIERS

Cattle, African buffalo, and water buffalo, but not suids, that have been infected with FMD can become carriers of the virus for several years. These hosts may excrete the virus after exposure and therefore can be the source of possible transmission of the disease to other susceptible animals. However, the likelihood that carriers transmit the infection under field conditions is rare. Particularly in semi-immune populations, animals have been recorded to "carry" FMDV without the appearance of lesions of the disease after recovery from undetected subclinical infection or on exposure of vaccinated animals to virus. Indeed, when considering the impact of disease on natural population dynamics, contact between carrier animals and unexposed individuals may be most important. Carrier animals can be identified by isolation of the virus from esophageal/pharyngeal (E/P) specimens, by the polymerase chain reaction (PCR), and by testing their sera for the presence of specific antibodies to FMDV types and/or antibody against nonstructural viral proteins. Consequently, it is very important in controlling livestock movement to be able to accurately detect the presence of specific antibodies in these animals and to distinguish them from those animals that have been vaccinated. Such identification has been carried out by screening for antibodies to virus-infection associated (VIAA) antigen and other bioengineered nonstructural viral proteins (2B, 2C, 3AB, and 3ABC), combined with assays to detect the agent itself, such as PCR or virus isolation. Isolation of the FMDVs from E/P fluid is the most specific method for identifying carrier animals, but it is an insensitive, inconsistent, and costly procedure.

Duration of the carrier state for long periods in the absence of observable clinical signs or lameness varies among species. For example, African buffalo may harbor the virus for a very long period (at least 5 years); water buffalo and cattle, for at least 3 years; and sheep and goats, for shorter periods (up to 9 months). Neither elephants nor suids (domestic and nondomestic wild pigs) have been shown to carry virus after infection, and therefore evidence of antibody against "virus-infection associated" (VIAA) antigen, a nonstructural viral protein, in these species would be especially valuable. In kudu, virus was detected for up to 136 days, and antibody to VIAA antigen was detected at least 494 days after experimental infection. Thus, it is likely that FMDV can persist as a carrier in kudu for longer periods, because high serum neutralization (SN) antibody titers persist as does antibody to VIAA antigen for a long period after experimental infection. Impala did not become virus carriers after infection, and their SN antibody titer is normally lower than that of other species, and antibody to VIAA antigen has been detected up to 136 days after experimental infection. Furthermore, warthog and bushpig also do not carry virus, and antibody to VIAA antigen was only detected up to 45 days after experimental infection with type SAT1 FMDV and serum neutralization titers being of short duration.

## FOOT-AND-MOUTH DISEASE IN WILDLIFE

As demonstrated by intensive surveys, nearly 60–70 species of wild animals were reported to be susceptible to either natural or experimental infection with FMD, and on several occasions FMDV was isolated from some in the natural state. Serological studies have shown significant SN antibody titers to one or more FMDV types, indicating that earlier infection probably occurs in a high proportion of the unvaccinated wild species. Therefore, it must be relevant to consider the susceptibility of a representative number of species of nondomestic animals that can readily be infected, particularly those that are occasionally in contact with susceptible domestic livestock. Many of theses species, as listed in TABLE 1, have been reported as affected and also implicated in the spread of the disease to farm livestock. In reviewing the literature, documented examples of the natural spread of FMD in wild animals have been recorded in several countries in several conditions and places. It may be of some particular interest to describe some specific accounts of FMD in wildlife. Since 1892 and before, particularly in several parts of South Africa, FMD infection in game was reported. In one case the disease was reported in cattle and sheep, but it was not serious, whereas two thirds of the antelope population died of the effects

**TABLE 1. Wild animal species naturally or experimentally susceptible to foot-and-mouth disease virus (FMDV)**

| Common name | Scientific name | Type of infection |
|---|---|---|
| Agouti-(Brazilian agouti) | *Dasyprocta leporina*<br>*Dasyprocta agouti* | Experimental |
| Armadillo (hair armadillo) | *Chaetophractus villosus* | Experimental |
| Babirusa | *Babyrousa babyrussa* | Natural/experimental |
| Bear | *Ursus horribilis* | Natural |
| Bear (Asiatic black bear) | *Ursus thibetanus* | Natural |
| Bear (Brown bear) | *Ursus arctos* | Natural |
| Bison | *Bos americanus* | Natural |
| Brown brocket | *Mazama gouzoubira* | Natural |
| Buffalo (African buffalo) | *Syncerus caffer* | Natural/experimental |
| Buffalo (Indian water buffalo) | *Bubalus bubalis* | Natural/experimental |
| Bushbuck | *Tragelaphus scriptus* | Natural |
| Bushpig | *Potamochoerus porcus* | Natural/experimental |
| Capybara | *Hydrochoerus hydrochaeris* | Experimental |
| Chamois (Alpine) | *Rupicapra rupicapra* | Natural/experimental |
| Collard peccary | *Tayassu tajacu* | Natural |
| Columbian deer | *Odocoileus columbicus* | Natural |
| Dromedary (Arabian camels) | *Camelus dromedarius* | Experimental |
| Duiker | *Sylvicapra grimmia* | Natural |
| Eland | *Taurotragus orys* | Natural/experimental |
| Elephant (African elephant) | *Loxodonta africana* | Natural/experimental |
| Elephant (Asian elephant) | *Elephas maximus* | Natural |
| Elk | *Alces machlis* | Natural/experimental |
| Fallow deer | *Dama dama* | Natural/experimental |
| Gaur | *Bos frontalis (Bos gaurus)* | Natural |
| Gayal (Mithun) | *Bos frontalis domesticus* | Natural |
| Gemsbok, Oryx | *Oryx oryx gazella* | Natural |
| Giraffe | *Giraffa camelopardalis* | Natural |
| Gnu (Blue Wildebeest) | *Connochaetes taurinus* | Natural/experimental |
| Grant gazella | *Gazella granti* | Natural |
| Guib | *Tragelaphus striptus* | Natural |
| Hedgehog (West European) | *Erinaceus europaeus* | Natural/experimental |
| Hedgehog (East African) | *Atelerix prurei hindu* | Experimental |
| Hippopotamus | *Hippopotamus amphibuis* | Experimental |
| Ibex | *Capra ibex* | Natural |

**TABLE 1.** **(*continued*) Wild animal species naturally or experimentally susceptible to foot-and-mouth disease virus (FMDV)**

| Common name | Scientific name | Type of infection |
|---|---|---|
| Impala | *Aepyceros melampus* | Natural/experimental |
| Kudu | *Tragelaphus strepsiceros* | Natural/experimental |
| Llama | *Lama guanicoeglama* | Experimental |
| Llama | *Lama glama* | Natural/experimental |
| Llama (Alpaca) | *Lama pacos* | Natural |
| Marsh deer | *Blastocerus dichotomus* | Experimental |
| Moose | *Alces alces* | Natural/experimental |
| Mountain gazelle | *Gazella gazella* | Natural |
| Mule deer | *Odocoileus hemionus* | Natural |
| Nilgai (antelope) | *Boselaphus tragocamelus* | Natural |
| Nyala | *Tragelaphus angasii* | Natural |
| Porcupine | *Hystrix galeata* | Experimental |
| Red brocket | *Mazama americana* | Experimental |
| Red deer | *Cervus elaphus* | Natural/experimental |
| Reedbuck | *Redunca arundinum* | Natural/experimental |
| Reindeer | *Rangifer tarandus* | Natural/experimental |
| Roe deer (western roe deer) | *Capreolus capreolus* | Natural/experimental |
| Sable antelope | *Hippotragus niger* | Natural |
| Sable antelope | *Ozanna grandicornis* | Natural |
| Saiga antelope | *Saiga tatarica* | Natural |
| Sambar deer | *Cervus unicolor* | Natural |
| Sika deer | *Cervus nippon* | Experimental |
| Southern pudu | *Pudu pudu* | Natural |
| Spotted deer | *Axis axis* | Natural |
| Tapir (Brazilian tapir) | *Tapirus terrestris* | Natural |
| Tapir (Malayan tapir) | *Tapirus indicus* | Natural |
| Thamin | *Cervus eldii* | Natural |
| Tsessebi | *Damaliscus lunatus* | Natural |
| Vicuna | *Vicugna vicugna* | Natural/experimental |
| Warthog | *Phacochoerus aethiopicus* | Natural/experimental |
| White-lipped peccary | *Tayassu pecari* | Natural |
| White-tailed deer | *Oedocoilleus virginianus* | Natural/experimental |
| Wildboar (*Sus scrofa*) | *European Sus scrofa* | Natural/experimental |
| Waterbuck | *Kobus ellipsiprymnus* | Natural |
| Yak | *Bos grunniens domesticus* | Natural |

of the disease (Annual Report of the colonial Veterinary Officer at Cape Town to the Colony of Cape of Good Hope). During the last outbreak of the disease in California, USA, from 1926–1929, 10% of mule deer were found to have FMD lesions. In Scandinavia, reindeer were affected with FMD when an outbreak occurred in 1927. Infection with FMDV was found in wildebeest, kudu, and wild pigs in Southern Rhodesia in 1937. The occurrence of FMD in wildlife was reported in an outbreak at the Paris Zoo in 1938 when cattle, gaur, African buffalo, gayal from India, bison, boars, warthogs, and tapirs were infected. At the zoo in Bern, Switzerland in 1940, bison and ibex were infected, but the disease did not spread to the deer, chamois, and wild boar population in adjacent pens. In the Zurich zoo in 1950, 28 mammals either died or had to be sacrified because of FMD. Three outbreaks of FMD occurred in the Buenos Aires zoo in Argentina in 1942, 1948, and 1955. In these outbreaks, it was reported that cattle, buffalo, suids, antelope, tapirs, deer, and bear were affected with FMDV. In Britain in 1946, during a series of outbreaks, European hedgehogs and deer with well-developed FMD lesions were found near infected premises. In 2001, during the outbreak of FMD in domestic animals, in Great Britain and the Netherlands, free-living and farmed deer were culled as part of a control program to eradicate the disease. The hedgehog was also found to have serious and fatal FMD lesions during the last outbreak of FMD in the UK. In the USSR, on several occasions, FMD infection spread from cattle was reported to infect vagrant Saiga antelope, which in turn transferred the disease back to cattle and to other animal species in places far from the original outbreak. A severe outbreak of FMD in a population of Mountain gazelles in Israel killed 50% of the population in the affected area. Once infection is established in a wild vagrant animal population, it could spread to other susceptible animal species. To reduce the chances of disease transmission, fenced areas designed for livestock production were erected in African countries to prevent direct contact between domestic stock and wild animals, particularly African buffalo, impala, and kudu. However, in certain cases it is difficult or impossible to provide adequate separation between feral wild animals and other livestock.

## INTERRELATION BETWEEN DISEASE IN DOMESTIC LIVESTOCK AND WILDLIFE

Literature on the interrelation between FMD in domestic livestock and in wild animals is still insufficient, particularly their role in the initiation and spread of the disease in domestic and indigenous fauna, and vice versa, under field conditions. In many cases, the potential role of some wildlife species in the epizootiology of FMD may no longer be relevant. On the basis of circunstantial evidence, several wildlife species have been implicated in FMDV transmission between, and among, wildlife and livestock. Free-living, wild or feral, African buffalo, impala, several deer species, Saiga antelope, and eland are the most frequently infected animals with FMDV under field conditions and are considered of particular importance in the epizootiology of the disease. Some have been recognized as a major source of infection among livestock. FMDV is known to circulate through free-ranging African buffalo populations, sometimes transmitting the disease from buffalo to other wild animal and to cattle. In this aspect, it had been suggested that transmission of the virus from carrier to susceptible animal may occur, but rarely. Thus, the lack of evidence of infection in other susceptible species in close contact with buffalo or cattle harboring FMDV suggests that initiation of infection in these species may require a high virus challenge as presented by clinically infected animals. FMD also spreads among tame water buffalo, with evidence of spread to in-contact cattle, pigs, and goats. Transmission of FMDV from infected cattle to susceptible water buffalo also occurs in the same way as from buffalo to buffalo. The possibility of transmission between species, for example, African buffalo to antelope, deer species, bushpig, and/or cattle, as occurs in African countries, and deer, feral pigs, tapir, capybara, water buffalo, and cow in the feral state, and other game in some South American countries, may be increased when they share grazing. In fact, it has been reported that wild animals straying into farms and coming into close contact with farm livestock, particularly when they share their grazing, as well cattle straying into Save Wildlife Conservancy and coming into contact with wildlife species, are factors that predispose to a high risk of FMD transmission. However, what has been said of wild species and is valid in African countries may not be of great significance in other regions, where the local wildlife species as a possible repository for virus between outbreaks in farm livestock, is not the same or does not exist. In South America, for example, free-living African buffalo, the principal source of FMDV in domestic and wild animals under field conditions, do not exist as in African countries. On the other hand, the systematic, repeated, and controlled vaccination program against FMD in domestic animals in conjunction with effective application of zoo-sanitary measures prevents the occurrence of the disease among the wildlife population. Therefore, the potential for a disease occurrence among the indigenous population is reduced, and obviously the wild animal population only will be infected if livestock was previously infected. Despite the fact that several South American wild animal species are suceptible to FMDV, no history of previous disease has been available or reported, under field conditions, in wildlife populations in South American countries.

## REFERENCES


1. Anderson, E.C., W.J. Doughty, J. Anderson & R. Paling. 1979. The pathogenesis of foot-and-mouth disease in the African buffalo (*Syncerus caffer*) and the role of this species in the epidemiology of the disease in Kenya. J. Comp. Pathol. **89:** 541–549.
2. Assistant Director, Wellcome Institute for Research and Foot-and-Mouth Disease, Veterinary Department, Kenya. 1963. Foot-and-mouth disease in non-domestic animals. Bull. Epiz. Dis. Afr. **II:** 143–146.
3. Condy, J.B., R.S. Hedger, C. Hamblin & I.T.R. Barnett. 1985. The duration of the foot-and-mouth disease virus carrier state in African buffalo (i) in the individual animal and (ii) in free-living herd. Comp. Immunol. Microbiol. Infect. Dis. **8:** 259–265.
4. CD-ROM Module. 2001. Foot-andMouth Disease. WildPro.Website. June 30.
5. FMD Wildlife, Zoos, Pets and Other Animals (Foot-and-Mouth Disease Control). Health & Management/Foot&Mouth Disease Module/List Hyperlinked Techniques & Protocols. <http://www.wildlife information.org>. Accessed September 2001.
6. Gomes, I., A.K. Ramalho & P. Augé de Mello. 1997. Infectivity assays of foot-and-mouth disease virus: contact transmission between cattle and buffalo (*Bubalus bubalis*) in the early stages of infection. Vet. Rec. **11:** 43–47.
7. Hedger, R.S., J.B. Condy & S.M. Golding. 1972. Infection of some species of African wild life with foot-and-mouth disease virus. J. Comp. Pathol. **82:** 455–461.

8. HEDGER, R.S. 1976. Foot-and-mouth disease in wildlife with particular reference to the African buffalo (*Syncerus caffer*). *In* Wildlife Diseases. L. Andrew Page, Ed.: 235–244. Plenum Publishing Corp. New York.
9. PINTO, A.A. & R.S. HEDGER. 1978. The detection of antibody to virus-infection associated (VIA) antigen in various species of African wildlife following natural and experimental infection with foot and mouth disease Virus. Arch. Virol. **57:** 307–314.
10. SUTMÖLER, P. 2001. Strategic foot-and-mouth disease in deer (*Cervidae*): its implications for policy of control of the disease in the Netherlands. <http://www.wildlife information.org>. Accessed September 2001.

# Foot-and-Mouth Disease in the Americas

## Epidemiology and Ecologic Changes Affecting Distribution

VICTOR SARAIVA

*Foot-and-Mouth Disease Centre/PAHO/WHO, Rio de Janeiro, RJ, Brazil*

**ABSTRACT: Foot-and-mouth disease(FMD) was first recorded in South America (SA) circa 1870, in Buenos Aires, Argentina, in Uruguay, and in southern Brazil as a result of the introduction of cattle from Europe during the early days of colonization. Livestock production to trade with neighboring countries was established in the La Plata Region, and the trade of livestock and products with Chile, northeastern and central western states of Brazil, to Peru, Bolivia, and Paraguay spread FMD, which reached Venezuela and Colombia in the 1950s and finally Ecuador in 1961. The traditional forms of livestock husbandry influence the diffusion and maintenance of the FMD virus (FMDV) in different areas. Cattle production in SA depends mainly on a strong relation between cattle-calf operations and fattening operations in a complementary cycle, revealing the vulnerability and susceptibility of these areas to FMDV. Understanding the relationship between time-space behavior of the disease and the forms of production defines the FMD ecosystems, a key concept to elaborating the control/eradication strategies of national FMD eradication programs, which must be modified when trade opportunities between zones of differing sanitary status change. The role of other susceptible species besides bovines, including wildlife, in maintaining and spreading FMDV has been the subject of several studies, but in SA, bovines are so far considered to determine disease presentation. Buffalo (*Bubalus bubalis*) have been implicated in the spread of the disease between farms in at least one case in Brazil. Sheep are almost on a par with bovine in terms of number, especially in the Southern Cone, but their role in the maintenance of infection is not considered important, possibly owing to rearing practices. Camelid populations in the Andean region do not play an important role in the maintenance of FMD, because of short persistence of infection and low population densities in these species. The importance of wildlife is not clear, but it is accepted that animals are mostly affected as a spinoff during outbreaks in domestic species. Experimentally infected capybaras (*Hydrochoerus hydrochoeris hydrochoeris*) showed clinical signs and infected other susceptible species, but their role in the maintenance of infection in nature is so far not clear.**

**KEYWORDS: foot-and-mouth disease; epidemiology; distribution; ecologic changes**

Address for correspondence: Dr. Victor Saraiva, Foot-and-Mouth Disease Centre/PAHO/WHO, Rio de Janeiro, RJ, Brazil. voice: 5521-3661-9022; fax: 5521-3661-9001.
vsaraiva@panaftosa.ops-oms.org; victor saraiva@bol.com.br; vsaraiva50@yahoo.com

**Ann. N.Y. Acad. Sci. 1026: 73–78 (2004). © 2004 New York Academy of Sciences.**
**doi: 10.1196/annals.1307.009**

## INTRODUCTION

The introduction of cattle into South America dates back to the very early days of colonization when trade of livestock and livestock products was a monopoly of Spain and Portugal. South American countries under their rule were protected from the disease because the Iberian Peninsula was not affected by foot-and-mouth disease (FMD) until the late nineteenth century. Livestock production for trade with neighboring countries was first established in the great plains of Argentina, southern Brazil, and Uruguay, using the abundant local cattle of European origin for the production of tallow, leather, and salted meats.

However, development of industrial cold storage brought time and efficiency factors of food conversion to the fore, prompting the importation of cattle with better genetics to improve productivity. It is thus accepted that the virus reached South America circa 1870, being recorded almost at the same time in Buenos Aires, Argentina, Uruguay, and southern Brazil. Its epidemiology was unknown, and no actions were then taken to control the disease allowing the virus to spread to Chile as well as to the northeastern and central western states of Brazil, to Peru, Bolivia, and Paraguay, being registered in Venezuela and Colombia in the 1950s and in Ecuador in 1961. North American countries (United States in 1921; Mexico in 1947 and 1954; and Canada in 1952) conducted successful eradication campaigns, but the countries of South America did not act effectively until the early 1960s to prevent the entry and spread of FMD.

The closing of the North American markets to products of animal origin after World War II and the aforementioned introduction of FMD into Venezuela and Colombia in the early 1950s may be considered the main reasons behind the decision to wage an organized, continent-wide fight on the disease that included the creation of PANAFTOSA in Brazil in 1951.

## EARLY NATIONAL FMD CONTROL PROGRAMS, ECOSYSTEMS, AND THE HEMISPHERIC FMD ERADICATION PLAN

The first plans for controlling the disease in South America were seen in the early 1960s in Argentina and the neighboring State of Rio Grande do Sul, in Brazil. Nationwide programs supported by international loans started by the mid-1970s and became the basis for many national animal health services in operation today. The FMD control method used by the national programs relied on a 3 cycles/year vaccination with aluminum hydroxide vaccines of the entire cattle herd of the country. This was supported by a transit control exerted without risk analysis and an outbreak control process that was not always effective. This approach did not rely on the epidemiologic studies about the ecosystems mentioned earlier, which set the basis for a regionalized approach. The ecologic changes affecting the distribution of the disease, mentioned in the title, are related to the productive system in place.

The traditional forms of raising, fattening, and processing livestock influence the diffusion and maintenance of infection in different areas of South America. There is a strong relation between range farming areas and fattening areas in a complementary cycle of production. Range farming areas, characterized by low rates of produc-

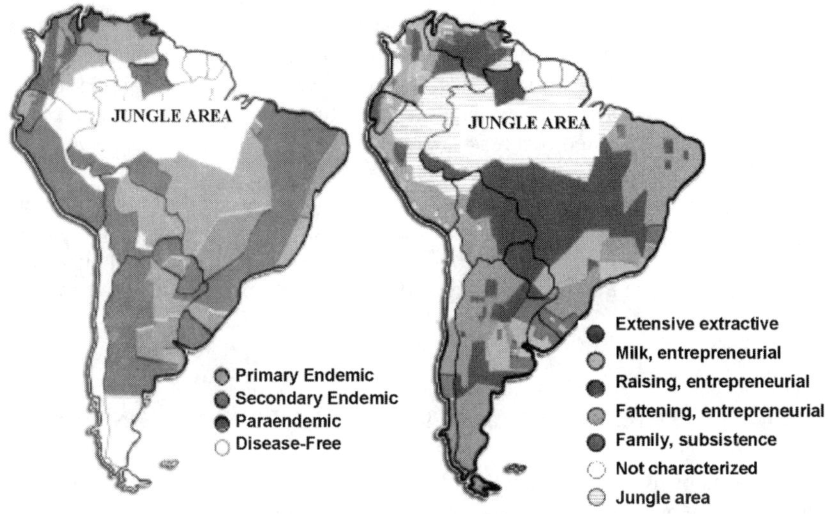


**FIGURE 1.** Forms of production and FMD ecosystems (1990).

tivity but high outputs, produce steers for fattening in areas with better productive infrastructure, closer to the centers of processing and consumption.

The seasonal flow of animals is responsible for a concentration of susceptible and infected animals in the fattening areas. The vulnerability and susceptibility of these areas, due to their peculiar livestock structure and levels of immunity of animal populations, might eventually spur epidemic FMD outbreaks, commonly seen in the early stages of the programs (FIG. 1). The development of knowledge about the time-space behavior of the disease, based on the traditional trade relationship, known in South America as FMD ecosystems, was basic to elaborate control/eradication strategies of the national programs.

The Hemispheric Plan for the Eradication of Foot-and-Mouth Disease (PHEFA) was developed late in the 1980s, with three main objectives: create and maintain FMD-free areas; increase availability of meat and milk by increasing efficiency of livestock operations; and improve the access of animals and products from originally affected countries to international markets.

Foot-and-mouth-disease eradication programs had their activities reoriented with a regional focus, and national animal health services were strengthened in nearly all the South American countries. Experiences in joint administration of the FMD programs between the government and the livestock sector were developed in some countries, usually to manage the vaccination cycles. This was effective in increasing vaccination coverage, resulting in better immunization due to closer control. Vaccination, outbreak control, and restricted animal movement resulted in a decrease in the number of FMD outbreaks from an average of 1,200 outbreaks in 1990 to about 130 in 1999.

The strategy of the Plan still relies today on regionalization of the countries with a special focus on the forms of production mentioned earlier. Different levels of risk require specific disease control/eradication strategies based on twice-a-year cyclic vaccination and emergency vaccinations, when needed, with oil-adjuvanted vaccines. Other species such as sheep, goats, and pigs usually are not vaccinated except when under serious risk. This policy proved helpful when seroepidemiologic surveys were carried out, as these species acted like negative controls. Strict animal movement control and expedited outbreak attention based on a surveillance system were also enforced.

To preserve activities from interruptions due to political interference, social and inter-sector participation was stressed in the administration of the campaigns as well as differentiated attention to small livestock owners, usually resistant to animal health programs. The Plan was created in 1987 and is expected to fulfill its goals by 2009.

## ROLE OF SPECIES IN THE MAINTENANCE AND DIFFUSION OF FMD

The forms of livestock production described previously are considered determinants of FMD endemism and causative of epidemic outburst in highly concentrated areas, as seen in the past. The participation of susceptible species in the maintenance and diffusion of FMDV in South America has been the subject of several studies, and bovines are accepted as the species mainly responsible. Even when the ovine population was au pair with the bovine one, specially in the Southern Cone of South America, its role in the maintenance of infection was not regarded as important, and the species were not compulsorily vaccinated. The importance of the small species in the maintenance of infection through longer periods has yet to be evaluated.

The importance of the camelid populations in the spread of the disease in the Andean region is also limited to small ranges, and persistence studies showed that infection is short-lived in camelids. The importance of wildlife in the maintenance of infection is yet undetermined. Nevertheless, it is known that wild species can be affected during outbreaks in South America, as a spinoff. There is, though, one case in which buffalo (*Bubalus bubalis*) seemed implicated in the spread of the disease in Brazil, when the introduction of this species to a controlled farm brought in FMDV A, which eventually spread to another farm in SA.

## FUTURE FOCUS

We have seen that eradication of FMD is indeed attainable in the countries of South America. Maintaining a disease-free status without vaccination for some countries while others in the region remain infected became a special challenge to the governments and veterinary authorities in the region.

Nevertheless, the recent resurgence of FMD in the Southern Cone of South America might reflect three situations: the urgency of the process adopted by some of the countries in the liberation of zones, which did not allow for the consolidation of the achievements in disease control, before advancing to the eradication step in the process. The second could be the existence of unnoticed "niches of infection," which, added to the end of vaccination, allowed diffusion of the agent within the herds. Another possibility is that the existing trade relations at border areas between countries with different sanitary status might have brought the agent to previously clear areas.

It may be argued that it resulted from the failure of authorities to recognize that continuing high levels of prevention and surveillance are cornerstones of the effort to maintain FMD-free status. Political and economic pressures to achieve disease-free status rapidly are constant and intense, but support and investment wither when FMD is declared eradicated in a country, and the government usually shifts resources to support other high-priority needs. The veterinary infrastructure collapses and the cooperative effort with the private sector ends when vaccination is suspended. There is little interest in maintaining FMD awareness in the minds of the producers, trade people, and the public.

The Plan of Action of PHEFA for 2003–2009 is being restructured and focuses on prevention, strengthening community participation, and concentrating efforts on border zones between the countries. In this aspect, a study carried out by PANAFTOSA and countries in South America described international zones where the agent can cross by means of transit of animals and products.

National programs must have their strategy reoriented under a regional focus based on the use of risk analysis method. Vaccination should be homogeneous in coverage and in timing. Primary prevention activities, such as surveillance for high-risk operations, should be included in the regional program. Border areas should be a priority for animal health programs with well-defined territorial coverage, joint programming, execution, and evaluation of control/eradication activities. This would create an "epidemiologic border" to protect against reintroduction of the agent in areas considered to be free.

Another area of interest is the development of new experiences in joint administration of programs between the public and private sectors. Strategic alliances should be developed with other sectors of the economy interested in the improvement of the sanitary situation of the livestock industry and its impact on international trade, bearing in mind cultural and economic idiosyncrasies.

The small producer in some areas became an important risk factor for the completion of the PHEFA objectives and must be brought into the program by the inclusion of FMD-related activities in packages of veterinary attention that takes into account the sanitary problems that cause the major impact on their production.

It is extremely unlikely that any country or zone in South America achieving free status will be able to "isolate" itself and thereby insulate its national herd from the effects of an FMD-infected neighbor or trading partner. FMD in South America is not any one country's challenge. It is a transboundary disease and demands that countries avoid focusing only on national programs and instead develop and implement *regional* eradication strategies.

## REFERENCES

1. Goic, R. 1971. Historia de la fiebre aftosa en América del Sur. *In* Primer Seminario sobre Sanidad Animal y Fiebre aftosa. Panamá, June 16–20. :32-39.
2. Astudillo, V.M. 1992. La fiebre aftosa en América del Sur. Hora Vet. **70:** 16–22.
3. Machado A. 1969. Aftosa. 182 pp. State University of New York Press. Albany, New York.
4. Rosenberg, F.J. & R. Goic. 1973. Programas de control y prevención de la fiebre aftosa en las Américas. Bol. Cent. Panam. Fiebre Aftosa **12:** 1–22.

5. ASTUDILLO, V. *et al.* 1985. Caracterización de los ecosistemas de la fiebre aftosa. *In* Programa de Adiestramiento de Profesionales Lationamericanos en Salud Animal – PROASA, Centro Panamericano de Fiebre Aftosa, 1985
6. CORREA MELO, E. & A. LÓPEZ. 2002. Control of foot and mouth disease: the experience of the Americas. Review Scientific et Technique, Office International des Epizooties, 21 (2).
7. ORGANIZACIÓN PANAMERICANA DE LA SALUD. 1988. Plan de Acción: Programa Hemisférico de Erradicación de la Fiebre Aftosa en América del Sur. Reunión del Comité Hemisférico de Erradicacón de la Fiebre Aftosa. Washington, June 6-7, Acta, 32 pp.
8. Situación de los Programas de Erradicación de la Fiebre Aftosa en América del Sur 2002. Centro Panamericano de Fiebre Aftosa, Rio de Janeiro, Brasil, 42 pp.
9. CORREA MELO, E. & V. SARAIVA. 2003. How to promote joint participation of the public and private sectors in the organisation of animal health programmes. Rev. Sci. Tech. Off. Int. Epiz. **22**(2): 517–522.
10. SARAIVA-VIEIRA, V. 2003. Regionalisation as an instrument for preventing the propagation of animal diseases, including those of camelids. *In* 71st General Session, OIE, May 18-23, 2003
11. LUBROTH, J., R. YEDLOUTSCHNIG, V. CULHANE & E. MIKICIUK. 1990. Foot and mouth disease virus in the llama (*Lama glama)*: diagnosis, transmission and susceptibility. J. Vet. Diagn. Invest. **2:** 197–203.
12. GOMES, I. & P. AUGÉ DE MELLO. 1992. Un episodio subclinico de fiebre aftosa en bovinos probablemente causado por la introducción de bufalos (*Bubalus bubalis*). [An episode of subclinical FMD in cattle probably caused by the introduction of water buffalo (*Bubalus bubalis*).] Seminario Internacional sobre Planes Locales y Zonales con Movilizacion de Recursos y Participación Communitaris para la Erradicación de la Fiebre Aftosa. Buenos Aires, Argentina, 30 de marzo al 1 de abril de 1992. Centro Panamericano de Fiebre Aftosa.

# Suspected Cases of *Neorickettsia*-Like Organisms in Brazilian Dogs

SELWYN A. HEADLEY,[a] ODILON VIDOTTO,[b] DIANA SCORPIO,[c]
J. STEPHEN DUMLER,[d] AND JOSEPH MANKOWSKI[c,d]

[a]*Laboratório de Patologia Veterinária, Hospital Veterinário, Centro Universitário de Maringá, Maringá, PR, Brazil*

[b]*Universidade Estadual de Londrina, Londrina, PR, Brazil*

[c]*Department of Comparative Pathology,* [d] *Department of Pathology, The Johns Hopkins University, Baltimore, Maryland, USA*

ABSTRACT: **Preliminary findings of gross and histopathological lesions consistent with salmon poisoning disease in 10 dogs from southern Brazil are described. Lesions were restricted to the spleen, lymph nodes, and intestinal lymphoid tissues. Grossly, there was marked hyperplasia of mesenteric lymph nodes and Peyer's patches. Microscopic alterations were characterized by diffuse hyperplasia of intestinal lymphoid tissues and Peyer's patches. Intracytoplasmic organisms consistent with *Neorickettsia helminthoeca* were demonstrated by Giemsa stain in reticuloendothelial cells of the intestine, spleen, Peyer's patches, and lymph nodes. We have named this organism *Neorickettsia helminthoeca*-like because of its marked similarity with the agent described in the United States.**

KEYWORDS: ***Neorickettsia helminthoeca*; pathology; parasitology; rickettsia; dogs**

## INTRODUCTION

Salmon poisoning disease (SPD) is a granulomatous enterocolitis of dogs and foxes caused by *Neorickettsia helminthoeca,* a coccoid or cocobacillary rickettsia that occurs in specific geographical locations of the United States.[1,2] The apparent limitation of SPD to the United States has been directly related to the restricted presence of the intermediate host, *Oxytrema silicula*, a pleurocerid snail.[3] However, cases of SPD have been described elsewhere.[4]

Animals infected by SPD are severely emaciated, anorexic, depressed with enlarged lymph nodes, and normally die within 14 days after infection.[3] Principal gross lesions are marked hypertrophy of the tonsils, mesenteric lymph nodes, Peyer's patches, intestinal lymphoid tissue, moderate splenomegaly, and hemorrhagic enteritis.[5,6] Microscopically, SPD is characterized by severe and diffuse

Address for correspondence: Selwyn A. Headley, Laboratório de Patologia Veterinária, Hospital Veterinário, Centro Universitário de Maringá, Av. Guedner, 1610, Jd. Aclimação, Maringá, PR, Brazil 87050-390. Voice/fax: +55 (044) 3027-6360.
headleysa@cesumar.br

**Ann. N.Y. Acad. Sci. 1026: 79–83 (2004). © 2004 New York Academy of Sciences.**
**doi: 10.1196/annals.1307.010**

hyperplasia of reticuloendothelial cells and lymphocytic depletion, principally in mesenteric lymph nodes.[1,5,6] Macrophages exhibit large quantities of elementary intracytoplasmic rickettsial organisms demonstrable by Giesma staining.[3,5,6]

This report describes the preliminary investigative results of 10 Brazilian dogs with intestinal and lymphoid alterations comparable to SPD.

## MATERIALS AND METHODS

Five mongrels (three males, two females) and five beagles (three males, two females), ages ranging between two and four years, were sacrificed. Routine necropsy was performed soon after death. Tissues were fixed in 10% formalin solution and routinely processed for histopathological evaluation. Selected specimens were stained by Giemsa to determine the presence or absence of intracytoplasmic organisms possibly consistent with *Neorickettsia helminthoeca*.

## RESULTS

All dogs were in good body condition and presented similar gross lesions. Grossly, prescapular, mesenteric, and axial lymph nodes were markedly enlarged. Sectioned surface revealed prominent white cortical follicles (FIG. 1); discrete to

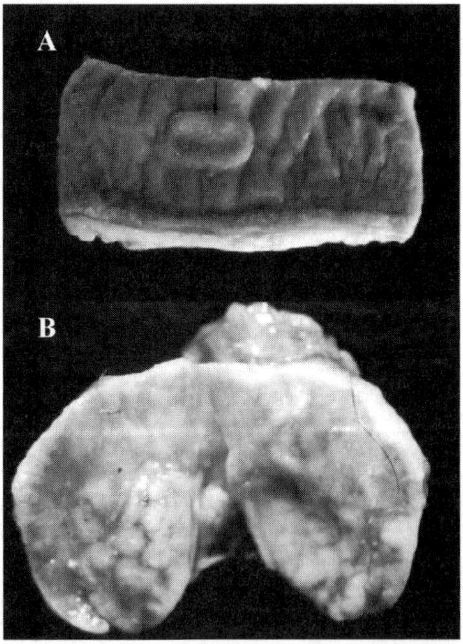


**FIGURE 1.** Gross lesions in the small intestine of a beagle. (**A**) Marked hypertrophy of Peyer's patches. (**B**) Sectioned mesenteric lymph node (observe the prominent white follicles, located principally at the cortex).

moderate edema was observed in some lymph nodes. All dogs demonstrated severe and multifocal hypertrophy of Peyer's patches and intestinal lymphoid tissue (FIG. 1). Moderate splenomegaly was observed in two mongrels; in these cases, prominent white follicles were observed at the sectioned splenic surface. Diffuse hemorrhagic colitis was observed in one mongrel.

Microscopic intestinal lesions were characterized by severe and diffuse hyperplasia of lymphoid tissue; Peyer's patches were markedly hyperplastic (FIG. 2). Mesenteric lymph nodes demonstrated marked cortical depletion, with sparing of few germinal centers. Similar microscopic alterations were observed in the spleen. Small intracytoplasmic organisms comparable to *Neorickettsia helminthoeca,*[1,5,6] demonstrated by Giemsa stain, were observed in macrophages within Peyer's patches, intestinal glands, germinal centers of mesenteric lymph nodes, and splenic corpuscles. Microscopic sections of the large intestine of one beagle revealed several thick-

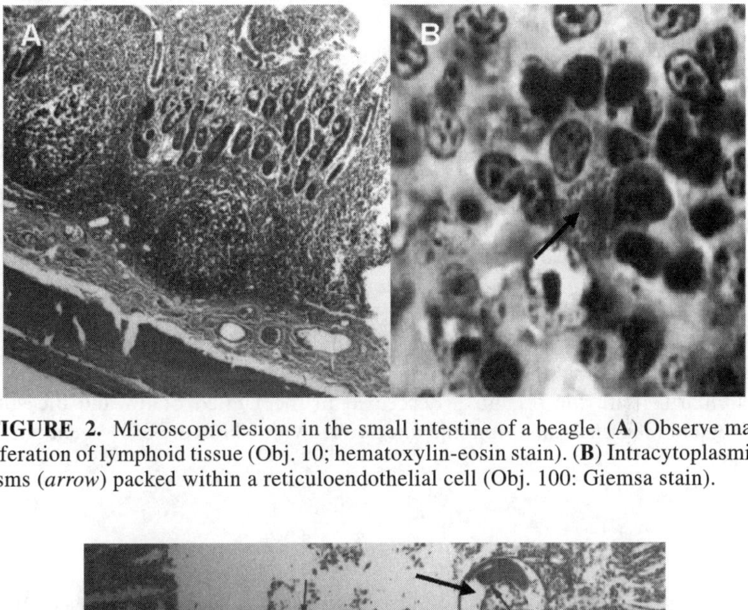


**FIGURE 2.** Microscopic lesions in the small intestine of a beagle. (**A**) Observe marked proliferation of lymphoid tissue (Obj. 10; hematoxylin-eosin stain). (**B**) Intracytoplasmic organisms (*arrow*) packed within a reticuloendothelial cell (Obj. 100: Giemsa stain).

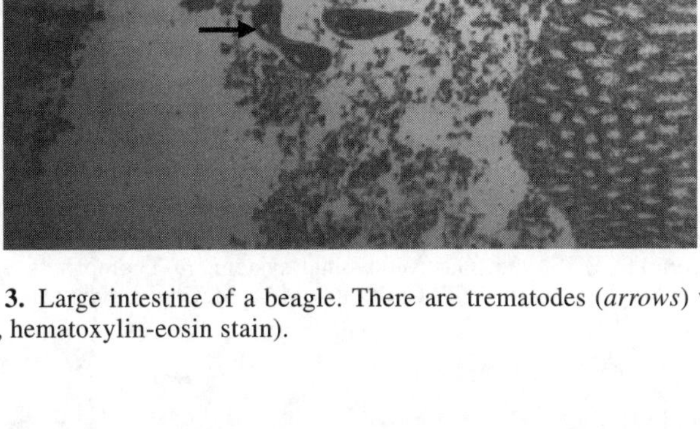

**FIGURE 3.** Large intestine of a beagle. There are trematodes (*arrows*) within the lumen (Obj. 10, hematoxylin-eosin stain).

walled trematodes (0.38 to 0.51 μm in length and 0.11 to 0.12 μm in width) that were comparable to *Ascocotyle* (*Phagicola*) *arnaldoi* (FIG. 3).[7]

## DISCUSSION AND CONCLUSIONS

The intracytoplasmic organisms observed in the present report are morphologically similar and produced pathological alterations characteristic of *Neorickettsia helminthoeca.*[1,3,5,6] Although SPD has been considered restricted to specific geographic locations of the United States,[1,3] cases have been diagnosed elsewhere,[4] so there exists a possibility of having SPD-like disease in other countries.

The feeding of crude fish to dogs is not a common practice in Brazil, so the biological cycle of *Neorickettsia helminthoeca* does not correspond exactly to the intracellular organism observed in this investigation. However, we believe that the intracytoplasmic organisms observed in the present report may be closely related to *Neorickettsia helminthoeca.* Three members (*Phagicola angrense, Phagicola angeloi*, and *Phagicola arnaldoi*) of the genus *Ascocotyle* (*Phagicola*) have been previously described in Brazil.[7] We strongly believe that the trematode observed in this report is *Ascocotyle* (*Phagicola*) *arnaldoi*; however, we were unable to obtain live specimens of the parasite to fully characterize the trematode. This trematode, like *Nanophyetus salmincola*, which is associated with SPD, requires two intermediate hosts (specific snails and fishes) to complete its life cycle in the intestine of the definitive host.[8]

Immunohistochemical and molecular biology techniques are being implemented to characterize the etiologic agent. Epidemiologic studies are being undertaken to determine the source of infection, the intermediate hosts, and the biological cycle of this *Neorickettsia helminthoeca*-like organism in Brazil. We have named the intracellular organism *Neorickettsia helminthoeca*-like because of its marked morphological similarity with the organism described in the United States and the similar pathological intestinal and lymphoid alterations with which the SPD agent has been associated.[1,3,5,6]

## REFERENCES

1. TIMONEY, J.F., J.H. GILLESPIE, F.W. SCOTT & J.F. BARLOUGH. 1992. Hagan and Bruner's Microbiology and Infectious Diseases of Domestic Animals. 8th edit.: 335–337. Cornell University Press. Ithaca, NY.
2. WALKER, D.J. & J.S. DUMLER. 1996. Emergence of the ehrlichioses as human health problems. Emerg. Infect. Dis. **2:** 18–29.
3. GORHAM, J.R. & W.J. FOREYT. 1998. Salmon poisoning disease. *In* Infectious Diseases of the Dog and Cat, 2nd edit. C.E. Green, Ed.: 135–139. W.B. Saunders. Philadelphia, PA.
4. BOOTH, A.J., L. STOGDALE & J.A. GRIGOR. 1984. Salmon poisoning disease in dogs in southern Vancouver Island. Can. Vet. J. **25:** 2–6.
5. JONES, T.C.H., R.D. HUNT & N.W. KING. 1997. Veterinary Pathology. 6th edit.: 388–390. Lippincott Williams & Wilkins. Baltimore, MD.
6. VAN KRUNINGEN, H.J. 1995. Gastrointestinal system. *In* Thompson's Special Veterinary Pathology. 2nd edit. W.W. Carlton & M.D. McGavin, Eds.: 61–62. Mosby. St. Louis, MO.

7. Travassos, L., J.F. Teixeira De Freitas & A. Kohn. 1969. Trematódeos do Brasil. Mem. Inst. Oswaldo Cruz **67:** 560–575.
8. Armitage, M. 1998. Complex life cycles in the heterophyid trematodes: structural and developmental design in the *Ascocotyle* complex of species. *In* Proceedings of the Fourth International Conference on Creationism. R.E. Walsh, Ed. Creation Science Fellowship. Pittsburgh, PA. <http://www.icr.org/research/ma/>

# Identification of IgG2-Specific Antigens in Mexican *Anaplasma marginale* Strains

R. BARIGYE,[a] M.A. GARCÍA-ORTIZ,[b] E.E. ROJAS RAMÍREZ,[b] AND S.D. RODRÍGUEZ[b]

[a]*Facultad de Medicina Veterinaria y Zootecnia, Universidad Nacional Autónoma de México, Col. Coyoacán, México D.F.*

[b]*CENID-Parasitología Veterinaria, INIFAP, SAGARPA, Km 11.5 Carr. Cuernvaca-Cuautla, Col. Progreso, Jiutepec, Morelos, México C.P.*

**ABSTRACT: To identify novel antigens with immunoglobulin G2 (IgG2) specificity and immunostimulant properties for bovine Th1 cells, humoral and cellular responses were studied in cattle inoculated with initial bodies from a Mexican isolate of *Anaplasma marginale* and challenged with a heterologous strain. Analysis of post-immunization sera by ELISA and assaying of *in vitro* cellular responses in peripheral blood mononuclear cells (PBMCs) cultured in the presence of protein extracts from three *Anaplasma marginale* strains showed positive values of optical density ELISA readings and stimulation indices in the immunized but not control cattle. Post-immunization and post-challenge sera recognized in Western blots several proteins with molecular weights ranging from 15 to 209 kDa, twelve of which were recognized by IgG2 in the three *Anaplasma marginale* strains. Seven of these are novel and have not been previously reported for their IgG2 specificity; three are confirmed to be major surface proteins (MSP-1a, MSP-2, and MSP-5); and the others correspond to other well-studied MSPs but were not confirmed. Partially purified fractions of protein extracts of the Mex-17 strain were tested against PBMCs cultured *in vitro*. One out of the seven novel proteins induced detectable lymphoproliferation (LP) of PBMCs, and interferon-γ was detected in supernatants of PBMC cultured in the presence of two protein fractions, including the one that caused LP. It is concluded that novel antigens, particularly the 28-kDa protein, played an additional role in the protection of immunized cattle and should be considered vaccine candidates after *in vivo* immunization experiments are concluded.**

**KEYWORDS: *Anaplasma marginale*; antigen identification; novel antigens; IgG2-specific antigens; vaccine candidate antigens; rickettsia**

## INTRODUCTION

Bovine anaplasmosis is an important disease of cattle caused by *Anaplasma marginale*, a rickettsia that invades and infects bovine erythrocytes, resulting in severe

Address for correspondence: S.D. Rodríguez, A.P. 206 CIVAC, Jiutepec, Morelos 62500, Mexico.
rodriguez.sergio@inifap.gob.mx

**Ann. N.Y. Acad. Sci. 1026: 84–94 (2004).** **© 2004 New York Academy of Sciences.**
**doi: 10.1196/annals.1307.011**

anemia associated with anorexia, fever, loss of weight, drop in milk production, and sometimes death. *A. marginale* is biologically transmitted by ixodid ticks. It is mechanically transmitted by some biting flies and by unsterilized instruments that have contacted blood infected by *A. marginale*. Very few cases of vertical transmission have been reported in this disease.[1,2]

The use of inactivated *A. marginale* immunogens to protect cattle has been reported, and attempts to make improved vaccines have been made by purifying and using specific surface membrane proteins[3,4] as immunizing agents. Immunization of cattle using killed whole organisms or purified outer membranes has been shown to induce partial protection against high-level rickettsemia and severe disease.[4–6] Like live vaccines, the killed organisms and purified outer membranes are associated with both specific antibody and cellular responses.[7–9] It has been postulated that protective immunity against anaplasmosis requires production of immunoglobulin G2 (IgG2),[10,11] which enhances phagocytosis and microbial killing by activated macrophages.[12] Interferon-$\gamma$ (IFN-$\gamma$) produced by activated $CD4^+$ helper cells stimulates isotype switching from IgG1 to IgG2 production by activated B cells[13] and synthesis of the rickettsicidal nitric oxide by activated macrophages.[14,15] On the basis of these mechanisms, specific T cell lines have been used for corroboration of potentially protective *A. marginale* major surface proteins (MSPs).[11] Also, these lines have shown that MSP-1a, MSP-2, and MSP-3 contain T cell epitopes conserved in various strains of *A. marginale* from different countries.[11] However, MSP-2 and MSP-3 are highly polymorphic and share about 55% of conserved amino acid sequences.[11] In vaccination experiments, MSP-1a, MSP-2, and MSP-3 from *A. marginale* (Florida strain) were used with saponin adjuvant as immunizing agents in calves and shown to independently induce complete and partial protection against homologous and heterologous challenge, respectively.[11,16]

While some workers have shown a number of immunoreactive proteins in conventional Western blots probed with the whole sera of immunized cattle,[17,18] to date no work has been done to identify other antigens that by specific immunoglobulin type (IgG1 or IgG2) are related to protection in vaccinated cattle.

The use of an initial-body-based immunogen in México has been shown to induce full protection against homologous challenge and a very high level of protection against heterologous challenge.[19] Despite these observations, no work involving the Mexican isolates of *A. marginale* has been done to establish the various antigens involved in protection of vaccinated cattle. The overall objective of this study was thus to identify novel antigens in addition to the known MSPs present in extracts of a Mexican *A. marginale* strain with the capacity to induce an IgG2 and a Th1 cellular response in fully susceptible adult cattle immunized and challenged with a heterologous *A. marginale* strain.

## MATERIALS AND METHODS

### Anaplasma marginale *Strains*

This study used three Mexican *A. marginale* strains: Mex-15, Mex-17, and Mex-30. These strains, originally isolated from the states of Mexico, Morelos, and Veracruz, respectively, have been discussed in detail elsewhere.[20] Infected blood for preparation of the experimental immunogen was produced by passaging *A. marginale* seed material through three splenectomized heifers as previously described.[21] The infected blood was washed in Puck's saline, pH 7.4, and cryopreserved in 10% plolyvinylpyrollidone at –70°C until use.

### *Preparation of Experimental Immunogen*

Experimental immunogen and antigen for *in vitro* studies were prepared from infected red blood cells by sonication and differential centrifugation in Puck's saline, pH 7.4.[21,22] To reduce erythrocyte membrane contamination, partially purified *A. marginale* initial bodies were centrifuged in a 30% isoosmotic Percoll density gradient formed *in situ* according to the manufacturer's instructions. Protein concentration of purified antigen preparation was determined by the Bradford method.[23]

### *Immunization*

Eight heifers purchased from a tuberculosis- and tick-free herd from northern Mexico and reared under tick-free conditions were used in the study. On days 0, 21, and 93, three animals were subcutaneously inoculated with 100 μg of *A. marginale* (Mex-17) protein suspended in 2 mL of Puck's saline containing 6 mg of Quil-A saponin adjuvant (Superfos Bio-sector A/S, Denmark), and three other animals were inoculated with the same amount of immunogen in oil-based IMS1313 adjuvant (SEPPIC, France), which forms water in oil emulsion. The immunogen for the first and second immunizations was inactivated in 1% glutaraldehyde at 37°C, whereas that for the third and last immunization was not. Two heifers used as controls were administered a placebo of either saponin or IMS1313 adjuvant. Following immunization, rectal temperatures (rT) and packed cell volume (PCV) values were recorded. Giemsa-stained thin blood smears were examined for the presence of *A. marginale* organisms, and serum samples were taken weekly throughout the time of the experiment for monitoring of antibody kinetics by ELISA.

### *Challenge*

Twenty-three weeks after the first immunization, experimental cattle were challenged by intramuscular injection of fresh $1 \times 10^8$ red blood cells (RBCs) infected with the virulent *A. marginale* Mex-30 strain. Post-challenge rT, PCV, and rickettsemias were evaluated as described above, and serum samples were taken for ELISA and immunoblot assays.

### *IgG Production and IgG1- and IgG2-Specific Antigens*

Pre-immunization, post-immunization, and post-challenge sera were tested by indirect ELISA, to evaluate IgG kinetics (as previously described[19]), and by Western blotting, to determine the antigenic pattern of the three Mexican *A. marginale* strains. For each lane, 2.5 μg of Percoll-purified *A. marginale* protein was fractionated in a 10% polyacrylamide-SDS gel, and the transfer of proteins to a polyvinylidene fluoride (PVDF) Immobilon-P membrane (Millipore, Mexico) was carried out as described by García and colleagues.[20] The membrane was blocked and probed with the sera of the immunized/protected heifers to detect IgG-specific antigens. The monoclonal antibodies (mAbs) Ana22B1 and AnaF16C1 were used to confirm the presence of MSP-1a and MSP-5, respectively; a polyclonal antibody specific for MSP-2/3 (kindly provided by Dr. Guy Palmer, Washington State University, Pullman, WA) was used to confirm the presence of MSP-2 in the Mex-17 strain. In a procedure that used mouse anti-bovine IgG1 and anti-IgG2 mAb, membrane strips containing *A. marginale* Mex-17 protein and previously incubated with post-immunization and post-challenge sera of a protectively immunized bovine were assayed for IgG1- and IgG2-specific antigens. (This procedure was a modified version of a procedure described by Brown and colleagues.[24]) Following incubation with test sera, and three washes in a 1% bovine serum albumin solution in phosphate-buffered saline-Tween 20 pH 7.4 (PBS-T-1%), the membranes were separately incubated overnight at 4°C with murine anti-bovine IgG1 or IgG2 mAb (Serotec, Oxford, UK) diluted 1:100 in PBS-T-1%. They were washed twice in Tris-0.05% Tween-20 pH 7.5 (T-T-1%), incubated for 1 hr at room temperature with goat anti-mouse IgG alkaline phosphatase conjugate (SIGMA) diluted 1:5000 in T-T-1%, washed three times in the same buffer solution, and developed with premixed BCIP/NBT substrate (product number B6404, Sigma Chemical Company, St. Louis, MO).

Two-dimensional electrophoresis was carried out according to the method described by O'Farrel.[25] This method used a 3–10 pH range in a Hoefer vertical electrophoresis cell at 700 V for 16 hr for the first dimension of the separation. For the second dimension of the separation, the gels were equilibrated for 15 min in SDS sample buffer (0.625M Tris-HCl pH 6.8, 10% v/v glycerol, 2% w/v SDS, 5% v/v 2-mercaptoethanol) and electrophoresed in a 10% polyacrylamide-SDS gel at 110 V for 2 hr. The proteins were transferred to a PVDF membrane and subjected to immunoblotting for IgG- and IgG2-specific antigens, as described above.

### *Continuous Flow Electrophoresis*

Six milligrams of *A. marginale* protein were treated with SDS sample buffer and fractionated in a cylindrical 10% polyacrlamide gel according to the method described by Van Kleef and colleagues.[26] Proteins in the fractions were precipitated at –20°C with cold acetone and washed in 70% ethanol, and the precipitate was suspended in sterile phosphate-buffered saline with 50 μg/mL of culture-grade gentamicin and kept at –70°C until use. Every tenth fraction was run by PAGE-SDS and selected for proteins of interest following silver staining (Bio-Rad, Hercules, CA) or immunoblotting, as described above.

### **In Vitro** *Lymphoproliferation Assays and Flow Cytometry*

Peripheral blood mononuclear cells (PBMCs) were isolated in Ficoll-Paque PLUS density gradients (Amersham Pharmacia Biotech AB, Uppsala, Sweden) and cultured in the presence of crude antigen or protein fractions in RPMI-1640 medium (GIBCO, Langley, OK ). This medium was supplemented with 2 mM L-glutamine, 50 μM 2-mercaptoethanol (2-ME) (Sigma-Aldrich, St Louis, MO), 10,000 IU of penicillin, and 50 μg/mL of streptomycin, and it was buffered by 2 g/L of sodium bicarbonate and 25 mM HEPES (Sigma-Aldrich). On the fifth day of incubation, the PBMCs were assayed for LP by cell proliferation ELISA and for $CD4^+$ activation by flow cytometry. Briefly, PBMCs were incubated in triplicate wells in 96 tissue culture plates (Corning Glass Works, Corning, NY) at $5 \times 10^5$ cells per well in a final volume of 200 μL. Crude *A. marginale* extracts of the three strains were used at 2.5 and 5 μg/mL, and fractions were used at 5% v/v. Triplicate wells with concanavalin-A (Con-A) (Sigma-Aldrich) and non-infected red blood cell (nRBC) membrane protein at the same concentrations were included in the plate as controls. In both cases, plates were left for a total of 5 days in a high humidity incubator at 37°C and 5% carbon dioxide ($CO_2$). On the fourth day of incubation, the cells meant for proliferation ELISA were labeled with 10 μM per well of 5-bromo-2-deoxyuridine (BrdU) reagent diluted in RPMI-1640 medium. After a further 12 hr of incubation, cell proliferation ELISA was carried out using a cell proliferation kit according to the manufacturer's instructions (Cell Proliferation ELISA System, Version 2, Code RPN 250, Amersham-Pharmacia). The reaction was read at 450 nm in a spectrophotometer (Labsystems Multiscan Plus, Fisher Scientific, Pittsburgh, PA), and the optical density readings were used to derive the stimulation index by dividing the average optical density reading of the test wells with that of the blank wells with only medium. A stimulation index of 2.0 or more was considered positive.

For the flow cytometry study, the cells were not labeled with BrdU as described above. On the fifth day of incubation, the PBMCs were washed three times in Hank's balanced salt solution pH 7.4 (HBSS) and labeled with mouse anti-bovine $CD4^+$ (Serotec, Oxford, UK) diluted 1:100 in HBSS. After three washes (as before), the cells were reacted with anti-mouse IgG conjugated to flourescein isothiocyanate diluted 1:100 in HBSS, washed three times (as before), fixed in 0.5% paraformaldehyde, and analyzed in a flow-activated cell sorter. Results were analyzed with Cell Quest (Largo, FL) software.

### *Immunoenzymatic Assay for IFN-γ*

Supernatants were taken before BrdU labeling of PBMCs cultured the in the presence of test fraction material or before doing flow cytometry. The determination of IFN-γ was carried out according to the method described in the BOVIGAM kit (Interferon Gamma Test, CSL, Parkville 3052, Victoria, Australia).

## RESULTS

The degree of protection in the immunized bovines was evaluated based on post-challenge clinical data, which included the percentage decrease in hematocrit, the level of rickettsemias, and the rT. The presence of clinical disease indicated that among vaccinated animals there were two types of response. In the first type, animals did not suffer any clinical changes. In the second type, animals presented clinical disease even in the presence of antibodies (FIG. 1). Analysis of sera from experimental cattle assayed by ELISA to follow the IgG kinetics in the post-immunization and post-challenge periods showed that all vaccinated cattle with either preparation induced the production of specific high IgG titers after vaccination, yet those titers came down to low levels some time after the last inoculation and did not increased after challenge (FIG. 1). Sera selected on the basis of bovine vaccination response were used in immunoblots for identification of IgG2-specific proteins.

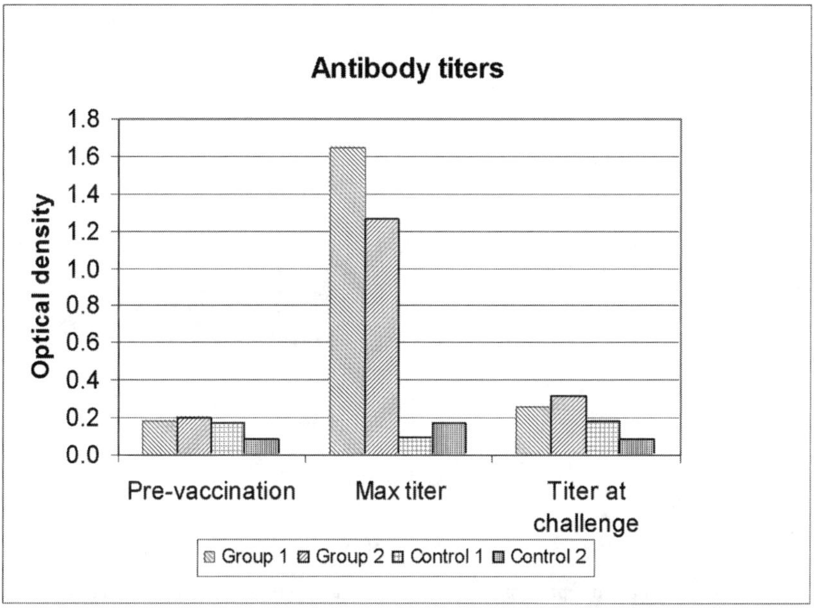


**FIGURE 1.** IgG titers at vaccination points, and maximal values after vaccination and challenge. Animals were inoculated three times with inactivated M-17 *Anaplasma marginale* initial bodies.

Analysis of the antigenic profiles of the three Mexican strains of *A. marginale* in one-dimensional Western blots showed various proteins that were recognized by post-immunization serum of a protectively immunized animal (FIG. 2a) where there appear eleven IgG2-positive antigens in the three study strains. MSP-1a, MSP-2, and MSP-5 are presented in FIGURE 2b as recognized by specific antibodies. Two-dimensional Western blotting of proteins from the immunizing Mex-17 strain showed nine of the eleven IgG2-specific proteins demonstrated in one-dimensional blots with an additional protein that co-migrated with MSP-2 (FIG. 3). Remarkably, none of the IgG2-positive proteins were recognized by sera of the two control animals; however, there was very minimal IgG2 recognition by serum of the immunized and unprotected animal (data not shown).

*A. marginale* proteins were separated by continuous flow electrophoresis, and fractions containing novel antigens were tested for their ability to induce a Th1-type response. Although the flow cytometry results were inconclusive (results not shown), studies with the protein fractions showed a positive LP response for a fraction in which a 28-kDa protein (FIG. 4) was the only protein recognized by bovine IgG2 present only in serum of a protectively immunized animal. IFN-γ was detected in supernatants of PBMCs cultured in the presence of this and another fraction containing a 37-kDa protein.

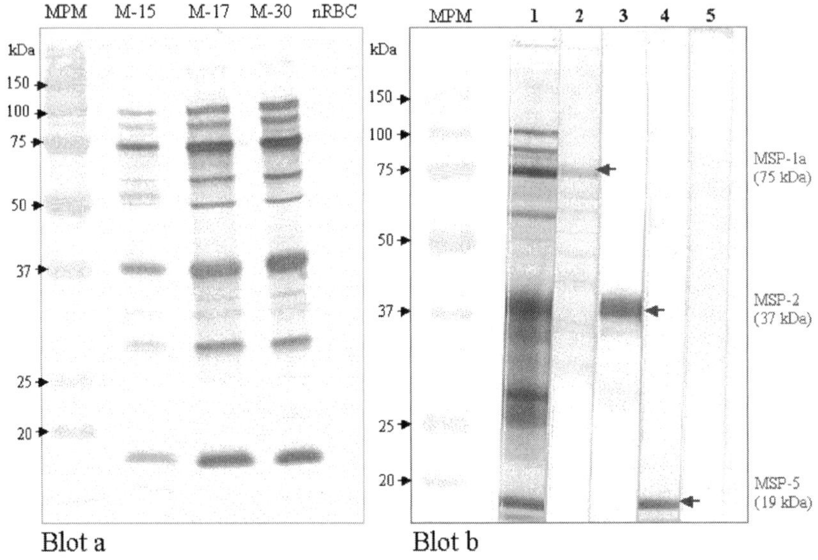


**FIGURE 2.** Antigenic pattern of IgG2-specific antigens in three Mexican strains of *A. marginale* and major surface proteins in the *A. marginale* Mex-17 strain.

## DISCUSSION

The identification of IgG2-specific antigens in three Mexican *A. marginale* strains is hereby reported in this study. These include novel antigens as well as three of the six previously reported MSPs.

As demonstrated in the immunoblot results, there was a remarkable similarity in the antigenic patterns of the three *A. marginale* study strains. In addition, *in vitro* study of the cellular response showed that antigen from all the study strains were capable of inducing a positive LP in PBMCs of cattle immunized with the Mex-17 strain. Phylogenetic analyses by de la Fuente and colleagues,[27] of four Mexican *A. marginale* isolates, including the three (Mex-15, Mex-17, and Mex-30) reported in this study, have shown great genetic similarity. On the basis of this similarity, these investigators grouped these isolates in one genetic clade. In consideration of this genetic relationship, the similarity in the antigenic pattern of the three strains would have been expected. However, the Mex-15 strain showed an additional IgG2-specific antigen of 52 kDa that might signify some subtle genetic differences between this and the other two strains.

In agreement with the result reported by Brown and colleagues,[24] an immunized bovine (no. 2624) failed to develop a substantial IgG2 response despite presence of a strong IgG1 response. Interestingly, this animal died following challenge despite detectable LP and a largely IgG1 humoral response during this period. This finding can be explained within the confines of the working protection model, in which it is hypothesized that immunity against anaplasmosis involves IgG2 switching as a result of production of IFN-$\gamma$ by $CD4^+$ T cells.

Eleven IgG2-specific proteins have been identified in the Mex-17 and Mex-30 strains, and twelve have been identified in the Mex-15 isolate. Remarkably, the apparent molecular weights of some of these proteins correspond to those of some of the MSPs identified and characterized in *A. marginale* strains elsewhere, of which MSP-1a, MSP-2, and MSP-5 are confirmed in this study. In agreement with various other studies,[28] a 31-kDa protein identified in this study and thought to be the non-variable MSP-4 has remarkably shown no IgG2 specificity.

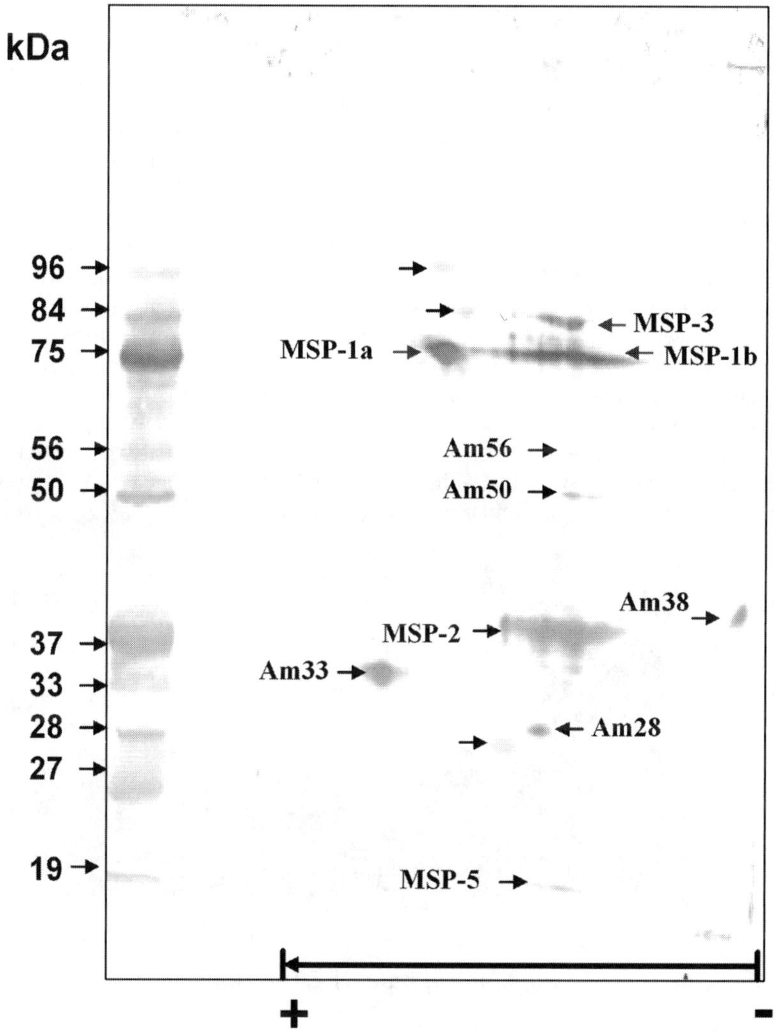


**FIGURE 3.** Western blot profile of IgG2-specific proteins. After two-dimensional electrophoresis, the membrane was probed with the serum of one animal that was protectively immunized.

On the basis of one-dimensional and two-dimensional blot results, molecular weight, and migration characteristics, the 75-kDa protein that focused into a single dot in the two-dimensional immunoblot is the single-copy MSP-1a, as confirmed in a one-dimensional blot with a specific mAb. The molecular weight of MSP-1a may vary between geographical *A. marginale* strains because of the presence of repeated tandem sequences of up to 29 amino acids that interestingly bear conserved neutralization epitopes in some of the *A. marginale* strains studied.[29]

Another IgG2-specific protein with an apparent molecular weight of 84 kDa and thought to be MSP-1b was shown to focus as two dots. MSP-1b is known to be coded by two different genes, *msp1β1* and *msp*1β2,[30] and would thus be expected to appear as two individual proteins comigrating close together. Immunization of cattle with purified MSP-1 complex (MSP-1a and MSP-1b) has shown protection against both homologous and heterologous challenge.[16] Thus, MSP-1b should have been expected to appear in the IgG2-specific recognition patterns reported in this study.

The 37-kDa protein showed the highest immunoreactivity among all antigens demonstrated in the Western blots, whereas isoelectric focusing has shown a number of proteins of the same molecular weight and nearly contiguous isoelectric points. Despite having the same molecular weight as MSP-2 (37 kDa) and thus comigrating with the later in one-dimensional gels, an IgG2-specific protein with a distribution pattern distinct from that of MSP-2 was shown in two-dimensional blots. Different polymorphic variants of MSP-2 with different isoelectric points are reported to occur in the same animal,[25,28] and these might migrate together in one-dimensional polyacrylamaide gels to give a prominent 37-kDa band that is resolved by two-dimensional electrophoresis to show the variant proteins. It is, however, concluded that the single IgG2-specific protein with a distinct distribution pattern is novel and not a variant of MSP-2. mAb typing in one dimension has confirmed the presence of MSP-2 migrating with a molecular weight of 37 kDa.

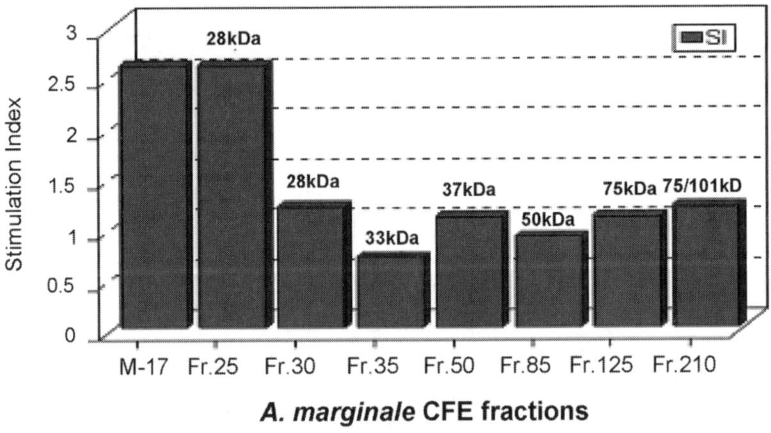


**FIGURE 4.** Continuous flow electrophoresis fractions were tested for their *in vitro* capacity to stimulate proliferation. PBMCs from a protectively immunized bovine were cultured in the presence of fractions collected from continuous flow electrophoresis. *In vitro* proliferation assays were carried out in triplicate wells as described in the text.

Testing of individual continuous flow electrophoresis fractions of *A. marginale* protein showed a positive LP response for the 28-kDa protein, and IFN-γ was detected for at least the fraction containing this protein and another containing a 37-kDa protein (FIG. 4). Of the novel proteins hereby reported, the 28-kDa protein should be considered a vaccine candidate. Further studies are being undertaken on the others before reliable conclusions can be made.

## ACKNOWLEDGMENTS

This research was supported by the CONACYT (a Mexican government agency) under Grant No. 30416B. Robert Barigye was supported by the Secretaría de Relaciones Exteriores (Foreign Affairs Ministry) of México and a complimentary scholarship from The National Autonomous University of México through the School of Veterinary Medicine.

## REFERENCES

1. POTGIETER, F.T. & VAN RENSBURG. 1987. The persistence of colostral antibodies and the incidence on *in utero* transmission of *Anaplasma* infections in calves under laboratory conditions. Onderstepoort J. Vet. Res. **54**(4): 557–560.
2. SALABARRIA, F.F. & R. PINO. 1988. Transmisión vertical de *Anaplasma marginale* en bovinos infectados durante el periodo final de la gestación. Rvta. Cub. Cienc. Vet. **19**(3): 179–182.
3. PALMER, A.H. 1989. Anaplasma vaccines. *In* Veterinary Protozoan and Hemoparasite Vaccines. I.G. Wright, Ed.: 1–29. CRC Press. Boca Raton, FL.
4. TEBELE, N., T.C. MCGUIRE & G.H. PALMER. 1991. Induction of protective immunity using *Anaplasma marginale* initial bodies membranes. Infect. Immun. **59:** 3199–3204.
5. BROCK, W.E., I.O. KLIEWER & C.C. PEARSON. 1965. A vaccine for anaplasmosis. J. Am. Vet. Med. Assoc. **147:** 948–951.
6. MONTENEGRO-JAMES, S., A.T. GUILLEN, S.J. MA, *et al.* 1990. Use of the dot enzyme-linked immunosorbent assay with isolated *Anaplasma marginale* initial bodies for serodiagnosis of anaplasmosis in cattle. Am. J. Vet. Res. **51:** 1518–1521.
7. BUENING, G.M. 1973. The role of cell-mediated immunity in anaplasmosis infection. *In* Proceedings of the Sixth National Anaplasmosis Conference, Las Vegas, Nevada, pp. 71–77. Heritage Press. Stillwater, OK.
8. BUENING, G.M. 1976. Cell-mediated immune response in anaplasmosis as measured by a micro cell-mediated cytotoxicity assay and leukocyte migration inhibition test. Am. J. Vet. Res. **34:** 1215–1218.
9. CARSON, C.A., D.M. SELLS & M. RISTIC. 1976. Cell-mediated immunity in bovine anaplasmosis and correlation with protection induced by vaccination. Vet. Parasitol. **2:** 75–81.
10. MCGUIRE, T.C., A.J. MUSOKE & T. KURTTI. 1979. Functional properties of bovine IgG1 and IgG2 interaction with complement, macrophages, neutrophils and skin. Immunology **38:** 249–256.
11. BROWN, W.C., D. ZHU, V. SHKAP, *et al.* 1998. The repertoire of *Anaplasma marginale* antigens recognized by CD4+ clones from protectively immunized cattle is diverse and includes major surface protein 2 (MSP-2) and MSP-3. Infect. Immun. **66:** 5414–5422.
12. PALMER, G.H. & T.F. MCELWAIN. 1995. Molecular basis for development of a vaccine against anaplasmosis and babesiosis. Vet. Parasitol. **57:** 233–253.
13. ESTES, D. M., N.M. CLOSER & G.K. ALLEN. 1994. IFN-γ stimulates IgG2 production from bovine B-cells co-stimulated with anti-μ and mitogen. Cell Immunol. **154:** 287–295.

14. Adler, H.E., E. Peterhans, J. Nicolet & T.W. Jungi. 1994. Inducible L-arginine-dependent nitric oxide synthatase activity in bone marrow-derived macrophages. Biochem. Biophys. Res. Commun. **198:** 510–515.
15. Stich, R.W., L.K.M. Shoda, M. Dreewes, *et al.* 1998. Stimulation of nitric oxide production by *Babesia bovis.* Infect. Immun. **66:** 4130–4136.
16. Palmer, A.H., A.F. Barbet, G.H. Cantor & T.C. McGuire. 1989. Immunization of cattle with the MSP-1 surface protein complex induces protection against a structurally variant *Anaplasma marginale* isolate. Infect. Immun. **57:** 3666–3669.
17. Montenegro-James, S., M.A. James, M. Toro Benitez, *et al.* 1991. Efficacy of purified *Anaplasma marginale* initial bodies as a vaccine against anaplasmosis. Parasitol. Res. **77:** 93–101.
18. Patarroyo, J.H.S., D.J. Henckel, A.A. Prates & C.L. Mafra. 1994. Antigenic profile of a pure isolate of *Anaplasma marginale* of Brazilian origin, using a Western blot technique. Vet. Parasitol. **52:** 129–137.
19. Rodríguez, S.D., M.A. García-Ortiz, G. J. Cantó Alarcón, *et al.* 1999. Ensayo de una vacuna experimental inactivada contra *Anaplasma marginale.* Tec. Pecu. Mex. **37**(1): 1–12.
20. García-Ortiz, M.A., O. Ángeles, S. Hernández, *et al.* 1998. Caracterización de la virulencia de un aislado mexicano de *Anaplasma marginale.* Téc. Pécu. Méx. **36**(3): 197–202.
21. Rodríguez, S.D., M.A. García-Ortiz, G. Hernández Salgado, *et al.* 2000. *Anaplasma marginale* inactivated vaccine: dose titration against a homologous challenge. Comp. Immunol. Microbial. Infect. Dis. **23:** 239–252.
22. Puck, T.T., S.J. Cieciura & A. Robinson. 1958. Genetics of somatic mammalian cells. III. Long-term cultivation of euploid cells from human and animal subjects. J. Exp. Med. **108:** 945–955.
23. Bradford, O.F. 1986. Adaptation of the protein assay to membrane-bound proteins by solubilizing in glucopyranoside detergents. Anal. Biochem. **162:** 11–17.
24. Brown, W.C., V. Shkap, D. Zhu, *et al.* 1998. CD4+ T-lymphocyte and immunoglobulin G2 responses in calves immunized with *Anaplasma marginale* outer membranes and protected against homologous challenge. Infect. Immun. **66:** 5406–5413.
25. O'Farrel, P.H. 1975. High resolution of two-dimensional electroforesis of proteins. J. Biol. Chem. **250:** 4007.
26. Van Kleef, M., N.J. Gunter, H. Macmillian, *et al.* 2000. Identification of *Cowdria ruminantium* antigens that stimulate proliferation of lymphocytes from cattle immunized by infection and treatment or with inactivated organisms. Infect. Immun. **68**(2): 603–614.
27. de la Fuente, J., R.A. Van den Bussche, J.C. García-García, *et al.* 2002. Phylogeography of New World isolates of *Anaplasma marginale* based on major surface protein sequences. Vet. Microbiol. **88:** 275–285.
28. Oberle, S.M. & A.F. Barbet. 1993. Derivation of the complete *msp4* gene sequence of *Anaplasma marginale* without cloning. Gene **136:** 291–294.
29. Allred, D.R. & A.F. Barbet. 1990. Molecular basis for surface antigen polymorphisms and conservation of a neutralization-sensitive epitope in *Anaplasma marginale.* Proc. Natl. Acad. Sci. USA **87:** 3220–3224.
30. Barbet, A.F., G.H. Palmer, P.J. Myler & T.C. McGuire. 1987. Characterization of an immunoprotective protein complex of *Anaplasma marginale* by cloning and expression of the gene coding for polypeptide Am105L. Infect. Immun. **55:** 2428–2435.

# Detection of *Anaplasma marginale* DNA in Larvae of *Boophilus microplus* Ticks by Polymerase Chain Reaction

MÁRCIA KIYOE SHIMADA,[a] MILTON HISSASHI YAMAMURA,[a] PAULA MIYUKI KAWASAKI,[a] KÁTIA TAMEKUNI,[a] MICHELLE IGARASHI,[a] ODILON VIDOTTO,[a] AND MARILDA CARLOS VIDOTTO[b]

[a]*Departamento Med. Vet. Preventiva, CCA, Universidade Estadual de Londrina, Campus Universitário, Londrina, PR, Brazil*

[b]*Departamento de Microbiologia, CCB, Universidade Estadual de Londrina, Londrina, PR, Brazil*

**ABSTRACT: *Boophilus microplus* larvae from two different sources were used for the detection of *Anaplasma marginale* DNA: larvae A, which were collected from a pasture of an endemic farm, and larvae B, which originated from engorged female ticks fed on calves with no clinical signs of disease and with low rickettsemia (approximately 0.01 to 1.0%). Larvae A were collected monthly, from January to May in 2001. Two hundred engorged female ticks fed on calves that provided larvae B were divided into groups of 10 and kept in a controlled environment at either 18°C or 28°C. Fifty larvae were used from each sample for DNA extraction, and 5 µL of DNA were submitted to amplification of the sequence of *msp5* gene of *A. marginale* by polymerase chain reaction (PCR). Seven out of 50 samples of larvae A (14%) were positive for the presence of DNA of *A. marginale* showing amplified product of 457 bp. Ten out of 91 samples of larvae B (11%) kept at 18°C were positive, and all larvae B at 28°C were negative. Thus, this study confirmed the presence of *A. marginale* DNA in *B. microplus* larvae by PCR. The *Eco*RI restriction enzyme analysis confirmed the specificity of the amplicon, which resulted in two fragments: 265 bp and 192 bp. The sequencing analysis of the amplicon from larvae demonstrated 98% homology with the *msp5* sequence from Florida *A. marginale* strain .**

**KEYWORDS: *Anaplasma marginale*; larvae; *Boophilus microplus*; polymerase chain reaction (PCR)**

## INTRODUCTION

Bovine anaplasmosis is a tick-borne disease of cattle caused by the obligate intraerythrocytic bacteria *Anaplasma marginale* (order *Rickettsiales*, family *Anaplasmataceae*).[1] This disease is endemic in many tropical and subtropical areas, and the

Address for correspondence: Odilon Vidotto, Departamento Med. Vet. Preventiva, CCA, Universidade Estadual de Londrina, Campus Universitário, C.P. 6001, CEP 86051-990, Londrina, PR, Brazil. Fax: 55-43-3371-4714.
vidotto@uel.br

**Ann. N.Y. Acad. Sci. 1026: 95–102 (2004).** **© 2004 New York Academy of Sciences.**
**doi: 10.1196/annals.1307.012**

acute phase causes anemia, abortion, weight loss, and eventually death, resulting in significant losses to meat and milk production.[2] *A. marginale* is usually transmitted biologically by feeding ticks, whereas mechanical transmission occurs when infected blood is transferred to susceptible animals by biting flies or blood-contaminated fomites. Approximately 20 species of ticks have been implicated as vectors worldwide.[3] In Brazil, *Boophilus microplus* is considered to be the main vector of *A. marginale*,[4] but the mechanism of transmission employed by this tick is still controversial.

*B. microplus* is a monoxenic tick, but adult stages can migrate between animals by physical contact.[4–6] Intrastadial transmission by male ticks may be important in maintaining the organism in enzootic areas. The transtadial transmission has been reported for many ticks including *B. microplus*.[7,8] Transovarial transmission of *A. marginale* has also been reported for some ticks, but has not been demonstrated for *B. microplus*.[4,5,8–11]

Ribeiro and Lima[12,13] demonstrated the multiplication of *A. marginale* in *B. microplus*, and verified the influence of temperature on the development of colonies of *A. marginale* in the midgut epithelial cells of experimentally infected *B. microplus* females. The colonies were observed in the midgut epithelial cells in 11.1% of females that were kept at environmental temperature by 19 days after detachment from donor calf, suggesting that transovarial transmission occurs in winter (temperatures ranged from 10 to 15°C and from 22 to 32°C), when the infection could be restricted to the last eggs laid.

This work shows the results of a polymerase chain reaction (PCR) study with larvae of *B. microplus* collected from pasture and from engorged female ticks infected with *A. marginale* incubated at different temperatures.

## MATERIALS AND METHODS

### Boophilus microplus *Ticks*

Larvae from two sources were studied: larvae A, from a pasture in a dairy farm with endemic anaplasmosis, located in Londrina municipality (23° 30′ 17″ S, 51° 07′ 28″ W), Paraná State, Brazil; larvae B, resulting from engorged female ticks fed on calves infected with *A. marginale* with rickettsemia varying from 0.01% to 1.0% (confirmed by PCR), without clinical sign of anaplasmosis.

Larvae A were collected monthly, from January to May 2001, in a pasture of *Brachiaria spp* and coast-cross grass. The minimum temperatures in this period ranged from 8.2 to 20.5°C, and the maximum temperatures ranged from 21.2 to 31.5°C. The humidity ranged from 59.6 to 79.7%. Each month the collected larvae were separated in 10 samples of 50 larvae each.

The engorged female ticks that provided larvae B were first transferred to a controlled environment under two temperatures. Two hundred female ticks fed on calves were divided into groups of 10 and kept in Petri plates. Ten plates were incubated at 18 ± 2°C with 80–83% humidity, and the other 10 plates were incubated at 27 ± 2°C with 86–89% humidity. The pre-oviposition and oviposition periods were monitored daily, and the eggs were collected approximately every 5 days. All the eggs from each period were collected and kept in sterile flasks at 28 ± 1°C with 87% relative humidity until the hatching of the larvae. Fifty larvae were separated for each sample. All larvae and tick controls were washed with ethanol 70% and ultra-pure $H_2O$ and kept at –20°C until further use.

*B. microplus* male ticks collected from a calf naturally infected with *A. marginale* (60% of rickettsemia) were used as positive control. The negative control consisted of larvae originating from female ticks fed on cattle negative for *A. marginale*, kindly provided by EMBRAPA/CNPGC/Campo Grande/MS.

### *DNA Extraction*

The DNA from ticks and larvae were extracted by the phenol-clorophorm/isoamilic-alcohol and silica/guanidium thiocyanate[14] methods with some modifications. Fifty larvae from each sample were used. The male ticks or larvae groups were triturated in 450 μL of TE (Tris-HCl 10 mM; EDTA 1 mM, pH 8.0). After cell lysis with 10% SDS and proteinase K (10 mg/mL) for 1 hr at 37°C, the same volume of phenol-clorophorm/isoamilic-alcohol (24:24:1) was added. The suspension was incubated at 56°C during 15 min and centrifuged at 10,000 × g 10 min. The aqueous phase was then processed in silica/guanidium thiocyanate (120 g guanidium thiocyanate; 100 mL Tris-HCl 10 mM pH 6.4; EDTA 20 mM pH 8.0; 2.6 g Triton X-100). The washes were realized twice with L2 solution (120 g guanidium thiocyanate; 100 mL Tris-HCl 0.1 M, pH 6.4), twice with ethanol 70%, and once with acetone. The pellet was kept at 56°C for 15 min to dry, and the DNA was eluted in 25 μL of ultra-pure autoclaved $H_2O$. This procedure was repeated, and 50 μL of DNA was obtained. All extracted DNA was stored at –20°C until further use.

### *PCR Procedure*

PCR was performed with primers obtained from a sequence of the *msp5* gene of Florida *A. marginale* (GenBank number M93392) according to a procedure described by de Echaide and colleagues.[15] These primers included external forward, 5′-GCATAGCCTCCGCGTCTTTC-3′ (position 254 a 273), and external reverse, 5′-TCCTCGCCTTGGCCCTCAGA-3′ (position 710 a 692). The analysis of these primers did not show genomic similarity with microorganisms from the *Ehrlichia* and *Wolbachia* genera. Amplification reactions were performed in a final volume of 25 μL containing 5 μL of DNA, 20 pmol of each *primer*, 0.2 mM of deoxynucleoside triphosphate, 20 mM Tris-HCl (pH 8.3), 50 mM KCl, 1.5 mM $MgCl_2$, and 1.25 U of Taq DNA polymerase (GIBCO-Life Technology, Rockville, MD) with thermocycler (MJ Research, Watertown, MA) for 5 min at 95°C, 35 cycles at 95°C for 1 min, 65°C for 2 min, and 72°C for 1 min with a final extension at 72°C for 10 min followed by cooling to 4°C. Ultra-pure $H_2O$ was used as a reaction control in the same conditions. After the amplification reaction, 10 μL of the amplicon was electrophoresed on polyacrylamide gel (10%) and stained by silver. A 123-bp ladder (GIBCO-Life Technology) was used as molecular size standard.

### *Specificity of PCR Product*

To determine the specificity of the amplicon, the bands with 457 bp were digested with *Eco*RI restriction endonuclease (Bethesda Research Laboratories, Rockville, MD), according to the manufacturer's instructions. The digests were subjected to eletrophoresis through polyacrylamide gel (10%) and stained by silver. 123-bp ladder was used as a molecular mass marker.

The sequence analysis of the amplicon from larvae was carried out with the nested PCR product obtained with external forward 5′-CACCATGAGAATTTTCAAGATTGTG-3′, external reverse 5′-TCCTCGCCTTGCCCCTCAGA-3′, and internal forward 5′-AGAATTAAGCATGTGACCGCTG-3′. The amplicon was submitted at sequencing (3100, Applied Biosystems, Foster City, CA), and the sequence analysis was done with BLAST (<http://www.ncbi.nih.gov/BLAST/>) program.

## RESULTS AND DISCUSSION

Seven out of the 50 samples (14%) of larvae A collected from pasture in different months and 10 of the 91 samples (11%) of larvae B from engorged female ticks fed on calves were positive to presence of *msp5* gene of *A. marginale* by PCR, amplifying the fragment of 457 bp (TABLE 1). The *Eco*RI restriction enzyme analysis of this amplicon showed its specificity, showing two fragments, one of 265 bp and one of 192 bp (FIG. 1). The nucleotide sequence of this amplicon was determined and is displayed in FIGURE 2. The sequencing analysis of the amplicon from larvae demonstrated 98% homology with the *msp5* sequence from Florida *A. marginale* strain.

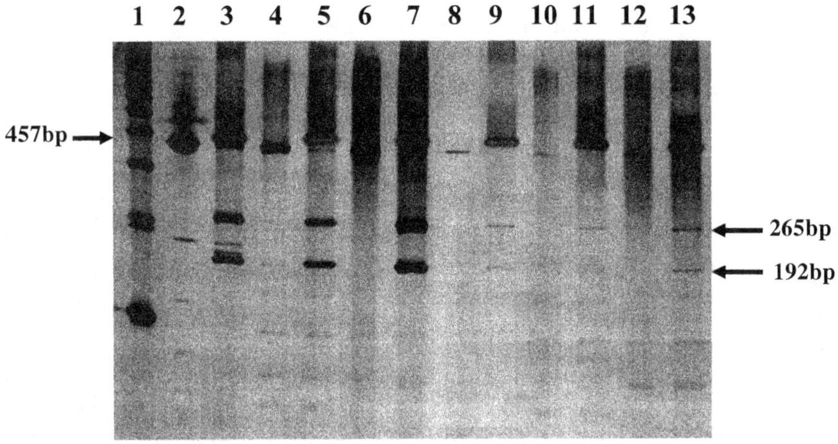


**FIGURE 1.** Polyacrylamide (10%) gel stained by silver of PCR product from *msp5* gene of *A. marginale*. *Lane 1:* molecular size markers (123-bp ladder). *Lane 2:* PCR product from infected blood of calf not cleaved (457 bp). *Lane 3:* PCR product from infected blood of calf cleaved with *Eco*RI. *Lanes 4 & 5: B. microplus* male ticks not cleaved and cleaved. *Lanes 6 & 7:* engorged female ticks not cleaved and cleaved. *Lanes 8 & 9:* larvae from pasture not cleaved and cleaved. Lanes 10–13: larvae from engorged female ticks not cleaved and cleaved.

The occurrence of positive larvae from pasture was higher in April (four samples), and in January all samples were negative for the presence of *A. marginale* DNA. The reported influence of environmental temperature on the multiplication of the *A marginale* in the *B. microplus* females[12] might explain the higher number of positive larvae samples in April (autumn) than in January (summer).

The period of incubation of engorged female ticks that provided larvae B at 18°C was longer than at 28°C. The pre-oviposition period of engorged female ticks that provide larvae B was 8–9 and 2–3 days at 18°C and 28°C, respectively. The oviposition period in larvae B at 18°C was 13–44 days, whereas the period at 28°C was

```
msp5/larva   :11  tgagaattttcaagattgtgtctaaccttctgctgttcgttgctgccgtgttcctgggg 70
                  ||||||||||||||||||||||||||||||||||||||||||||||||||||||||||||
msp5/Genbank:118  tgagaattttcaagattgtgtctaaccttctgctgttcgttgctgccgtgttcctgggg 177
                  ************************************************************

msp5/larva   :71  tactcctatgtgaacaagaaaggcatttcagcaaaatcggcgagaggtttaccacttcc 130
                  ||||||||||||||||||||||||||||||||||||||||||||||||||||||||||||
msp5/Genbank:178  tactcctatgtgaacaagaaaggcatttcagcaaaatcggcgagaggtttaccacttcc 237
                  ************************************************************

msp5/larva   :131 gaagttgtaagtgagggcatagcctccgcgtctttcaacaatttggttaatcacgagggg 190
                  ||||||||||||||||||||||||||||||||||||||||||||||||||||||||||||
msp5/Genbank:238  gaagttgtaagtgagggcatagcctccgcgtctttcaacaatttggttaatcacgagggg 297
                  ************************************************************

msp5/larva   :191 gtcaccgtcagtagcggcgatttggcggcaagcacatgttggtaatattcggcttctca 250
                  ||||||||||||||||||||||||||||||||||||||||||||||||||||||||||||
msp5/Genbank:298  gtcaccgtcagtagcggcgatttggcggcaagcacatgttggtaatattcggcttctca 357
                  ************************************************************

msp5/larva   :251 gcctgtaagtacacgtgccctaccgagttaggcatggcttctcagctcctaagtaaacta 310
                  ||||||||||||||||||||||||||||||||||||||||||||||||||||||||||||
msp5/Genbank:358  gcctgtaagtacacgtgccctaccgagttaggcatggcttctcagctcctaagtaaacta 417
                  ************************************************************

msp5/larva   :311 ggcgaccatgccgataagttgcaagttgtgttcataactgttgatccgaaaaatgacacc 370
                  ||||||||||||||||||||||||||||||||||||||||||||||||||||||||||||
msp5/Genbank:418  ggcgaccatgccgataagttgcaagttgtgttcataactgttgatccgaaaaatgacacc 477
                  ************************************************************

msp5/larva   :371 gtagccaagcttaaagagtaccacaagtcttttgatgcgagaattcagatgctcacaggc 430
                  ||||||||||||||||||||||||||||||||||||||||||||||||||||||||||||
msp5/Genbank:478  gtagccaagcttaaagagtaccacaagtcttttgatgcgagaattcagatgctcacaggc 537
                  ************************************************************

msp5/larva   :431 gaagaagcagacataaagagcctggttgaaaactacaaggtgtacgtaggcgataaaaag 490
                  ||||||||||||||||||||| |||||||||||||||||||||| || ||||| || |||
msp5/Genbank:538  gaagaagcagacataaagagcgtggttgaaaactacaaggtgtatgtgggcgacaagaag 597
                  ********************* ********************** ** ***** ** ***

msp5/larva   :491 gcaagcgatggtgatattgatcactcaacgttcatgtacctcatcaatgggaaaggcagg 550
                   |||| ||||||||||| || |||||||||||||||||||||||||||||||||||||||
msp5/Genbank:598  ccaagtgatggtgatatcgaccactcaacgttcatgtacctcatcaatgggaaaggcagg 657
                   **** *********** ** ***************************************

msp5/larva   :551 tat 553
                  |||
msp5/Genbank:658  tat 660
                  ***
```

**FIGURE 2.** Comparison of sequence obtained GenBank (number M93392) with sequence of *msp5* of *A. marginale* from positive larvae of *B. microplus* ticks, showing 98% homology.

**TABLE 1. Positive samples of *Boophilus microplus* larvae collected from pasture and engorged females by collection period using polymerase chain reaction (PCR)**

| Larvae from pasture (ambient T °C) (Larvae A) | | | Larvae from females incubated in chambers (Larvae B) | | | | |
|---|---|---|---|---|---|---|---|
| | | | | 28°C | | 18°C | |
| Larvae collection month | Samples/ collection[a] | Positive samples | Egg collection days | Total number of samples | Number of positive samples | Total number of samples | Number of positive samples |
| January | 10 | 0 | 7–8 | 20 | 0 | 0 | 0 |
| February | 10 | 1 | 12–14 | 20[b] | 0 | 20 | 2 |
| March | 10 | 1 | 18–19 | | | 20 | 2 |
| April | 10 | 4 | 23–24 | | | 20 | 0 |
| May | 10 | 1 | 28–29 | | | 20 | 3 |
| | | | 33–34 | | | 10 | 3 |
| | | | 43–44 | | | 01[b] | 0 |
| Total | 50 | 7 (14%) | | 40 | 0 | 91 | |

[a]Fifty larvae each sample.
[b]End of oviposition period.

7–13 days. All larvae B at 28°C were negative for the presence of *A. marginale* DNA. Ten samples of larvae B were positive between 13 and 34 days of egg collection (TABLE 1).

These results show that longer pre-oviposition and oviposition periods, which occur at lower temperatures, contributed to an increase in the number of positive larvae, a result that is in agreement with Ribeiro *et al.*[11] and Ribeiro and Lima,[12] who observed colonies of *A. marginale* in the midgut of *B. microplus* engorged females only at low environmental temperatures, when the pre-oviposition was 6–11 days and the oviposition period was 28–41 days. Also, these authors observed colonies only in 11% of female ticks by the 19th day after detachment from the donor calf, which increased to 33% of infected vectors in the following weeks, suggesting that winter is the most probable period for the occurrence of vertical transmission of this agent in nature, and that it is limited to the last eggs laid.

The increased number of positive larvae obtained in this work, when the tick females were incubated at 18°C, can be explained because oviposition periods longer than 14 days can enable the migration of *A. marginale* from midgut to ovaries, whereas in optimal conditions the oviposition in *B. microplus* is bigger (99.9%) from 10th to the 13th days after detachment of engorged female ticks,[16,17] limiting the passage of *Anaplasma* to ovaries. Most of the unsuccessful attempts of transovarial transfer of *A. marginale* infection were performed with larvae originated from *B. microplus* females incubated under high temperatures (28 to 30°C).[8–10] Such conditions of engorged tick female incubations could explain the failure of these attempts of transovarial transfer of *A. marginale* infection.

The rickettsemia level of persistently infected animals influences *A. marginale* transmission by *Dermacentor andersoni* and *B. microplus*.[18,19] Rickettsemia levels during persistence are cyclical and vary from $10^2$ to $10^7$ $mL^{-1}$. Although the tick infection rate correlates with the level of rickettsemia during acquisition feeding, a substantial percentage of ticks (27%) become infected even while feeding at the low levels of persistent rickettsemia that occur between the cyclic peaks.[20] Our results show an increase in the number of positive sample in larvae from engorged female ticks fed on calves with low parasitemia.

This study determined the presence of *A. marginale* DNA in *B. microplus* larvae by PCR. These larvae originated from a long oviposition period of engorged female ticks. Even though these data are not conclusive, they indicate that *A. marginale* can migrate from the midgut of *B. microplus* engorged female ticks to ovaries and consequently infect larvae.

## ACKNOWLEDGMENTS

This research was supported by Conselho Nacional de Desenvolvimento Cientifico e Tecnológico, CAPES, and CPG-UEL. We thank the Instituto de Biologia Molecular do Parana/Fiocruz for the DNA sequencing analysis.

## REFERENCES

1. DUMLER, J.S., A.F. BARBET, C.P.J. BEKKER, *et al.* 2001. Reorganization of genera in families *Rickettsiaceae* and *Anaplasmataceae* in the order *Rickettsiales*: unification of some species of *Ehrlichia* with *Anaplasma*, *Cowdria* with *Ehrlichia* with *Neorickettsia*, descriptions of six new species combinations and designation of *Ehrlichia equi* and "EGH agent" as subjective synonyms of *Ehrlichia phagocytophila*. Int. J. Syst. Evol. Microbiol. **51:** 2145–2165.
2. RISTIC, M. 1968. Anaplasmosis. *In* Infectious Blood Diseases of Man and Animals. D. Weinman & M. Ristic, Ed.: 478–572. Academic Press. New York.
3. EWING, S.A. 1981. Transmission of *Anaplasma marginale* by arthropods. *In* Proceedings of the Seventh National Anaplasmosis Conference. pp. 395–423. Mississippi State University. Mississippi State, MI.
4. KESSLER, R.H. 2001. Considerações sobre a transmissão de *Anaplasma marginale*. Pesq. Vet. Bras. **21**(4): 177–179.
5. CONNELL, M.L. & W.T.K. HALL. 1972. Transmission of *Anaplasma marginale* by the cattle tick *Boophilus microplus*. Aust. Vet. J. **48:** 477.
6. MASON, C.A.& R.A.I. NOVAL. 1981. The transfer of *Boophilus microplus* (Acarina: Ixodidae) from infested to uninfested cattle under field conditions. Vet. Parasitol. **8:** 185–188.
7. STICH, R.W., K.M. KOCAN, G.H. PALMER, *et al.* 1989. Transtadial and attempted transovarial transmission of *Anaplasma marginale* by *Dermacentor variabilis*. Am. J. Vet. Res. **50**(8): 1377–1385.
8. CONNELL, M.L. 1974. Transmission of *Anaplasma marginale* by the cattle tick *Boophilus microplus*. Queensland J. Agric. Anim. Sci. **31**(3): 185–195.
9. LEATCH, G. 1973. Preliminary studies on the transmission of *Anaplasma marginale* by *Boophilus microplus*. Aust. Vet. J. **49**(1): 16–19.
10. THOMPSON, K.C. & J.C. ROA. 1978 Transmission de *Anaplasma marginale* por la garrapata *Boophilus microplus*. Rev. Inst. Colomb. Agropecu. **13**(1): 131–134.
11. RIBEIRO, M.F.B., J.D. LIMA & J.H.P. SALCEDO. 1996. Attempted transmission of *Anaplasma marginale* by infected *Boophilus microplus*. Arq. Bras. Med. Vet. Zootec. **48**(4): 396–401.

12. RIBEIRO, M.F.B. & J.D. LIMA. 1995. Influence of temperature on the development of *Anaplasma marginale* in *Boophilus microplus*. Arq. Bras. Med. Vet. Zootec. **47**(4): 525–533.
13. RIBEIRO, M.F.B. & J.D. LIMA. 1996. Morphology and development of *Anaplasma marginale* in midgut of engorged female ticks of *Boophilus microplus*. Vet. Parasitol. **61:** 31–39.
14. BOOM, R., C.J.A. SOL, M.M.M. SALIMANS, *et al.* 1990. Rapid and simple method for purification of nucleic acids. J. Clin. Microbiol. **28**(3): 495–503.
15. TORIONI DE ECHAIDE, S., D.P. KNOWLES, T.C. MCGUIRE, *et al.* 1998. Detection of cattle naturally infected with *Anaplasma marginale* in a region of endemicity by nested PCR and a competitive enzyme-linked immunosorbent assay using recombinant major surface protein 5. J. Clin. Microbiol. **36**(3): 777–782.
16. VEGA, R. 1976. Contribuición al estudio de la biologia de *Boophilus microplus* (Canestrini, 1887) en Acad. Cien. Cuba. Ser. Biol. **64:** 1–8.
17. ALVARADO, R.U. & J.C. GONZALES. 1979. A postura e a viabilidade do *Boophilus microplus* (Canestrini, 1887) (Acarina: Ixodidae) em condições de laboratório. Revista Latinoamericana de Microbiologia **21:** 31–36.
18. KOCAN, K.M., D. HOLBERT, S.A. EWING, *et al.* 1983. Influence of parasitemia level at feeding on development of *Anaplasma marginale* Theiler in *Dermacentor andersoni* Stiles. Am. J. Vet. Res. **44**(4): 554–557.
19. AGUIRRE, P.H., A.B. GAIDO & A.E. VIÑABAL. 1994. Transmission of *Anaplasma marginale* with adult *Boophilus microplus* ticks fed as nymphs on calves with different levels of rickettsaemia. Parasite **1**(4): 405–407.
20. ERIKS, I.S., D. STILLER & G.H. PALMER. 1993. Impact of persistent *Anaplasma marginale* rickettsemia on tick infection and transmission. J. Clin. Microbiol. **31**(8): 2091–2096.

# Assessment of Feline Ehrlichiosis in Central Spain Using Serology and a Polymerase Chain Reaction Technique

ENARA AGUIRRE,[a] MIGUEL A. TESOURO,[b] INMACULADA AMUSATEGUI,[a] FERNANDO RODRÍGUEZ-FRANCO,[a] AND ANGEL SAINZ[a]

[a]*College of Veterinary Medicine, Universidad Complutense de Madrid, Spain*

[b]*College of Veterinary Medicine, Universidad de León, Spain*

**ABSTRACT: Antibodies to *Ehrlichia* spp. and inclusion bodies compatible with *Ehrlichia* spp. in feline blood cells have been previously detected in Spain. The aim of this study was to assess the presence of antibodies to *E. canis*, *N. risticii*, and *A. phagocytophilum* in 122 feline serum samples from Madrid (central Spain). In addition, *Ehrlichia* genus-specific, one-tube, nested polymerase chain reaction (PCR) was performed from blood samples from these cats. Of the cats, 10.6% were seropositive for *E. canis*, 2.4% were positive for *N. risticii*, and 4.9% were seropositive for *A. phagocytophilum*. Two *N. risticii*–positive cats and one animal seropositive to *A. phagocytophilum* were also seropositive for *E. canis*. Despite these seropositive results, all the blood samples analyzed by PCR were negative. Our results demonstrate reactivity against agents implicated in feline ehrlichiosis in Spain. Further studies should be performed in order to clarify the significance of serology and PCR in the diagnosis of feline ehrlichiosis.**

**KEYWORDS: feline ehrlichiosis; *Ehrlichia*; Spain; PCR; serology; cat**

## INTRODUCTION

Feline ehrlichiosis was first described in 1986 in France,[1] and subsequently in Kenya,[2] the United States,[3] France,[4] Brazil,[5] and Sweden.[6] *Ehrlichia* spp.–seropositive cats and inclusion bodies compatible with *Ehrlichia* spp. in feline mononuclear cells have also been reported in Spain.[7]

Different studies have demonstrated the presence of antibodies to *Ehrlichia canis*, *Neorickettsia risticii*, and *Anaplasma phagocytophilum* in feline sera.[8–10] Culture of *Ehrlichia* spp. has not yet been achieved from feline samples, but *Ehrlichia equi*-like[6] and *Ehrlichia canis*-like DNA have been amplified and sequenced from feline blood samples.[11]

Address for correspondence: Enara Aguirre, Departamento de Medicina y Cirugía Animal, Facultad de Veterinaria, Universidad Complutense de Madrid, Avda. Puerta de Hierro s/n, 28040, Madrid, Spain. Voice: 34-91-3943820; fax: 34-91-3943811.
angelehr@vet.ucm.es

**Ann. N.Y. Acad. Sci. 1026: 103–105 (2004). © 2004 New York Academy of Sciences.
doi: 10.1196/annals.1307.013**

The aim of this study was to assess the presence of antibodies to *Ehrlichia canis*, *Neorickettsia risticii*, and *Anaplasma phagocytophilum*, using the immune-fluorescent antibody (IFA) test, in feline serum samples from Madrid (central Spain). In addition, *Ehrlichia* genus-specific, one-tube, nested polymerase chain reaction (PCR) was performed from EDTA blood samples.

## MATERIALS AND METHODS

This study involved 122 cats treated at the Hospital of the College of Veterinary Medicine in Madrid for diverse medical or surgical diseases between April 2002 and February 2003. Serum and EDTA anticoagulated blood samples were taken from every cat.

Each serum sample was tested against *E. canis*, *A. phagocytophilum*, and *N. risticii* antigens using the IFA test. Positive and negative canine serum samples were used as controls, and the cutoff was established at an antibody titer of 1:40.

With a commercially available Ultraclean DNA Blood Spin Kit (Mobio Laboratories, Madrid, Spain), DNA was extracted from 200 μL of EDTA whole-blood samples that had been stored at –20°C.

One-tube, nested PCR was performed with *Ehrlichia* genus-specific primers as previously described.[12,13] To increase the sensitivity of the technique, concentrations of the primers were modified. In the case of the outer primers EHR-OUT1 and EHR-OUT2, 0.05 μmol was used. For the inner primers GE2f and EHRL3-IP2, we used 12.5 μmol.

## RESULTS AND DISCUSSION

Thirteen cats (10.6%) were seropositive for *E. canis*, three (2.4%) were positive for *N. risticii*, and six (4.9%) were seropositive for *A. phagocytophilum*. Two of the animals positive for *N. risticii* and one *A. phagocytophilum*–positive cat were also positive for *E. canis*. The possibility of cross-reactions among these agents should be considered.

A relationship between the antibody titer and the presence of clinical signs could not be established. Seropositive cats were admitted to the Hospital of the College of Veterinary Medicine with different diseases. One cat suffered from traumatism, three from renal failure, one from struvite urolithiasis, two from respiratory distress, one from gingivitis, one from pyometra, three from neoplastic disorders, and one from anterior uveitis. In addition, four cats presented with vomiting due to unknown causes, and another had fever of unknown origin. One of the seropositive cats had been taken to the Hospital for voluntary ovariohysterectomy.

In relation to laboratory findings, five (26.3%) of the 19 cats seropositive for at least one of the three agents tested presented with anemia, 3 (15.8%) had leukopenia, 3 (15.8%) displayed leukocytosis, 5 (26.3%) had hyperproteinemia, and 3 (15.8%) demonstrated altered renal biochemistry. The rest of the hematological and biochemical parameters remained normal.

Some studies have associated feline ehrlichiosis with fever, vomiting, polyarthritis, lameness, anorexia, weight loss, ocular discharge, anemia, thrombocytopenia, and hyperglobulinemia.[10] Our results show the presence of some of these clinical signs in several cats.

Despite these seropositive results, all the blood samples analyzed by PCR were negative. PCR amplification is considered a very specific and sensitive test. The negative PCR results in our seropositive cats may have been caused by the absence of these agents in peripheral blood because of their sequestration in other tissues. In any case, few data regarding the significance of seropositivity against these agents are available.

Still, all of the seronegative cats in our study were shown by PCR to be negative for *Ehrlichia* genus. It is important to be aware, however, that recent PCR studies on seronegative cats have shown that the cats were positive for *Ehrlichia*.[11] Until isolation, culture, and sequencing of *Ehrlichia* spp. from cats can be performed, molecular evidence of *Ehrlichia*-like DNA detected in these animals must be interpreted with caution.

Our results demonstrate reactivity against agents implicated in feline ehrlichiosis in Spain. The present study suggests the difficulty of finding *Ehrlichia*-positive feline blood samples in our area using PCR. Further studies should be performed to clarify the significance of serology and PCR in the diagnosis of feline ehrlichiosis.

## REFERENCES


1. CHARPENTIER, F. & P. GROULADE. 1986. Probable case of ehrlichiosis in a cat. Bull. Acad. Vet. France **59:** 287–290.
2. BUORO, I.B.J., R.B. ATWELL, J. KIPTOON, *et al.* 1989. Feline anemia associated with *Ehrlichia*-like bodies in three domestic shorthaired cats. Vet. Rec. **125:** 434–436.
3. BOULOY, R.P., M.R. LAPPIN, C.H. HOLLAND, *et al.* 1994. Clinical ehrlichiosis in a cat. J. Am. Vet. Med. Assoc. **204:** 1475–1478.
4. BEAUFILS, J.P., J. MARIN-GRANEL & P. JUMELLE. 1995. *Ehrlichia* infection in cats: a review of three cases. Prat. Med. Chir. Anim. Comp. **30:** 397–402.
5. ALMOSNY, N.R.P. & C.L. MASSARD. 1999. Feline ehrlichiosis. Rev. Clin. Vet. **4:** 30–32.
6. BJOERSDORFF, A., L. SVENDENIUS, J.H. OWENS, *et al.* 1999. Feline granulocytic ehrlichiosis: a report of a new clinical entity and characterization of the new infectious agent. J. Small Anim. Pract. **40:** 20–24.
7. TESOURO, M.A., I. AMUSATEGUI, F. RODRÍGUEZ, *et al.* Detección de anticuerpos frente a *Ehrlichia* canis en sangre de gatos de la zona centro de España. Paper presented at the IV Simposium Ibérico sobre Ixodoidea y Enfermedades Transmitidas. Setúbal, Portugal, September 24–26, 2000.
8. MATTHEMAN, L.A., P.J. KELLY, K.WRAY, *et al.* 1996. Antibodies in cat sera from southern Africa react with antigens of *Ehrlichia canis*. Vet. Rec. **138:** 364–365.
9. PEAVY, G.M., C.J. HOLLAND, S.K. DULTA, *et al.* 1997. Suspected ehrlichial infection in five cats from a household. J. Am. Vet. Med. Assoc. **210:** 231–234.
10. STUBBS, C.J., C.J. HOLLAND, J.S. REIF, *et al.* 2000. Feline ehrlichiosis: literature review and serologic survey. Compend. Cont. Educ. Pract. Vet. **22:** 307–317.
11. BREITSCHWERDT, E.B., A.C.G. ABRAMS-OGG, M.R. LAPPIN, *et al.* 2002. Molecular evidence supporting *Ehrlichia canis*-like infection in cats. J. Vet. Intern. Med. **16:** 642–649.
12. BREITSCHWERDT, E.B., B.C. HEGARTY & S.I. HANCOCK. 1998. Sequential evaluation of dogs naturally infected with *Ehrlichia canis, Ehrlichia chaffensis, Ehrlichia equi, Ehrlichia ewingii*, or *Bartonella vinsonii*. J. Clin. Microbiol. **36:** 2645–2651.
13. KORDICK, S.K., E.B. BREITSCHWERDT, B.C. HEGARTY, *et al.* 1999. Coinfection with multiple tick-borne pathogens in a Walker Hound kennel in North Carolina. J. Clin. Microbiol. **37:** 2631–2638.

# Nested PCR for Detection and Genotyping of *Ehrlichia ruminantium*

## Use in Genetic Diversity Analysis

DOMINIQUE MARTINEZ,[a] NATHALIE VACHIÉRY,[a] FREDERIC STACHURSKI,[b] YANE KANDASSAMY,[a] MODESTINE RALINIAINA,[c] ROSALIE APRELON,[a] AND ARONA GUEYE[d]

[a]*CIRAD, Guadeloupe, France*

[b]*CIRDES, Bobo Dioulasso, Burkina Faso*

[c]*FOFIFA, Antananarivo, Madagascar*

[d]*ISRA, Dakar, Senegal*

**ABSTRACT: *Ehrlichia ruminantium*, the agent of cowdriosis transmitted by *Amblyomma* ticks, presents an extensive genetic and antigenic diversity of key importance for vaccine formulation. Two means of nested polymerase chain reaction (PCR) targeting were developed to conduct molecular epidemiology studies in the Caribbean and Africa. The first used a conserved DNA fragment for detection of the pathogen in animals and vectors, and the second relied on the polymorphic *map1* gene for genotyping. As compared to a PCR, the nested PCR showed a 2-Log10 improvement of sensitivity and allowed amplification from ticks, blood, brain, and lungs from infected animals, providing a more accurate picture of the tick infection rate. In Guadeloupe, this rate reached 36% (*N* = 212) instead of 1.7% (*N* = 224), as previously estimated. Genetic typing was done by restriction fragment length polymorphism or sequencing of *map1* amplification products. Molecular epidemiology studies conducted in field sites selected for vaccination trials with inactivated vaccine, revealed the circulation of genetically divergent strains in limited geographical areas. It is known, then, that genetic clustering based on *map1* has no predictive value regarding the protective value of a given strain against a new strain. However, tracing the strains by this technique revealed the extent of *E. ruminantium* diversity that one can expect in a given region, and the method allows differentiation between an inadequate immune response and the challenge by a breakthrough strain on animals dying despite vaccination. Up to now, genetic typing does not avoid cross-protection studies, which were conducted in parallel, although on a more limited scale. The importance of pathogen diversity studies for optimization of vaccine design is discussed as well as the research for new polymorphic genes. These genes may allow better predictions on cross-protection, given the recent completion of the sequence of the full genome of two *E. ruminantium* strains.**

**KEYWORDS: *Ehrlichia ruminantium*; *Amblyomma* ticks; vaccine development; pCS20; *map-1***

Address for correspondence: Dominique Martinez, CIRAD-EMVT, Domaine Duclos, 97170 Petit Bourg, Guadeloupe. Voice: 590-590-25-59-95; fax: 590-590-94-03-96.
dominique.martinez@cirad.fr

**Ann. N.Y. Acad. Sci. 1026: 106–113 (2004).** **© 2004 New York Academy of Sciences.**
**doi: 10.1196/annals.1307.014**

## INTRODUCTION

*Ehrlichia ruminantium*, the causative agent of heartwater, is an obligate intracellular bacterium transmitted by *Amblyomma* ticks and infecting wild and domestic ruminants.[1] It is distributed all over sub-Saharan Africa, Madagascar, and some Caribbean islands. The parasite has a high degree of genetic diversity as shown by random amplified polymorphic DNA (RAPD)[2] and sequencing of the polymorphic *map1* gene.[3] The observation of various levels of cross-protection induced between different isolates reveals an antigenic diversity with implications for the development of vaccines.[4–6]

Besides the identification of protective antigens for vaccination, the evaluation of strain diversity in the field is essential to formulate an appropriate cocktail vaccine. The recent evaluation of an inactivated cowdriosis vaccine in selected field sites in Africa has emphasized the need to conduct such studies on diversity (unpublished data). To do so, it appeared necessary to develop molecular diagnostic tests of improved sensitivity to detect and type *E. ruminantium* in hosts and vectors. On the basis of an existing PCR detection method,[7] a nested PCR targeting the same portion of gene, pCS20, has been developed. A nested PCR to amplify the polymorphic *map1* gene was also developed for typing by restriction fragment length polymorphism (RFLP) or sequencing of the amplification fragment. The techniques were evaluated on blood, brain, and lungs of hosts as well as tick vectors before being used in molecular epidemiologic studies.

## MATERIAL AND METHODS

### *Samples and DNA Extraction*

The PCR methods were primarily developed using *E. ruminantium* DNA purified from *in vitro* cultures and then evaluated using ticks or biological samples from the ruminant hosts. In a first step, *Amblyomma variegatum* ticks fed as nymphs on febrile goats experimentally infected with *E. ruminantium* were used. The blood, brain, and lung samples from the same animals were also processed after death. After this first validation step, field ticks collected during epidemiologic studies on domestic ruminants in Guadeloupe (French West Indies) and Burkina Faso were tested. All samples, ticks and organs, were preserved either frozen at –20°C or in 70% ethanol.

DNA was extracted from ticks and host samples using the QiaAmp DNA minikit (Qiagen, Courtaboeuf, France) according to the manufacturer's instructions.

### *pCS20 Nested PCR*

The current PCR detection test using AB128 and AB129 primers has already proved to be highly specific.[8] To avoid an extensive retesting of specificity, AB128 and AB129 were used as internal primers of the nested PCR. AB 128 was also used as the external forward primer, whereas a new primer, named AB130, was selected as reverse external primer. It was selected in a region without single nucleotide polymorphism (SNP) polymorphism after alignment of the corresponding genome fragment of three different strains. After preheating the DNA at 94°C for 3 min, the first round of PCR with AB128 and AB130 primers was conducted using the following conditions: 35 cycles of a 45-sec denaturation at 94°C, a 45-sec annealing at 50°C, a 45-sec elongation at 72°C, and a final 10-min extension at 72°C. One microliter of pure or 1/10 dilution of the PCR product from the first round was submitted to a second round of PCR with AB128 and AB129 primers consisting of 35 cycles of a 45-sec denaturation at 94°C, a 45-sec annealing at 55°C, and 45 sec elongation at 72°C, followed by a 10-min extension step at 72°C.

### **map1** *Nested PCR*

The *map1* was amplified with the following primers:

- external forward (Map1NT) 5′-CTCGTAAGAAGTGCGTTAAT-3′
- external reverse (MapCT1) 5′-TTAAAATACAAACCTTCCTCC-3′
- internal forward (Map1LP) 5′-CTTGGTGTGTCCTTTTCTGA-3′
- internal reverse (Map1CT2) 5′-CCTTCCTCCAATTTCTATACC-3′

The conditions for the first round of PCR using Map1NT and Map1CT1 primers were identical to those of the first round of the pCS20 nested PCR. One microliter of pure or 1/10 dilution of the PCR product from the first round was used as template in the second round of PCR consisting of 40 cycles of a 1-min denaturation at 94°C, a 1-min annealing at 55°C, and 1 min elongation at 72°C, followed by a 10-min extension at 72°C. Amplification products from all PCR reactions were visualized in 1.5% agarose gels.

### *RFLP of PCR Amplified* **map1** *Genes*

The amplification products obtained from the *map1* nested PCR procedures were digested with the restriction enzymes RsaI, NdeI, XbaI, and PstI. The fragments were separated on agarose and visualized under ultraviolet illumination.

### *Sequencing of* **map1** *PCR Products*

PCR products were extracted from low-melting agarose gels using the "gel extraction kit" (Eurobio, Les Ulis, France) following the manufacturer's recommendations and sent for sequencing to a commercial company (Genome Express, Grenoble, France).

### *Epidemiologic Studies*

The epidemiologic studies aimed at describing the extent of strain diversity. This was done by the isolation of a range of strains originating from a limited geographical area for testing in cross-protection and genotyping studies. In Burkina Faso, the work was conducted in four villages that were separated by a distance of 15 to 40 km (Sara, Bekuy, Lamba and Bankouma). The ticks were collected on animals between October and December during the seasonal peak of *Amblyomma variegatum* nymphs. The resulting adults were fed on sheep, which developed hyperthermia. Blood from reacting sheep was used to isolate *E. ruminantium* in endothelial cells.

Afterwards, cross-protection studies were carried out with these isolates. After feeding on sheep during hyperthermia, engorged nymphs were also collected for typing *E. ruminantium* stocks. In Guadeloupe, 212 ticks (mainly adults) were collected on different herds all around the island and tested for infection by *E. ruminantium*. Positive ones were subsequently typed.

## RESULTS

### *Sensitivity of the Nested PCR*

Serial 10-fold dilutions of purified DNA showed a 2-log10 improvement of the sensitivity for both pCS20 and *map1* nested PCR as compared to PCR (FIG. 1). This corresponds to an average detection limit of 6 and 60 *E. ruminantium* elementary bodies respectively for pCS20 and *map1* nested PCR. This difference of sensitivity was confirmed by the comparison of both nested PCR, which showed that 91% of 113 pCS20-positive DNA samples were also positive for *map1*. Conversely, a *map1*-positive reaction was never observed together with a pCS20-negative result. Thus, some weak pCS20-positive samples could not be amplified for *map1* and subsequent typing.

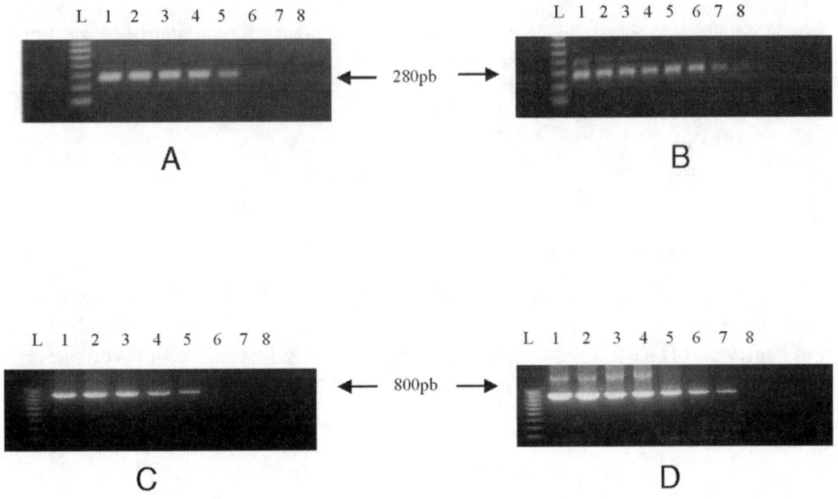


**FIGURE 1.** Detection limit of pCS20 and *map1* nested PCR. Nested PCR and PCR products from reactions with 10-fold serial dilutions of purified *E. ruminantium* DNA after agarose gel electrophoresis and ethidium bromide staining. The DNA ladder of 100 bp is shown in *lane L. Lanes 1 to 8*: serial tenfold dilution of *ER* DNA $6\times10^7$ to 0.6 estimated *E. ruminantium* elementary bodies. (**A**) pCS20 PCR. (**B**) pCS20 nested PCR. (**C**) *map1* PCR. (**D**) *map1* nested PCR.

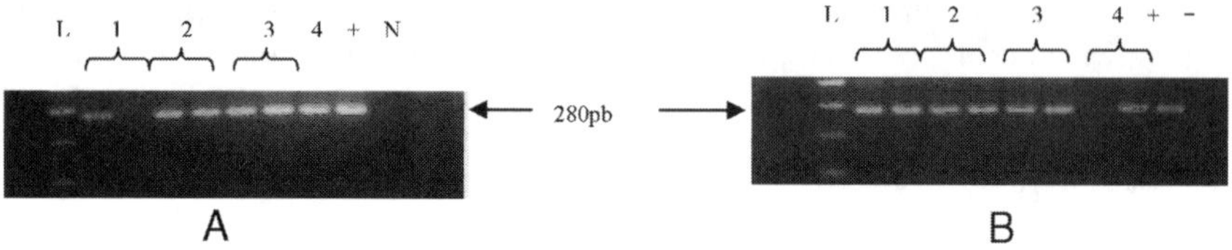


**FIGURE 2.** pCS20 nested PCR detecton of *E. ruminantium* in fresh and frozen blood, lung, and brain samples from goats. (**A**) Agarose gel electrophoresis of pCS20 nested PCR products representative of reactions with fresh (*left lane*) and frozen blood (*right lane*) obtained from febrile reacting goats 1, 2, and 3. Only fresh blood was processed for the goat 4. N: fresh blood from a naive goat. (**B**) Nested PCR products from lung (*left lane*) and brain (*right lane*) samples from goats 1, 2, 3, and 4. +: Nested PCR products of positive *E. ruminantium* DNA. –: Nested PCR products of water. The DNA ladder of 100 bp is shown in *lane L.*

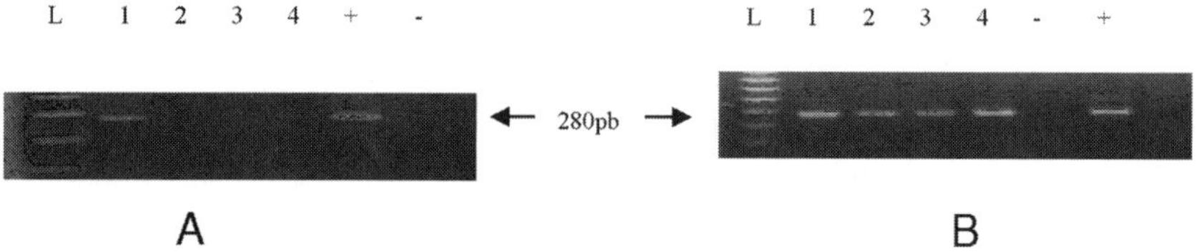


**FIGURE 3.** Detection of *E. ruminantium* in field ticks by pCS20 PCR and nested PCR. Agarose gel electrophoresis of PCR and nested PCR products representative of reactions. Samples from the same ticks (1, 2, 3, and 4) were processed by (**A**) PCR or (**B**) nested PCR. +: Positive reaction control PCR products. –: Negative reaction control PCR products.

### *Detection of* E. ruminantium *in the Ruminant Host*

The nested PCR techniques were successfully used to detect *E. ruminantium* in the blood of reacting animals as well as brain and lung samples from animals dying from heartwater (FIG. 2). The nested PCR techniques significantly improved the sensitivity of detection in all organs tested, and detection was possible in frozen samples as well as samples preserved in 70% ethanol.

### *Detection of* E. ruminantium *in Ticks*

The nested PCR techniques were first evaluated on ticks infected by engorgement on animals reacting after experimental inoculation. Ticks collected in the fields were processed afterwards. Here again, the sensitivity was increased (FIG. 3), and detection was possible both in fresh and ethanol-preserved ticks. The improvement of the detection limit by nested PCR provided a more accurate description of the epidemiologic situation. Indeed, the tick infection rate reached 36% ($N = 212$) by nested PCR, whereas it was evaluated around 1.7% ($N = 224$) by PCR.

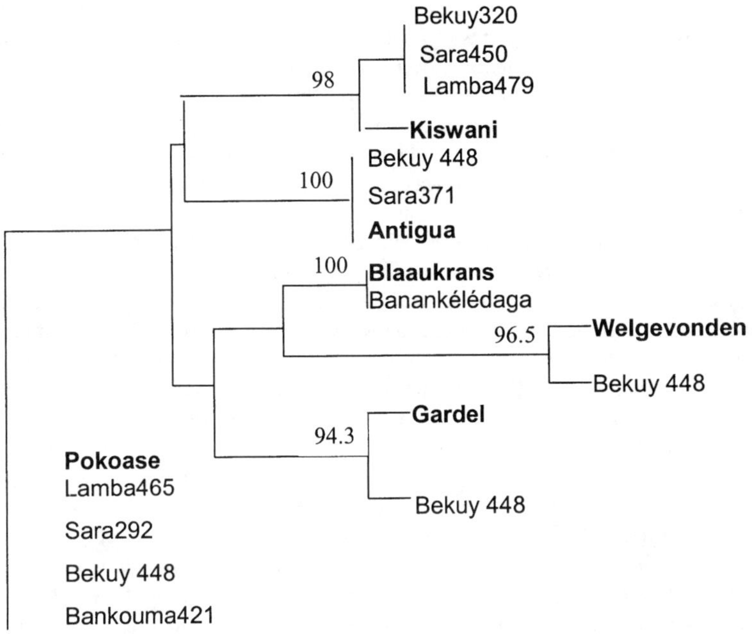


**FIGURE 4.** The *map1* genetic tree of Burkina Faso isolates. Numbers represente percentage of homology between isolates. Reference strains are indicated in **bold**.

### *Genetic Diversity in Burkina Faso and Guadeloupe*

In Burkina Faso, 12 *E. ruminantium* stocks, from eight different herds distributed among four villages, were sequenced after *map1* nested PCR on *E. ruminantium*-positive tick DNA. As shown in FIGURE 4, which shows the *map1* genetic tree, there was a wide genetic diversity between Burkina Faso isolates. The *map1* genotypes present in Burkina Faso were identical to different strains distributed in all infected Africa and Caribbean countries, such as Pokoase (Ghana), Kiswani (Kenya), and Antigua (Caribbean island). One out of 12 stocks was a mix of four different strains, whereas all the other stocks had a single genotype. In Guadeloupe, DNA from nine *E. ruminantium*-positive ticks was sequenced. As in Burkina Faso, stocks from Guadeloupe were genotypically identical to several African and Caribbean strains. Moreover, only 3 out of 35 *E. ruminantium*-positive ticks had a mix of different genotypes as tested by RFLP or sequencing.

## DISCUSSION

In this study, a nested PCR technique for the detection and typing (*map1* gene) of *E. ruminantium* was validated. A 2-Log10 improvement of the sensitivity was obtained for each PCR target (pCS20 and *map1*) as compared to PCR. The detection limit, found to be around six organisms for the pCS20 nested PCR, is similar to that obtained by another method based on PCR followed by Southern blotting of the pCS20 probe on the amplification fragment.[7] However, the advantage of a nested PCR is that it does not necessitate the laborious hybridization step and can be easily automated for high-throughput analysis in epidemiologic studies. The *map1* nested PCR was 1-Log10 less sensitive (a detection limit of around 60 organisms) than the pCS20 one. A positive *map1* result was always associated to a positive pCS20 result, confirming the high specificity of the latter. The *map1*-negative results were only observed associated with weak positive pCS20 signals from some field tick samples. However, 91% of pCS20 field-positive ticks gave a *map1* amplification product that could be used for genotyping, which is sufficient in molecular epidemiologic studies in which many samples are analyzed. Moreover, fresh samples, frozen samples, or samples preserved in 70% ethanol could be processed by nested PCR without any loss of signal intensity, confirming the possibility that *E. ruminantium* nested PCR could be used as a new molecular tool for epidemiologic studies.

With these methods, recent studies associated with large-scale field vaccination of the inactivated heartwater vaccine have shown that in a very limited geographical area at least five different *map1* genotypes were present. The breakthrough of immunity could be explained by the diversity as confirmed by cross-protection trials. However, although *map1* is a stable molecular marker useful for tracing strains, clusters obtained in the *map1* genetic distance tree are not predictors of cross-protection between isolates. This emphasizes the need to carry out epidemiologic studies on vaccination trial sites. Moreover, extensive mapping of genotype diversity associated with additional cross-protection studies appears essential to formulate appropriate vaccines.

The recent completion of two full sequences of *E. ruminantium* genomes (Gardel and Welgevonden strains) giving access to polymorphic genes of the pathogen open the way to the identification of predictors of cross-protection and, ultimately, to the development of protective vaccine candidate genes.

## ACKNOWLEDGMENTS

This research was supported by the European Union fund of the International Scientific Cooperation Projects (INCO) project contract number ICA4-CT-2000-30026.

## REFERENCES

1. Oberem, P.T. & J.D. Bezuindenhout. 1987. Heartwater in hosts other than domestic ruminants. Onderstepoort J. Vet. Res. **54:** 271–275.
2. Perez, J.M., D. Martinez, A. Debus, *et al.* 1997. Detection of genomic polymorphisms among isolates of the intracellular bacterium *Cowdria ruminantium* by random amplified polymorphic DNA and Southern blotting. FEMS Microbiol. Lett. **154:** 73–79.
3. Reddy, G.R., C.R. Sulsona, R.H. Harrison, *et al.* 1996. Sequence heterogeneity of the major antigenic protein 1 genes from *Cowdria ruminantium* isolates from different geographical areas. Clin. Diagn. Lab. Immunol. **3:** 417–422.
4. Du Plessis, J.L. & L. Van Gas. 1989.Immunity of tick-exposed seronegative and seropositive small stock challenged with two stocks of *Cowdria ruminantium.* Onderstepoort J. Vet. Res. **56:** 185–188.

5. UILENBERG, G., E. CAMUS & N. BARRÉ. 1985. Quelques observations sur une souche de *Cowdria ruminantium* isolée en Guadeloupe (Antilles Françaises). Rev. Elev. Med. Vet. Trop. 38(**1**): 34–42.
6. JONGEJAN, F., & M.J.C. THIELEMANS. 1991.Antigenic diversity of *Cowdria ruminantium* isolates determined by cross-immunity. Res. Vet. Sci. **51:** 24–28.
7. PETER, T.F., A.F. BARBET, A.R.R ALLEMAN, *et al.* 2000. Detection of the agent of heartwater, *Cowdria ruminantium*, in *Amblyomma* ticks by PCR: validation and application of the assay to field ticks. J. Clin. Microbiol. **38:** 1539–1544.
8. PETER, T.F., S.L. DEEM, A.F. BARBET, *et al.* 1995. Development and evaluation of PCR assay for detection of low levels of *Cowdria ruminantium* infection in *Amblyomma* ticks not detected by DNA probe. J. Clin. Microbiol. **33:** 166–172.

# Recent Studies on the Characterization of *Anaplasma marginale* Isolated from North American Bison

KATHERINE M. KOCAN,[a] JOSÉ DE LA FUENTE,[a]
ELIZABETH J. GOLSTEYN THOMAS,[b] RONALD A. VAN DEN BUSSCHE,[c]
ROBERT G. HAMILTON,[d] ELAINE E. TANAKA,[b] AND SUSAN E. DRUHAN[b]

[a]*Department of Veterinary Pathobiology, College of Veterinary Medicine,* [c]*Department of Zoology and Collection of Vertebrates, Oklahoma State University, Stillwater, Oklahoma, USA*

[b]*Canadian Food Inspection Agency, Lethbridge Laboratory, Animal Diseases Research Institute, Lethbridge, Alberta, Canada,*

[d]*The Nature Conservancy, Tallgrass Prairie Preserve, Pawhuska, Oklahoma, USA*

**ABSTRACT: *Anaplasma marginale* (Rickettsiales: Anaplasmataceae), a tick-borne pathogen of cattle, is endemic in tropical and subtropical regions of the world. Many geographic isolates of *A. marginale* occur worldwide that have been identified by major surface protein (MSP) 1a, which varies in sequence and molecular weight owing to different numbers of tandem 28-29 amino acid repeats. Although serologic tests have identified American bison, *Bison bison*, as being infected with *A. marginale*, the present studies were undertaken to confirm *A. marginale* infection in bison, to characterize bison isolates, and to compare the phylogenetic relationship of the bison isolates with other *A. marginale* isolates from North America. Nine *A. marginale* isolates derived from Canadian bison possessed identical *msp4* sequences with one characteristic silent nucleotide change. The sequence of MSP1a was determined for one Canadian and two U.S. bison isolates of *A. marginale*, and these isolates contained 4 and 5 tandem repeats, respectively. One U.S. bison isolate tested for infectivity proved to be infective for cattle and transmitted by *Dermacentor variabilis* ticks. the results of this study demonstrated that these *A. marginale* isolates obtained from bison were similar to ones derived from naturally infected cattle.**

**KEYWORDS: *Anaplasma marginale*; bison; ticks; cattle; major surface protein**

*Anaplasma marginale* (Rickettsiales: Anaplasmataceae) is a rickettsial pathogen that causes the disease anaplasmosis in cattle (see review by Kocan *et al.*[1]). Feeding ticks effect biological transmission of this obligate intraerythrocytic organism, whereas mechanical transmission occurs when infected blood is transferred to sus-

Address for correspondence: Katherine M. Kocan, Department of Veterinary Pathobiology, College of Veterinary Medicine, 250 McElroy Hall, Oklahoma State University, Stillwater, OK 74078-2007, USA. Voice: (405) 744-7271; fax: (405) 744-5275.
kmk285@cvm.okstate.edu

**Ann. N.Y. Acad. Sci. 1026: 114–117 (2004). © 2004 New York Academy of Sciences.
doi: 10.1196/annals.1307.015**

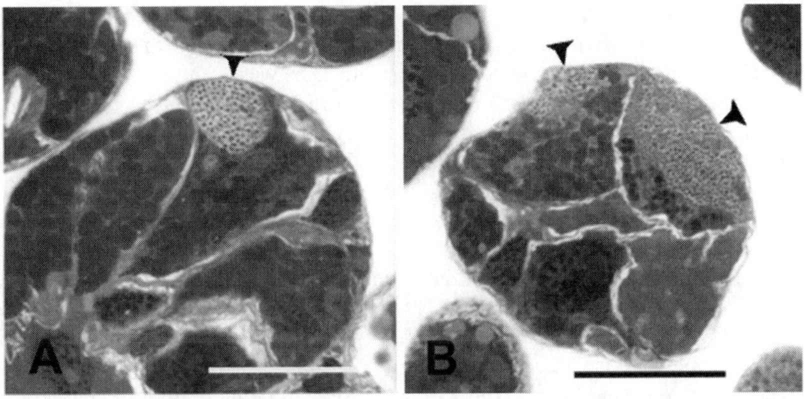


**FIGURE 1.** Light micrographs of salivary glands of *D. variabilis* that fed on a calf infected with the U.S. bison isolate of *A. marginale*. (**A**) One large colony of *A. marginale* at the margin of the salivary gland acinar cell (*arrowhead*). (**B**) Two salivary gland acinar cells with *A. marginale* colonies (*arrowheads*). *Bar*: 5 μm.

ceptible cattle by biting flies or blood-contaminated fomites (see reviews by Ewing[2] and Kocan *et al.*[3]). Many geographic isolates of *A. marginale* occur worldwide that differ in biology, morphology, protein sequence, and antigenic characteristics. These isolates have been identified by the characterization of major surface protein (MSP) 1a, which varies in sequence and molecular weight because of different numbers of tandem 28–29 amino acid repeats (see Allred *et al.*[4] and the review by de la Fuente *et al.*[5]). The *msp4* gene of *A. marginale* isolates was found in previous studies to provide phylogeographic information.[6–8]

American bison, *Bison bison*, were reported previously to be infected with *A. marginale* on the basis of serologic surveys of naturally infected animals and to be susceptible to *A. marginale* by experimental infection studies.[9–10] Recent studies demonstrated *A. marginale* infection in two herds of naturally infected bison from two widely separated geographic areas, the United States (Nature Conservancy, Tallgrass Prairie Preserve, Pawhuska, Oklahoma) and Canada (Saskatchewan).[11] In the Canadian herd, 10 animals proved to be infected with *A. marginale* on the basis of serology and polymerase chain reaction assays, whereas 42 of 50 bison culled from the Tallgrass Prairie Preserve (TGPP) proved to be persistent carriers of *A. marginale* on the basis of serology. The U.S. bison isolate of *A. marginale* was found to be infective when inoculated into susceptible splenectomized calves.

The *A. marginale* bison isolate proved to be infective for *Dermacentor* spp. ticks.[11] In two trials, male *D. andersoni* and *D. variabilis* ticks were allowed to feed for 7 days on a splenectomized calf that was experimentally infected with the bison isolate of *A. marginale* obtained from one of the TGPP bison. The ticks were then removed from the infected calf and held in a humidity chamber for 7 days, after which they were allowed to feed on a susceptible, splenectomized calf. The ticks transmitted *A. marginale* to the susceptible calf, and the gut and salivary glands dissected from the ticks were confirmed to be infected by use of light and electron microscopy (FIG. 1) and a polymerase chain reaction assay.

Phylogenetic analysis of *A. marginale* MSP1a and MSP4 sequences, shown previously to provide phylogeographic information, demonstrated that the *msp4* sequence for bison isolates grouped together with U.S. west-central cattle isolates of *A. marginale*. Although the *msp4* sequences from U.S. and Canadian bison isolates differed, the *msp4* sequences from the two U.S. bison isolates (an additional one obtained after publication of de la Fuente *et al.*[11]) were identical, as were the nine bison isolates from Canadian bison. The region for the *msp1* gene was amplified from one Canadian bison sample and one *A. marginale* isolate from U.S. bison after inoculation into a susceptible calf. The sequences demonstrated the presence of four and five tandem repeats, respectively, in the amino terminal portion of the protein as described for other *A. marginale* isolates.[5,11] However, the MSP1a sequence of bison isolates of *A. marginale* had three repeat forms (Q, R, and W) that are not present in any other sequenced isolate of *A. marginale*.

The results reported herein confirm that bison can be naturally infected with *A. marginale*. Sequence information of *A. marginale* isolates from bison in Canada and the United States suggests that these isolates are closely related to isolates of *A. marginale* from U.S. cattle. Furthermore, these studies documented that bison isolates of *A. marginale* are transmissible by *Dermacentor* spp. ticks, the main vector for *A. marginale* in the United States.

## ACKNOWLEDGMENTS

This research was supported by the Oklahoma Agricultural Experiment Station (project 1669); the Endowed Chair for Food Animal Research (K.M. Kocan, College of Veterinary Medicine, Oklahoma State University); the National Institutes of Health, Centers for Biomedical Research Excellence (through a subcontract to J. de la Fuente from the Oklahoma Medical Research Foundation), and the Oklahoma Center for the Advancement of Science and Technology (applied research grant AR021-037). Dollie Clawson and Joy Yoshioka (Department of Veterinary Pathobiology, Oklahoma State University) are acknowledged for technical assistance; Sue Ann Hudiburg and Janet J. Rogers (Core Sequencing Facility, Department of Biochemistry and Molecular Biology, Noble Research Center, Oklahoma State University) are acknowledged for oligonucleotide synthesis and DNA sequencing, respectively; Terry Carroll (CFIA Lethbridge Laboratory) is acknowledged for performing the anaplasmosis complement fixation test.

## REFERENCES

1. Kocan, K.M., E.F. Blouin & A.F. Barbet. 2000. Anaplasmosis control: past, present and future. Ann. N.Y. Acad. Sci. **916:** 501–509.
2. Ewing, S. A. 1981. Transmission of *Anaplasma marginale* by arthropods. *In* Proceedings of the Seventh National Anaplasmosis Conference: 395–423. Mississippi State University. Mississippi State, MS.
3. Kocan, K.M., J. de la Fuente, A.A. Guglielmone & R.D. Melendéz. 2004. Antigens and alternatives for control of *Anaplasma marginale* infection in cattle. Clin. Microbiol. Rev. **16:** 698–712.
4. Allred, D.R., T.C. McGuire, G.H. Palmer, *et al.* 1990. Molecular basis for surface antigen size polymorphisms and conservation of a neutralization-sensitive epitope in *Anaplasma marginale*. Proc. Natl. Acad. Sci. USA **87:** 3220–3224.

5. DE LA FUENTE, J., J.C. GARCIA-GARCIA, E.F. BLOUIN, *et al.* 2001. Evolution and function of tandem repeats in the major surface protein 1a of the ehrlichial pathogen *Anaplasma marginale*. Anim. Health Res. Rev. **2:** 163–173.
6. DE LA FUENTE, J., R.A. VAN DEN BUSSCHE, J.C. GARCIA-GARCIA, *et al.* 2002. Phylogeography of New World isolates of *Anaplasma marginale* (Rickettsiaceae: Anaplasmataceae) based on major surface protein sequences. Vet. Microbiol. **88:** 275–285.
7. DE LA FUENTE, J., R.A. VAN DEN BUSSCHE & K.M. KOCAN. 2001. Molecular phylogeny and biogeography of North American isolates of *Anaplasma marg*inale (Rickettsiaceae: Ehrlichieae). Vet. Parasitol. **97:** 65–76.
8. DE LA FUENTE, J., R.A. VAN DEN BUSSCHE, T.M. PRADO & K.M. KOCAN. 2003. *Anaplasma marginale* major surface protein 1α genotypes evolved under positive selection pressure but are not a marker for geographic isolates. J. Clin. Microbiol. **41:** 1609–1616.
9. TAYLOR, S.K., V.M. LANE, D.L. HUNTER, *et al.* 1997. Serologic survey for infectious pathogens in free-ranging American bison. J. Wildl. Dis. **33**: 308–311.
10. ZAUGG, J.L. & K.L. KUTTLER. 1985. *Anaplasma marginale* infections in American bison: experimental infection and serologic study. Am. J. Vet. Res. **46:** 438–441.
11. DE LA FUENTE, J., E.J. GOLSTEYN THOMAS, R.A. VAN DEN BUSSCHE, *et al.* 2003. Characterization of *Anaplasma marginale* isolated from North American bison. Appl. Environ. Microbiol. **69:** 5001–5005.

# Comparative Study of Three Surgical Treatments for Two Forms of the Clinical Presentation of Bovine Pododermatitis

L.A.F. SILVA,[a] I.B. ATAYDE,[a] M.C.S. FIORAVANTI,[a] D. EURIDES,[b] K.S. OLIVEIRA,[a] C.A. SILVA,[c] D. VIEIRA,[a] AND E.G. ARAÚJO[a]

[a]*Department of Veterinary Medicine, School of Veterinary Medicine, UFG, Goiânia, Goiás, Brazil*

[b]*Department of Veterinary Medicine, FMV/UFU, Uberlândia, Minas Gerais, Brazil*

[c]*Agriculture Science Center, UFG, Jataí, Goiás, Brazil*

**ABSTRACT: In this study, 1013 animals showing signs of clinical pododermatitis were examined and divided into five unevenly numbered groups. Affected animals in Groups I and II showed only signs of vegetative interdigital pododermatitis. The lesions were surgically removed and either protected with bandages (in Group I) or cauterized with incandescent iron (Group II). The animals in Groups III, IV, and V, showed signs of necrotic pododermatitis. These were treated with different protocols after the necrotic tissue was surgically removed: in Group III, the lesion was cauterized; in Group IV, the wound was protected with bandages; and in Group V, both the second and the third phalanges were amputated. There was a statistically significant relapse difference between Group III and Group V, as well as a difference among Group IV and Group V animals, and there were fewer relapses among the latter. The treatment used in Groups II and III proved to be efficient and inexpensive. Amputation of the phalanges was the treatment that resulted in fewer relapses among all protocols, despite its mutilating effect. The association of a local and parenteral treatment with an antibiotic agent, as well as the use of foot baths, contributed greatly to a fast recovery.**

**KEYWORDS: vegetative interdigital pododermatitis; necrotic pododermatitis**

## INTRODUCTION

Pododermatitis is an infectious disease that occurs in acute and chronic forms, and its clinical manifestations vary according to the stage of the disease. The lesion is characterized as vegetative interdigital pododermatitis when there is a proliferation of skin and subcutaneous tissue in the interdigital space,[1] or as necrotic infectious pododermatitis when necrosis is present. Necrosis can reach the circulation of the second or third phalanx, and affect the bones, tendons, and sheaths.[2–4]

Address for correspondence: L.A.F. Silva, Rua 18-A, 591 Apt. 502, Ed. Acauã. Setor Aeroporto, 74 070-060 Goiânia, Goiás, Brazil. Voice: 55-62-521-1572; fax: 55-62-521-1566. lafranco@vet.ufg.br.

**Ann. N.Y. Acad. Sci. 1026: 118–124 (2004). © 2004 New York Academy of Sciences. doi: 10.1196/annals.1307.016**

Acute pododermatitis presents as a hard tumefaction in the middle of the posterior and/or anterior region of the interdigital space of the affected limb.[5] This tumefaction may be followed by a chronic inflammatory process causing proliferation of tissue in the interdigital space. Necrotic lesions are attributed to *Bacterioides nodosus* and *Fusobacteriun necrophorum,* both Gram-negative anaerobes that do not release spores.[1,3,4,6–9] This disease affects cattle of all ages and breeds; however, animals that originate from *Bos taurus* breeds seem more susceptible than those from *Bos indicus*.[4]

Several treatments have been proposed. They can be local, systemic, or surgical, or (less commonly) they can involve vaccination.[6,10–13] Local or systemic treatments with antibiotics or sulfonamides have been reported to slow the course of the disease.[11,14–16] Further recommendations include thorough cleaning of the hooves, removal of the interdigital necrotic tissue, binding the lesion, applying protective caps and a wood block on the healthy nail, and the use of foot baths. The amputation of the phalanx in those cases of necrotic and purulent inflammation of the distal or proximal interphalanx articulation is recommended by others.[12,13,17]

The aim of this paper is to compare the results of different surgical treatments for bovine pododermatitis.

## MATERIALS AND METHODS

This study was carried in 105 rural properties of the State of Goiás, Brazil, from 1985 to 2002, in different months of the year. A total of 1013 bovines of different age, sex, and weight presenting clinical signs of pododermatitis were treated. These were from an estimated total of 28,000 animals from both meat and milking herds.

The animals were divided into five groups according to previously established clinical criteria.[18,19] Animals with vegetative interdigital pododermatitis showed a growth of fibrous tissue in the interdigital space but no ulceration of the soft tissue. Animals with ulceration of the soft tissue, fissures in the horn tissue, fibrosis, and a necrotic smell in the distal extremity of the limbs were diagnosed as having necrotic infectious pododermatitis. The degree of the pathology was not taken into consideration in establishing the type of treatment applied.

After clinical examination and characterization of the morbid process, the animals were divided as follows:

- Group I: 230 animals with vegetative interdigital pododermatitis whose lesions were surgically removed. Postoperative care included the protection of the wound with bandages.
- Group II: 199 animals with the same pathology as the animals in Group I. After surgical removal of the lesion, the wound was submitted to cauterization with an incandescent iron.
- Group III: 297 animals with necrotic pododermatitis. After the necrotic areas were surgically removed, the region was cauterized with an incandescent iron.
- Group IV: 124 animals with necrotic pododermatitis whose lesions were surgically removed. Postoperative care included protecting the wound with bandages;

- Group V: 163 animals with the same pathology as the animals in Groups III and IV. These animals, however, had their second and third phalanges amputated, and the wound was protected with bandages.

The preoperative care of all the animals included complete fastening for 12–18 hr. All the animals were constrained in adequate tie stalls. The surgical field was first cleaned with soap and water. Group V animals that had the distal forth of the phalanges amputated underwent trichotomy. Antisepsis was completed with local application of an iodophor solution (Biocid, Pfizer, Brazil). Lidocaine hydrochloride (Anestésico L. Pearson, Pearson, Brazil) was used as anesthetic, and it was injected through intravenous local-regional infiltration.

The animals that had vegetative interdigital pododermatitis were surgically treated according to a previously suggested technique.[14] In brief, fibrous, superficial tissue was removed, and the hoof was toileted. The animals that had septic pododermatitis were treated by having the deep, necrotic tissue removed followed by hoof toilet, according to another technique also previously reported.[20,21] Cauterization was accomplished with an incandescent iron, which was applied deeply enough to attain hemostasia, but carefully enough not to burn the healthy tissue.

During the postoperative period, all animals were given 10 mg/kg of intramuscular oxitetracycline (Oxitrat-LA, Vallée, Brazil), four applications every other day. The clinical evaluation of healing was performed daily for 45 days after the surgical procedure.

The Groups II and III animals received one application of impermeable divynil pirrolidone dihydrochloride, nitrofenil-fosfotioate, and vegetal-tar-associated ointment (Miostal, Minerthal, Brazil) directly on the just-cauterized wound. From day 2 onward, this treatment was replaced by daily 1% sodium hypochlorite foot baths.

After the surgical procedure, the animals from Groups I and IV had their wounds covered by an ointment combining gentamycin, sulphanilamide, sulphadiazine, urea, vitamin A, and palmitate (Vetaglós pomada, Univet S.A., Brazil), and bandaging was made impermeable by the same product used in the wounds of the animals from Groups II and III. The bandages were removed after 7 days, and the animals subjected to daily 1% sodium hypochlorite foot baths.

The animals of Group V had their wounds covered after the surgery with Vetaglós-saturated gauze (with the same product described in the above paragraph) and bandages, also made impermeable by Vetaglós. The dressings were changed weekly, and the stitches were removed after 21 days.

The results of the different protocols were compared through the $\chi^2$ test.[22]

## RESULTS

The prevalence for pododermatitis was 3.62%, emphasizing the fact that the majority of the affected animals were either cross-bred *Bos indicus* males or dairy females. Lameness was the most frequently observed clinical sign. In 79.9% of the cases, only one limb was affected; in 18.5%, two limbs were affected; in 1.4%, three limbs were affected; and in 0.2%, all four limbs were affected.

The results are shown in TABLE 1. In spite of using different treatments, the results were similar for animals in Groups I and II, with respective relapse rates of 23.48%

**TABLE 1. Recoveries and relapses of animals presenting bovine pododermatitis grouped according to treatment**

| | Treatment | | | | |
|---|---|---|---|---|---|
| Group | A[a] | B[b] | C[c] | Recoveries | Relapses |
| I[d] | 230 | | | 176 | 54 (23.48%) |
| II[d] | | 199 | | 167 | 32 (16.08%) |
| III[e] | | 297 | | 201 | 96 (32.32%)[f] |
| IV[e] | 124 | | | 74 | 50 (40.32%)[f] |
| V[e] | | | 163 | 153 | 10 (6.13%)[f] |
| Total | | | | 771 | 242 |

[a]Surgical treatment plus bandaging.
[b]Surgical treatment plus cauterization.
[c]Amputation of 2nd and 3rd phalanges plus bandaging.
[d]Animals presenting vegetative interdigital pododermatitis.
[e]Animals presenting septic pododermatitis.
[f]$\chi^2$ was significant for Group V when compared to Groups III ($\chi^2$ = 40.69) and IV ($\chi^2$ = 49.79).

and 16.08%, and there was no significant difference. Similarly, in Groups III and IV, which had respective relapse rates of 32.32% and 40.32%, there was no significant difference, in spite of the initial impression of cauterization being a more aggressive and effective treatment.

There is a significant difference in relapse rates between Group V to Group IV and between Group V to Group III with respect to postoperative complications (32.32% and 6.13%, respectively). In spite of the mutilation and postoperative complications in Group V, amputation appeared to be more effective in controlling the disease.

## DISCUSSION

The diagnoses of the cases were mainly made upon the clinical sign of lameness. This was related to the pain and edema on the talon of the affected side resulting from accumulation of exudate under pressure between the sensitive layers of the hoof and sole.[17] Some of the cases showed other factors that were thought to contribute to lameness; these factors included tissue necrosis reaching nerve endings, interdigital tyloma, and interphalangeal arthritis. Most of the animals changed the support point of their limbs, leading to exaggerated nail growth.

The animals in the early stage of disease showed signs similar to laminitis. The occurrence of laminitis is related to many factors, such as systemic disease, nutrition, lactation, season of the year, hard ground, age, and weight.[23–26] Considering the conditions under which the animals were managed, and also considering that in most cases only one limb was affected, it was possible to be sure that none of the cases were of laminitis. The similarity between laminitis and the initial stage of pododermatitis is not recognized by all, but the latter is considered a complication of the first.[27,28]

The tie stalls were a good choice for containment, making it easier to diagnose and treat the animals. Some animals with aggressive temperament or little previous human contact were, at first, a little uncomfortable; however, in a few days, these animals became used to the tie stalls.

Proliferation of fibrous tissue in the interdigital space of these animals was most likely a consequence of a chronic inflammatory process. The necrotic lesions, on the other hand, were attributed to *Fusobacterium necrophorum* and *Bacterioides nodosus*, recognizable by their repulsive odor, according to similar observations reported elsewhere.[6,8,19]

The characterization of the morbid process after clinical examination and subsequent classification in two categories (vegetative interdigital pododermatitis and necrotic pododermatitis), as well as the allocation of animals into five different groups, permitted the evaluation of the five surgical treatments efficiently without taking the degree of the pathology into consideration. The decision to amputate the phalanx was made on the basis of the severity of compromise of the limb, and was related mostly to the presence of necrosis of the third phalanx. Previous reports use a similar classification, but these reports were not concerned with establishing different treatments for this disease.[3,4] The variety of breeds and different temperaments of our animals, which are usually raised in large, open fields, proves the need for different therapeutic approaches suitable for different situations.

The use of local-regional intravenous anesthesia employing lidocaine hydrochloride avoided pain, making it possible to perform the surgery, a procedure that has already been recommended in literature.[12] In this study, analgesia was achieved after approximately 10 min of application.

Concerning cauterization, great care was taken not to damage healthy tissue, but to achieve total hemostasis and removal of necrotic tissue left after surgery. That was a potential contributing factor for the success of the treatments.

In spite of being effective, the bandaging treatment is expensive, since the need to change the dressing at regular intervals requires the presence of a technician. Additional difficulties in using this kind of treatment are due to the size of the properties, their distance from towns, and the raising system adopted by most of the farmers. These dressings can be successfully substituted by footbaths, which are less expensive and easier to manage when treating a greater number of animals.

Moreover, inadequate use of bandages may lead to worse complications, such as contamination of the wound in case of bandage loss, or ischemia leading to necrosis of the healthy phalange if the bandage is wrapped too tightly around the limb. Because the limb may become a culture medium for bacteria in wet areas, it is recommended that animals using bandages should be kept in dry and clean environments.

The use of footbaths not only contributed to a better healing of the wounds, but also acts as a prophylactic procedure for healthy animals.

The treatment that showed the best results was the one used in Group V. In this treatment, the amputation removed all apparent foci of necrosis, which favored the recovery from necrotic pododermatitis. In some cases, the complication was such that it was necessary to remove part of the hoof wall, even reaching the coronary band. Despite the invasiveness of the surgery, we recommend removal of the phalanges in severe necrotic pododermatitis, even considering the mutilating effect. This procedure favors recovery, but it is not supported by some.[29,30]

It is important to preserve the crown of the hoof and the greatest amount of healthy tissue as possible in order to have the phalangeal stump protected by corneal formation (rudimentary hoof) after complete healing.[20,21,31] This was taken into account in all the surgical operations and in strict postoperative care. This contributed to the low number of complications.

From the results of this study it is possible to conclude:

(1) The local-regional intravenous anesthesia technique employed was a satisfactory pain-avoiding agent, allowing safe surgical procedure.
(2) The animals were efficiently contained, making the procedure safe for the animals and the surgeon.
(3) The treatments employed in Groups II and III were proven to be efficient and inexpensive.
(4) The amputation of the second and third phalanges was the most successful treatment when compared to the rest; however, because of its mutilating effect, it is recommended only after more conservative approaches have failed.
(5) Incandescent iron cauterization allowed great hemostasia of the surgical wound.

## REFERENCES


1. Weaver, A.D. 1987. Cattle foot problems part 1: introduction and interdigital skin disease. Agri-Practice **9**(1): 34–38.
2. Weaver, A.D. 1987. Cattle foot problems part 2: diseases of the horn and corium. Agri-Practice **9**(2): 35–40.
3. Fraser, C.M. Manual Merck de Veterinária. 6th edit: 553–565. Roca. São Paulo.
4. Smith, B.P. 1994. Tratado de Medicina Interna do Grandes Animais: 1164–1167. Manole. São Paulo.
5. Raven, E.T. & J.L. Cornelisse. 1971. Dermatitis interdigital contagiosa de los bóvideos denominada em este articulo: pata hedionda. Not. Med. Vet. **2/3:** 219–243.
6. Ribeiro, L.A.O. 1980. Foot-rot dos ovinos: etiologia, patogenia e controle. Bol. IPVDF Porto Alegre **7:** 41–45.
7. Holdeman, L.V., R.W. Kelley & W.E.C. Moore. 1986. *In* Bergey's Manual of Systematic Bacteriology, 9th edit., vol. 1: 631–637. Williams & Wilkins. Baltimore, MD.
8. Fajardo, M.R.C., G.F. Larios, E.G. Valero & V. Tenorio. 1987. Estudio de um brote de pododermatite infecciosa severa bovina. Tecnica Pecuária Mexicana **25:** 99–102.
9. Baron, E.J., L.R. Peterson & S.M. Finegold. 1994. Diagnostic Microbiology: 524–548. Mosby. St. Louis, MO.
10. Mondino, H.C. 1981. Tratamiento de afecciones podales infectadas en el bovino. Revista Militar Veterinária **28:** 119–121.
11. Weaver, A.D. 1987. Cattle foot problems part 3: Surgical techniques. Agri-Practice **9**(3): 14–17.
12. Nuss, K. & M.P. Weaver. 1991. Resection of the distal interphalangeal joint in cattle: an alternative to amputation. Vet. Rec. **128:** 540–543.
13. Pesja, T.G., G.S. Jean, G.F. Hoffsis & J.M.B. Musser. 1993. Digit amputation in cattle: 85 cases (1971–1990). J. Am. Vet. Med. Assoc. **202:** 981–984.
14. Arkins, S. 1981. Lameness in dairy cows. Irish Vet. J. **35:** 163–170.
15. Greenough, P.R. 1986. Pododermatitis circunscripta (ulceration of the sole) in cattle. Agri-Practice **7:** 17–22.
16. Corrêa, W.M. & C.N.M. Corrêa. 1992. Enfermidades infecciosas dos mamiferos domesticos. 2nd edit.: 158. Medsi. Rio de Janeiro.

17. Kersjes, A.W., F. Németh & L.J.E. Rutgers. 1986. Atlas de cirurgia dos grandes animais: 94–99. Manole. São Paulo.
18. Leão, M.A. 1997. Contribuição ao estudo da pododermatite bovina — dados hematológicos e bioquímicos, Thesis MV, Universidade Federal de Goiás, Escola de Veterinária, Goiânia, Goiás, Brazil.
19. Silva, C.A. 1997. Identificação e isolamento do Dichelobacter nodosus e do Fusubacterium necrophorum de bovinos portadores de pododermatite, relação com a etiopatogenia, dados edafoclimáticos e avaliação do tratamento, Thesis MV, Universidade Federal de Goiás, Escola de Veterinária, Goiânia, Goiás, Brasil.
20. Berge, E. & M. Westhues. 1975. Técnica Operatória Veterinária: 444–448. Labor. Barcelona.
21. Hickeman, J. & R.G. Walker. 1983. Atlas de cirurgia veterinária. 2nd edit.: 208–209. Guanabara Koogan. Rio de Janeiro.
22. Cury, P.R. 1997. Metodologia e análise da pesquisa em ciências biológicas. Gráfica e Editora Tipomic: 263. Botucatu, Sao Paulo, Brazil.
23. Ebeid, M. 1993. Bovine laminitis: a review. Vet. Bull. **63:** 205–213.
24. Singh, S.S., W.R. Ward & R.D. Murray. 1993. Etiology and pathogenesis of sole lesions causing lameness in cattle: a review. Vet. Bull. **63:** 303–315.
25. Vemut, J.J. & P.B. Greenough. 1994. Predisposing factors of laminitis in catle. Br. Vet. J. **150:** 151–164.
26. Nocek, J.E. 1997. Bovine acidosis: implications on laminitis. J. Dairy Sci. **80:** 1005–1028.
27. Hoblet, K.H. & W. Weiss. 2001. Metabolic hoof horn disease. Vet. Clin. North Am. Food Anim. Pract. **17**(1): 111–127.
28. Lischer, C.J. & P. Ossent. 2002. Pathogenesis of sole lesions attributed to laminitis in cattle. Proceedings of the 12th International Symposium on Disorders of Ruminant Digit: 82–89. Orlando, FL.
29. Greenough, P.R. 2000. Diseases of the feet of dairy cows. *In* Congresso Brasileiro de Cirurgia e Anestesia. J.H Stringhini, Ed.: 1–21. Editora de UFG. Goiânia, Brazil.
30. Van Amstel, S.R. & J.K. Shearer. 2001. Abnormalities of hoof growth and development. Vet. Clin. North Am. Food Anim. Pract. **17**(1): 73–91.
31. Birgel, E.H. 1974. Amputação do casco, uma técnica a ser considerada. Informações veterinárias. Laboratório Bayer **8:** 25–30.

# Identification of a Coronin-Like Protein in *Babesia* Species

JULIO V. FIGUEROA,[a] ERIC PRECIGOUT,[b] BERNARD CARCY,[b] AND ANDRÉ GORENFLOT[b]

[a]*CENID-PAVET, INIFAP, Jiutepec, Morelos, Mexico*

[b]*Laboratoire de Biologie Cellulaire et Moleculaire, Fac de Pharmacie, Montpellier, France*

**ABSTRACT: The present study was designed to immunochemically identify a coronin-like protein in *Babesia bovis*, *B. bigemina*, *B. divergens*, and *B. canis*. A 2-kbp cDNA insert of *B. bovis* carried by plasmid BvN9 was sequenced by the dideoxichain-termination method on both strands. The cDNA insert contained a 1719-bp long open reading frame coding for a deduced protein sequence of 61.7 kDa. Sequence analysis using the PSI-BLAST program revealed about 30% protein sequence identity with a coronin-like protein of *Plasmodium falciparum*. The encoding sequence of the cDNA insert lacking 70 amino acids at the N-terminal was subcloned in frame into pGEX 4T-3 to produce a recombinant glutathione *S*-transferase (GST)-pBv fusion protein. Polyclonal antibodies prepared in rabbits immunized with the purified GST-fusion protein recognized a *Babesia*-specific component of approximately 60 kDa by immunoprecipitation with [$^{35}$S]methionine-labeled parasites. However, two molecules with relative sizes of 60 and 70 kDa were recognized in *Babesia*-infected erythrocyte extracts by immunobloting analysis. The 70-kDa component was apparently of host erythrocyte origin. In an indirect fluorescent antibody test, the rabbit serum strongly reacted with the merozoite stage of the four *Babesia* species, but also, although weakly, with the host erythrocyte. A cosedimentation assay performed with GST-pBv fusion protein and exogenous actin from rabbit liver showed that the GST-pBv fusion protein, but not the GST protein, was associated to actin. From these results, we conclude that the protein present in the four *Babesia* species analyzed here may be considered as a novel coronin-like, actin-binding protein.**

**KEYWORDS: *Babesia bovis*, *B. bigemina*, *B. divergens*, *B. canis*; WD repeat; coronin-like protein; actin-binding protein.**

## INTRODUCTION

Babesiosis is a tick-transmitted disease of mammalian hosts, caused by the intraerythrocytic protozoan parasites of the genus *Babesia*, which is manifested by anemia, fever, occasional hemoglobinuria, and (often) death.[1] More than 70 species

Address for correspondence: Dr. Julio V. Figueroa, CENID-PAVET, INIFAP, Apartado Postal 206, CIVAC, Morelos, 62500 Mexico. Voice: 777-3 21 13 60; fax: 777-3 20 55 44.
figueroa.julio@inifap.gob.mx

**Ann. N.Y. Acad. Sci. 1026: 125–138 (2004). © 2004 New York Academy of Sciences.
doi: 10.1196/annals.1307.017**

within the genus *Babesia* have been listed. Eighteen of those cause the disease in domestic animals.[2] *Babesia* is included in the phylum Apicomplexa,[2,3] and several species of *Babesia* of cattle have been reported in the literature: *B. bovis*, *B. bigemina*, *B. divergens*, and *B. major*.[4] Two *Babesia* species have been described in dogs: *B. canis* and *B. gibsoni*. These species are distributed worldwide.[5]

On a global basis, bovine babesiosis is considered one of the most important tick vector-borne infections of ruminants.[6] Babesiosis persists in tropical and semitropical regions of the world between the 32nd parallel south and the 40th parallel north of the equator.[7] Bovine babesiosis with its wide geographic distribution exerts an important economic impact on livestock industries, especially in developing countries. Up to half a billion cattle throughout the world may be endangered by the disease caused by species of *Babesia*.[7] The economic impact of the disease includes losses due to death or production loss in terms of milk or meat production, hide damage,[1] etc. Immunization of susceptible cattle with either living or inactivated organisms or components derived from ticks, infected animals, or *in vitro* cultures has been attempted.[7,8] The ideal vaccine, however, has not yet been developed. The delay is due, in part, to the lack of or little understanding of the processes underlying the parasite-host relationship at the cellular and molecular levels.

The sporozoites (infective forms) of tick trans-ovarially transmitted *Babesia* are reported to enter the erythrocyte directly.[9,10] The mechanism of sporozoite entry into erythrocytes is not completely understood. It is hypothesized that penetration is an active process, similar to that utilized by merozoites,[11,12] involving five steps[9]: 1) contact between merozoite and erythrocyte; 2) the orientation of the merozoite apical pole to the erythrocyte surface; 3) membrane fusion between merozoite and erythrocyte; 4) discharge of rhoptry contents; and 5) invagination of the erythrocyte membrane and entry of the merozoite.[9] The merozoite undergoes differentiation inside the erythrocyte (trophozoite) and multiplies by asexual division (merogony) in erythrocytes. This process occurs in most species by binary fission, leading to the characteristic appearance of paired merozoites inside the erythrocyte.[10]

Intraerythrocytic merozoites have common features, aside from variations in size and shape, listed as follows: 1) covered by a complete pellicle consisting of three membranes; 2) an apical and a posterior polar ring, rhoptries, micronemes, and subpellicular microtubules; and 3) a membrane-bound nucleus during division.[10] *Babesia bigemina* and *B. bovis* have spheroid bodies of unknown origin and function, adjacent to the nucleus. It is hypothesized that rhoptries and spherical bodies accumulate extranuclear proteins. The extranuclear proteins may play a role in cell penetration.[13]

Merozoites are produced in the erythrocytes, leave the host cell and immediately invade more erythrocytes. The asexual merogony will continue indefinitely until the host dies or eliminates the parasite.[9] Some *Babesia* parasites formed inside the erythrocyte acquire an unusual shape which is considered to represent gametocytes. Such stages will transform into gamonts. In some species, such as *B. canis*,[14] the morphological transformation into gamonts will begin inside the erythrocytes. Once ticks feed on an infected animal, trans-ovarial transmission occurs, and—after a complicated biological process that involves several types of cells, tissues, and organs of the tick—thousands of sporozoites are formed in the acinar cells of the salivary gland. The sporozoites among *Babesia* species aside from size differences are typically pyriform with a broad apical end and a small, pointed posterior pole.

Sporozoites contain the typical nucleus and mitochondria, as well as micronemes and five to seven rhoptries. Rhoptries apparently have a role in the invasion of erythrocytes of the vertebrate host when transmitted by the tick's saliva,[9] indirectly evidenced by *in vitro* inhibition assays.[11] Finally, the new progeny of sporozoites initiate the development in the vertebrate host to complete the cycle.[9,10,13]

*Babesia* parasites, as most members of the Apicomplexa, are motile. In apicomplexan parasites, locomotion is intimately associated with host cell invasion and, most probably, *Babesia* employs the same underlying cellular machinery as that identified in *Toxoplasma*, *Eimeria*, or *Plasmodium* species.[15,16] Among the similarities in the sequential events leading to the internalization of some parasite stages, the driving force underlying parasite motility and host cell invasion has been shown to be parasite actin- and myosin-based mechanisms.[16] Actin dynamics is tightly regulated by actin-associated proteins that promote or inhibit actin nucleation or polymer elongation, and efforts have been made to identify and functionally characterize parasite molecules that potentially control actin dynamics *in vivo*.[15] In this study, we report the identification of a single-copy *Babesia bovis* gene, the sequence of which encodes a protein with homology to the *Plasmodium falciparum* actin-binding, coronin-like protein.

## MATERIALS AND METHODS

### Babesia bovis *cDNA Clones and Sequencing*

Five *Escherichia coli* XL-1 blue clones containing recombinant PBluescript SK$^-$ plasmids (pBvO1, pBvO4, pBvO5, pBvN9, and pBvN11) derived from a *B. bovis* cDNA library[17] (library aliquot kindly donated by Guy H. Palmer, Washington State University, Pullman, WA) constructed in the λZAP vector (Stratagene, La Jolla, CA) were originally utilized. Clones were subjected to the alkaline lysis method for plasmid purification.[18] An aliquot of the purified plasmids was sent out for sequencing to Genome Express S.A. (Grenoble, France). The sequencing reaction was performed by the dideoxichain-termination method[19] on both strands of the cDNA insert. The clone containing the largest *B. bovis* cDNA insert (pBvN9) was selected for further characterization in this study.

### *Southern Blotting*

*Babesia bovis* genomic DNA was digested with *Hind*III, *Xho*I, *Hind*III/*Bam*HI, and *Xho*I/*Bam*HI restriction enzymes as recommended by the manufacturer (Life Technologies, Paisley, United Kingdom). Digested DNA was separated by agarose gel electrophoresis, transferred to a nylon membrane (Boeheringer-Mannheim Biochemicals, Indianapolis, IN), and fixed by ultraviolet cross-linking. The membrane was hybridized overnight at 42°C with a *Bam*HI/*Eco*RI-digested, purified, and dUTP-digoxigenin-labeled pBvN9 cDNA insert, prepared by the random-primed DNA probe-labeling system as suggested by the supplier (Boeheringer-Mannheim Biochemicals). The membrane was washed, reacted with anti-Digoxigenin-alkaline phosphatase conjugate, and treated with Lumiphos (Boeheringer-Mannheim), and the luminiscent signal was detected by exposure of membrane to X-ray film.

### *Expression of Glutathione S-Transferase-pBvN9 Fusion Protein*

A *Bam*HI/*Eco*RI fragment lacking 70 N-terminal residues of the encoding amino acid sequence of the *B. bovis* cDNA was excised from the pBvN9 recombinant plasmid and subcloned into the *Bam*HI/*Eco*RI-digested, phosphatase-treated pGEX-4T3 expression vector (Pharmacia Biotech, Uppsala, Sweden). Positive clones were selected from transformed *E. coli* (Sure strain) by using *Bam*HI/*Eco*RI restriction enzyme digestion analysis. Glutathione *S*-transferase (GST)-pBvN9 fusion protein was over-expressed in BL 21 cells transformed with the recombinant plasmid according to the manufacturer's protocols (Pharmacia Biotech). Transformed *E. coli* BL 21 cells were cultured overnight in 100 mL of LB medium and diluted 10-fold in fresh medium, grown for 2 h and then induced with 0.2 mM isopropylthio-β-D-galactoside. After 3 h of culture, the bacteria were pelleted, resuspended in 100 mL of lysis buffer (50 mM Tris, pH 8.0, 150 mM NaCl, 1 mM EDTA, and 1.0% Triton X-100), and incubated with 0.2 mg/mL lysozyme for 20 min on ice. Suspensions were then sonicated, insoluble material was pelleted at 10,000 × *g* for 10 min, and the supernatants were incubated with 2.5 mL of glutathione-agarose beads (Sigma, France) for 30 min at 4°C. The beads were washed with washing buffer (50 mM Tris, pH 8.0, 100 mM NaCl, 1 mM EDTA, and 1.0% Triton X-100) three times. The bound GST-pBVN9 fusion protein was eluted with 5 mM reduced glutathione in 50 mM Tris, pH 8.0. BL 21 cells carrying the expression vector alone were processed in the same way to obtain GST protein as a control. To control for the over-expression, and to determine the molecular mass and purity of the GST-pBvN9 fusion protein, eluates were analyzed in 12% sodium dodecyl sulfate polyacrilamide gel electrophoresis (SDS-PAGE) and Coomassie blue staining.[18] For some experiments, such as the rabbit immunization with the recombinante GST-pBvN9 fusion proteins, after preparative SDS-PAGE of a total bacterial lysate with 2× sample buffer, the band of interest was identified by short Coomassie blue staining, the piece of gel was placed in a dialysis bag filled with electrophoresis buffer, and the recombinant fusion protein was eluted from the gel during 1 h at 100 V.

### *Production of Polyclonal Antibodies to GST-pBvN9 Fusion Protein*

The production of polyclonal antibodies to the GST-pBvN9 (*B. bovis*-derived cDNA) fusion protein was carried out by immunizing rabbits three times, 3 weeks apart, with 50 µg of fusion protein emulsified with complete Freund's adjuvant (first immunization) and incomplete Freund's adjuvant (second and third immunization).

### *Source of Parasite Antigen*

*Babesia bovis* and *B. bigemina* parasites from Mexico were cultivated *in vitro* under standard culture conditions.[20,21] *Babesia divergens* was cultured in human $O^+$ erythrocytes essentially as described before.[22] *B. canis* parasites were grown in canine erythrocytes as previously described.[5]

### *Indirect Fluorescent Antibody Test*

The indirect fluorescent antibody test (IFAT) on acetone-fixed, parasite-infected erythrocytes was carried out as described before.[23] Infected erythrocytes smears made out of *in vitro*-cultured parasites were reacted with 1:50 dilution of serum collected from the rabbit immunized with GST fusion protein. Preimmune rabbit serum diluted 1:50 in phosphate-buffered saline was used as a negative control. In addition, a monoclonal antibody (Mab) directed to a surface protein of *B. bovis*[24] was included in the test as positive control. The antigen slides were allowed to incubate at 37°C for 30 min. Washing of parasites with phosphate buffered saline was carried out three times for 5 min. The antigen slides were reacted with fluorescein isothiocyanate-conjugated sheep anti-rabbit IgG, or rhodamine-conjugated goat anti-mouse IgG (Sigma) respectively, and incubated for a further 30 min. After three washes in phosphate-buffered saline, smears were examined on an epifluorescent light microscope.

### *Biosynthetic Labeling of Parasite*

*B. bovis* and *B. canis* cultures were expanded to 25 $cm^2$ culture flasks, and [$^{35}$S]methionine metabolic labeling of parasite polypeptides was accomplished as follows: At 24 h post initiation of cultures, the culture supernatant was replaced with complete culture medium supplemented with [$^{35}$S]L-methionine to a final concentration of 50 μCi/mL of culture (specific activity > $3.7 \times 10^4$ GBq/mmol$^{-1}$, Amersham, France). Cultures were returned to the incubator and incubated at 37°C for 12 hours.

### *Antigen Solubilization*

For immunoprecipitation and/or immunoblotting analysis of parasite components, cultured *Babesia* parasites were extracted for 1 h at 0–4°C in nine volumes of extraction buffer (2% Triton X-100, 0.6 M KCl, 0.15 M NaCl, 5 mM EDTA, 0.01 M Tris-HCl, pH 7.8, 1 mM phenylmethylsulfonyl fluoride (PMSF), and 0.1 mM p-tosyl-l-lysine chloromethyl (TLCK)). The extract was then centrifuged at 15,600 × *g* for 15 min 4°C to remove the insoluble material. The supernatant was removed and kept at –70°C. Preparations of normal bovine erythrocytes cultures were solubilized, and the components extracted in a similar manner.

### *Immunoprecipitation*

Approximately $10^6$ cpm of radiolabeled antigen were incubated overnight at 4°C with 5 μL of rabbit serum. The antigen/antibody complex was precipitated by adding 75 μL of Protein G-Sepharose (Pharmacia Biotech) during 1 h at room temperature. Reaction tubes were then washed twice by centrifugation (13,600 × *g*, for 2 min at 4°C) with washing buffer A (20 mM Tris HCl, pH 7.5, 5 mM EDTA, 100 mM NaCl (TEN), 1% Nonidet-P40, 1% BSA); four times with washing buffer B (TEN, 1% NP-40, 2 mM NaCl); and twice with washing buffer C (TEN, 1% NP-40). Bound [$^{35}$S]methionine-labeled antigens were eluted with SDS-sample buffer for electrophoretic analysis.

### *SDS-PAGE*

The [$^{35}$S]methionine-labeled parasite antigens immunoprecipitated by the procedure described above were boiled for 5 min in sample buffer containing 62.5 mM tris-HCl, pH 6.8, 3% SDS, 20% glycerol, and 5% 2-mercaptoethanol. Samples were analyzed by separation of components in 10 cm × 16 cm slab gels using a 12% discontinuous gel system.[25] The separating gel from the immunoprecipitation experiment was processed for fluorography, before drying for exposure to X-OMAT X-AR5 film (Eastman Kodak, Rochester, NY).

### *Immunoblotting*

The electrophoresed proteins from unlabeled material were electrotransferred from the gel onto a sheet of nitrocellulose as described before.[26] The electrophoretic transfer of polypeptide bands was performed at room temperature at 100 mA for 1 h using a commercial semi-dry transfer unit (Transfor, Biorad, France). Once blotting was completed, nitrocellulose sheets were submerged for 2 h in blocking solution Tris-buffered saline (TBS, 25 mM Tris-HC1, 0.15 M NaCl, pH 7.2) containing 5% (w/v) non-fat dried milk and 0.05% Tween 20. For analysis of multiple-antigen reactions, nitrocellulose sheets containing several lanes of separated antigenic components (normal red blood cells, infected red blood cells of the various *Babesia* species) were reacted with a 1:200 dilution of primary antibody calculated to contain at least 2.5 mL per lane of antigenic material. Reaction with primary antibody was allowed to continue for 1 h at room temperature. The nitrocellulose sheets were washed three times with TBS for 10 min each, after which an appropriate dilution (1:1000) of anti-rabbit IgG goat antiserum conjugated to alkaline phosphatase was added and incubated for 1 h at room temperature. Further washing of the nitrocellulose paper was done as indicated, three times, 10 min each time. This was followed by the addition of a commercially available substrate, BCIP/NBT (5-bromo-4-chloro-3-indolyl-phosphate/nitroblue tetrazolium), which was used according the supplier.

### *Cosedimentation Assay of GST-pBvN9 Fusion Protein with F-Actin*

The assay was performed following the conditions described earlier.[27] Briefly, G-actin (15 μg/reaction) (Sigma, France) and GST protein, GST-pBvN9 fusion protein, and fusion protein alone were incubated in F-actin buffer (20 mM Tris, pH 8.0, 160 mM KCl, 0.2 mM adenosine 5′-triphosphate) for 1 h at 25°C. The reaction mixtures were then ultracentrifuged at 100 000 × *g* for 1 h. The supernatants, pellets, and total reaction mixture (sample before ultracentrifugation) were resolved by discontinuous 12% SDS-PAGE, and gels were stained with Coomassie blue. In addition, gels were immunoblotted for anti-GST-pBvN9 fusion protein analysis.

## RESULTS AND DISCUSION

Sequence analysis of the approximately 2-kbp long cDNA insert of *B. bovis* carried by plasmid BvN9 revealed a nucleotide stretch of 1955 residues in length (nucleotide sequence deposited in the GenBank database under accession number AY324186). The cDNA insert contained a 1719-bp long open-reading frame, between nucleotide position number 219 (corresponding to a initiation codon ATG) and 1940 (corresponding to a stop codon TAG). This unique open-reading frame would code for a 573-translated aminoacid sequence of a protein with a predicted size of 61.7 kDa (Protean module, Lasegene software, DNAStar, Madison, WI). Secondary structure prediction analysis by the Chou-Fasman (Protean) probability method reveals that the encoded protein seems to be principally constituted of β re-

MSTVKLKNLFGEPFKQVYCDLKINPKPTAFSGGMAASPTYVAFPWEVGGG
GLVSLIGLDKLGRNSGAEKIDLR**GHAGSLQDMVFNDFDYSVLATGSDDCSV**
**RVWR**VGNNEGSALCNLA**GHTKKTTNVVWNASTDYVLLSGSMDNTVKVWD**
VKHGSAVSTIPIEGNYSYCNWSYDGNTVLVSTKESYVAFADPRDGKVKLAF
K**AHDSNKATSVQWLGGNYGGDYLATTGYVGNQTRQIRVWD**ARNTDKPVV
SKDIDSAPGPLIPYWDSDTGLLTVVGKGDLTVRIFQYLEGDLNRAGEFKCNG
TIKSFCFLPNSACDKSRCELGRLLYNCTSKEINPISIVVLRRNSQAAMGEIYG
NVEQRRRTLAEEWHGCDLGAPQKSISATFDETAPPSMQSSFNSQVSDSAK
ARTMLSVASPSADWESTATGKAFIEIVGHVNHLSIRYQPAFNSADMLEHLEN
LEKEVVTMINLVKKEHGISTPSMTSSGRNASQVSDRSSSHQPMNSSATPNM
AAVNTAAKVTSVIPKPKEETTSNNASDGQAAVESATKGLTGVKAAIAAMEAR
RAQAKGDASGRKM

**FIGURE 1.** Deduced amino acid sequence of *Babesia bovis* pN9. Underlined characters denote the WD repeats.

gions, more so than α-helix regions. The latter were localized mainly at the carboxy-terminal region of the molecule. Predicted sites or motifs identified by homology analysis[28] of the pBvN9 sequence with the data collection of GenBank (NCBI) indicate the presence of at least three WD domain sequences. WD proteins are made up of highly conserved repeating units usually ending with aminoacid residues Trp-Asp (WD).[29] WD proteins are found in all eukaryotes but not in prokaryotes. They regulate functions such as cell division, cell-fate determination, gene transcription, transmembrane signaling, mRNA modification, and vesicle fusion.[29]

Sequence analysis using the PSI-BLAST application program[28] revealed strong homology (about 50% protein sequence similarity and 30% protein sequence identity) with a coronin-like protein of *Plasmodium falciparum* (and *P. yoelli*), which include mainly the WD repeat region, but also extends towards the carboxy-terminus region (FIG. 1). A single-copy *P. falciparum* gene encoding a coronin-like protein has been reported.[30] This gene sequence encodes a protein that displays strong homology with the *Dictyostelium discoideum* actin-binding protein originally described as coronin.[31] In our BLAST search, the pBvN9 cDNA-encoded protein also shared similarities and identities, although at a lesser extent, with all members of the large family of coronin proteins reported so far, including that from *D. discoideum* (25% identity).

Analysis of the DNA sequence showed no restriction enzyme sites for *Hind* III and *Xho* I. Thus, southern blot analysis of restriction enzyme-digested *B. bovis* genomic DNA probed with the digoxigenin-labeled insert cDNA, demonstrated hybridization to one single band, both in the *Hind*III- and *Xho*I-single digested genomic DNA samples (FIG. 2, relative sizes of 6.5 and 7.0 kbp, respectively). PBvN9 cDNA contains an internal *Bam*HI site at position 510–515 of the sequence. Thus, it was expected that the digoxigenin-labeled probe would hybridize with at least two genomic *B. bovis* DNA fragments. This was not the case, however, as only one hybridized band could be detected in the double-digestion reactions (*Hind*III/*Bam*HI; *Xho*I/BamHI) in which fragments of approximately 2 and 2.5 kbp, respectively, were observed (FIG. 2). Presumably, this was because the probe utilized was not the complete labeled insert; rather, it is a probe that missed precisely 470 bp up-

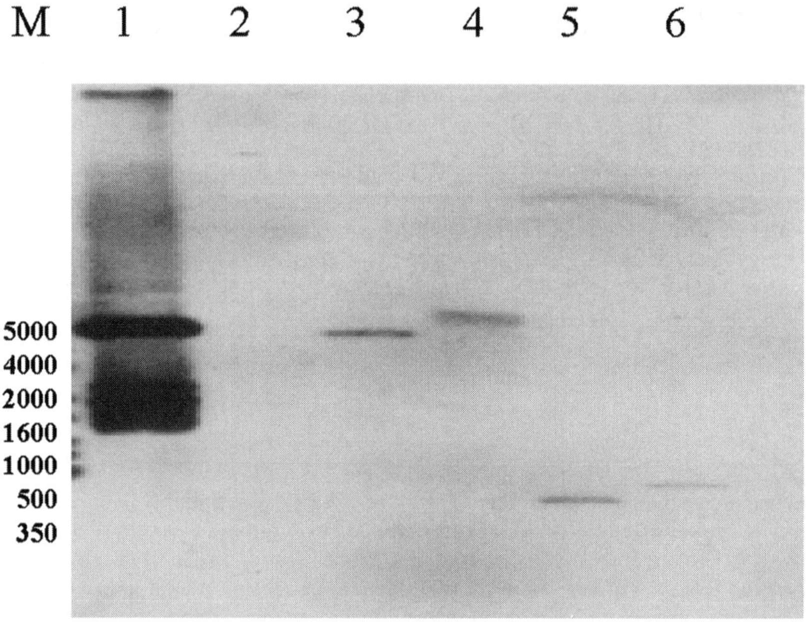


**FIGURE 2.** Southern blot hybridization of *Babesia bovis* genomic DNA using dUTP-digoxigenin-labeled pgBvN9 cDNA insert as probe. *Lane 1*: pgBvN9 DNA digested with restriction enzyme *Hind*III. *Lane 2*: *B. bovis* genomic DNA undigested. *Lane 3*: *B. bovis* genomic DNA digested with *Hind*III. *Lane 4*: *B. bovis* genomic DNA *Xho*I. *Lane 5*: *B. bovis* genomic DNA *Hind*III/*Bam*HI (5). *Lane 6*: *B. bovis* genomic DNA *Xho*I/*Bam*HI.

stream from the internal *Bam*HI site. Therefore, the genomic DNA sequences positioned 5′ upstream between the *Bam*HI and either the *Hind*III or *Xho*I sites detected in the single digests, would not be detected in the southern blot with the truncated probe used in this study. Whether the DNA sequence hybridized with the insert probe derived from pBvN9 represents a single-copy gene, as discovered in *P. falciparum*, needs further clarification.

The encoding sequence of the cDNA insert lacking 70 amino acids at the N-terminal was subcloned in frame into pGEX 4T-3 to produce a recombinant GST-pBv fusion protein. The GST-pBvN9 fusion protein was resolved by SDS-PAGE, in which it was shown that a recombinant protein with a calculated molecular mass of about 80 kDa (54.7 kDa pGEX-BvN9 + 25 kDa GST) was expressed (FIG. 3).

To define whether *B. bovis* actually expressed the putative coronin-like protein, three different immunoassays were performed. The IFAT was utilized to determine the distribution of the BvN9 protein. As demonstrated in FIGURE 4A, the serum from a rabbit immunized with the GST-BvN9 fusion protein and tested against acetone-fixed *B. bovis*-infected erythrocytes contained antibodies that reacted primarily, although in a diffuse pattern, with merozoite-stage parasite bodies. However, the antibodies in the rabbit serum also reacted, albeit weakly, with the red cell ghosts present in the preparation.

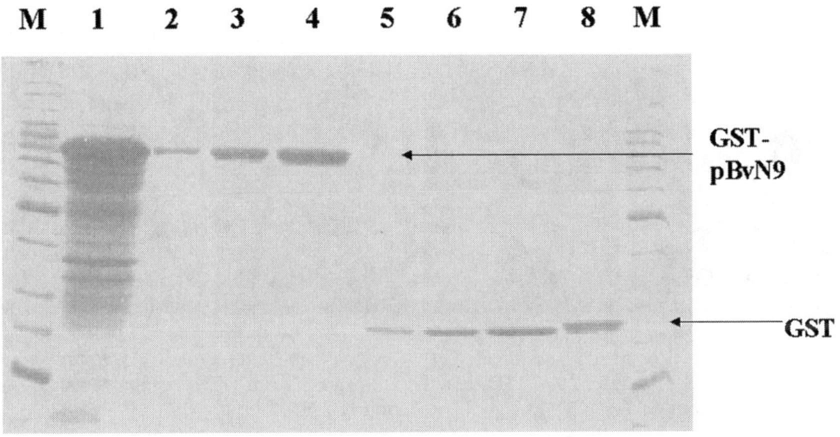


**FIGURE 3.** SDS-PAGE analysis of protein extracts from *E. coli* BL 21 induced by isopropylthio-β-D-galactoside and transformed with pGEX-BvN9. Purification of recombinant fusion protein GST-BvN9 (*lanes 2–4*) and GST (*lanes 5–8*) with reduced glutathion/agarose beads from the supernatant of bacterial lysate. Molecular weight standards are from the BechMark Protein Ladder (Life Technologies, Paisley, UK).

The rabbit preimmune serum did not contain antibodies that would bind to the parasite or red cell ghosts (not shown). That the fluorescent staining of *B. bovis* merozoites is an authentic reaction was demonstrated when an anti-*B. bovis* Mab was reacted against the same fixed parasite preparation by using rhodamine-conjugate antimouse IgG in a double-labeled IFAT-test (not shown). There seemed to be a particular compartmentalization or accumulation of pBvN9 protein towards the anterior end of the merozoite (FIG. 4A) as evidenced by the parasite staining pattern observed after incubation with the rabbit immune serum. However, the reaction was not stage-specific; that is, both mature and immature parasites were fluorescence labeled. In addition, the expression of a parasite component similar to BvN9 protein was identified in the other three species of *Babesia* tested, that is, *B. divergens*, *B. bigemina*, and *B. canis* (FIGS. 4B–4D).

To confirm the parasite expression of the BvN9 protein (or its homolog) in the *Babesia* species analyzed in this study, electrophoresed *Babesia* proteins were transferred to nitrocellulose, and the membrane reacted with the rabbit immune serum. FIGURE 5 shows the results after the immunoassay. Except for the *B. divergens* sample, in which only one component (approximate calculated molecular mass: 60 kDa) was recognized by the polyclonal antibodies present in the immunized rabbit serum, two components (60 kDa and 67 kDa) were detected in the other three species analyzed. As evidenced by the reaction of rabbit immune serum with the uninfected erythrocyte sample, it was demonstrated that the 67-kDa component is of bovine host origin (FIG. 5).

To precisely define and confirm the parasitic nature of components other than the 67-kDa band detected in the immunoblots, [$^{35}$S]methionine polypeptides were immunoprecipitated and separated by SDS-PAGE. FIGURE 6 shows the parasite

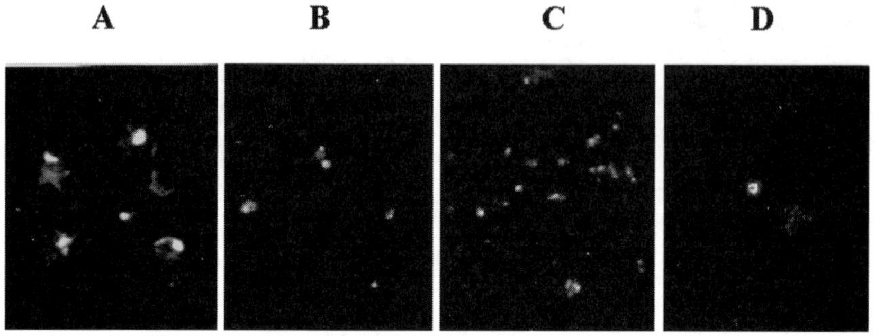


**FIGURE 4.** Blood smears of infected erythrocytes with *Babesia bovis* (**A**), *B. bigemina* (**B**), *B. divergens* (**C**), and *B. canis* (**D**) reacted with serum from rabbit immunized with recombinant protein GST-pBvN9 and goat anti-rabbit IgG/FITC conjugate.

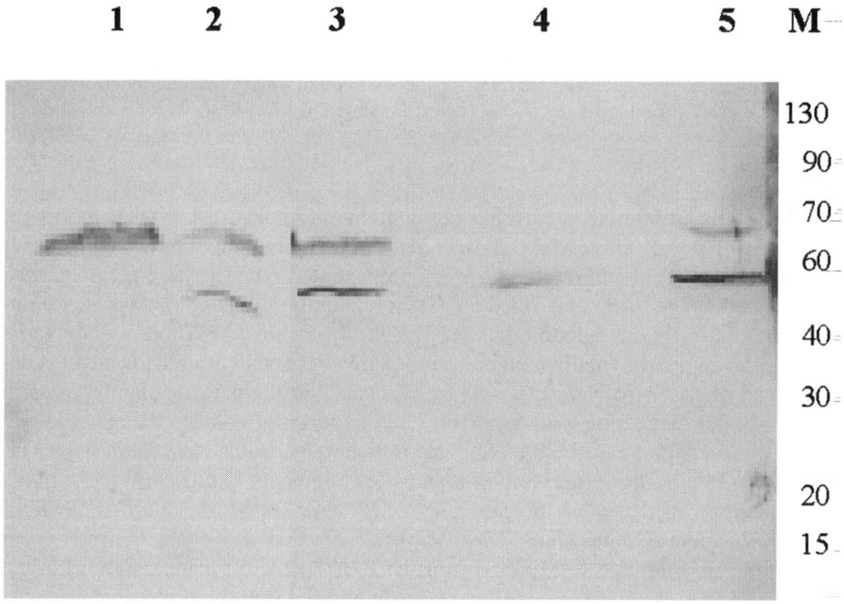


**FIGURE 5.** Immunoblot analysis with serum from rabbit immunized with recombinant protein GST-pBvN9. *Lane 1*: uninfected bovine erythrocytes protein extract. *Lane 2*: *B. bovis*-infected erythrocyte extract. *Lane 3*: *B. bigemina*-infected bovine erythrocytes. *Lane 4*: *B. divergens*-infected human erythrocytes. *Lane 5*: *B. canis*-infected canine erythrocytes.

polypeptides precipitated by the immunized rabbit serum. A radiolabeled polypeptide banding at a relative size of 60 was immunoprecipitated by the immunized rabbit serum from the *B. bovis* protein extract. Although it is difficult to discern from the high background present in the immunoprecipitation test, a *B. canis* component of the same size as that of *B. bovis* was precipitated by the rabbit serum.

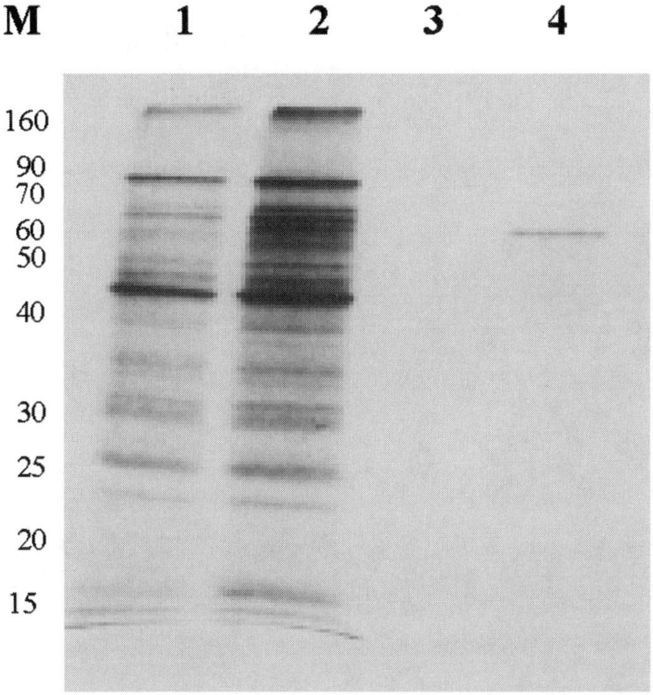


**FIGURE 6.** Immunoprecipitation assay with [$^{35}$S]methionine-labeled parasites. Lanes 1 & 2: *B. canis* antigen. Lanes 3 & 4: *B. bovis* antigen. *Lanes 1 & 3*: Antigen immunoprecipitated with rabbit preimmune serum. *Lanes 2 & 4*: Antigens immunoprecipitated with serum from rabbit immunized with GST-pBvN9.

Out of the components (the 60- and 67-kDa components) recognized by the rabbit immune serum in the immunoblot technique, the smaller one was demonstrated, unequivocally, to belong to the parasite, as assessed by the immunoprecipitation and SDS-PAGE of metabolically labeled *B. bovis* culture extracts. It is known that *in vitro* cultures of *Babesia* spp. (if deprived of bovine leukocytes and platelets) added with [$^{35}$S]methionine would incorporate the radioactive amino acid exclusively into the parasite polypeptides.

The similarity between pBvN9 protein and *P falciparum* coronin suggested that the *B. bovis* component could also be an actin-binding protein. Thus, a cosedimentation assay of the GST-fused BvN9 with F-actin was performed. When GST was mixed with F-actin in the co-sedimention assay, it was not able to bind to F-actin (FIG. 7). On the other hand, GST-pBvN9 was detected, although very faintly, along with F-actin in the pellet portion of the cosedimentation assay, but not detected in the pellet portion of the cosedimentation assay in the absence of F-actin (FIG. 7). The result was confirmed by immunoblotting analysis with rabbit serum anti-GSTpBvN9 protein (FIG. 8). This result suggests that pBvN9 was responsible for the actin-binding activity, and thus can be considered an actin-binding protein.

In conclusion, a parasite protein present in the four *Babesia* species analyzed here may be considered as a novel coronin-like, actin-binding protein.

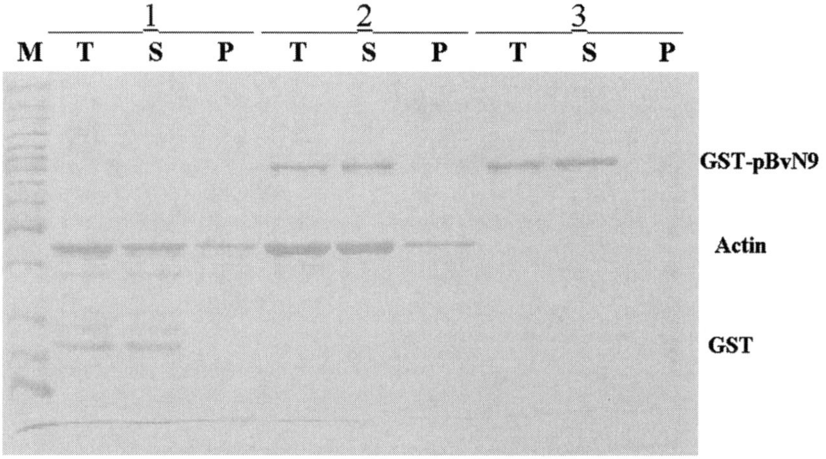


**FIGURE 7.** SDS-PAGE analysis of protein samples utilized in the actin cosedimentation assay. Sample designations: 1: actin + GST; 2: actin + GST-pBvN9; 3: GST-pBvN9. Lane designations: M: molecular weight markers; T: aliquot of sample without centrifugation; S: aliquot of supernatant after 2 h centrifugation at 100,000 × *g*; P: aliquot of pellet after centrifugation.

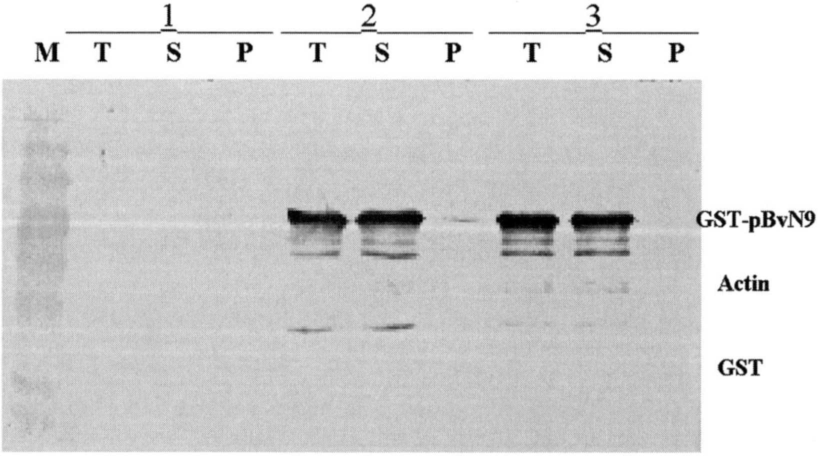


**FIGURE 8.** Immunoblotting analysis of a replica of FIGURE 7 with serum from rabbit immunized with recombinant protein GST-pBvN9. The sample and lane designations are identical to those in FIGURE 7.

## ACKNOWLEDGMENTS

Financial support for this study was received in part from the Ministére d'Education Nationale et la Researche, France and Conacyt, Mexico, Project No. 34473-B.

## REFERENCES

1. McCosker, P.J. 1981. The global importance of babesiosis. *In* Babesiosis. M. Ristic & J.P. Kreier, Eds.: 1–24. Academic Press. New York, NY.
2. Levine, N.D. 1971. Taxonomy of the piroplasmas. Trans. Am. Microsc. Soc. **90:** 2–33.
3. Levine, N.D. *et al.* 1980. A newly revised classification of the protozoa. J. Protozool. **27:** 37–58.
4. Hoyte, H.M.D. 1976. The tick fever parasites of cattle. Proc. R. Soc. Queensland **87:** v–xiii.
5. Carret C. *et al.* 1999. *Babesia canis canis*, *Babesia canis vogeli*, *Babesia canis rossi*: differentiation of the three subspecies by a restriction fragment length polymorphism analysis on amplified small subunit ribosomal RNA genes. J. Eukaryot. Microbiol. **46:** 298–303.
6. Uilenberg, G. 1980. Ticks and tick-borne diseases of veterinary interest in Europe and Africa: prospects for control and eradication. *In* Ticks and Tick-Borne Diseases: Proceedings of the 56th Annual Conference of the Australian Veterinary Association. L.A.Y. Johnston & M.G. Cooper, Eds.: 1–3. Townsville, Queensland, Australia.
7. Ristic, M. 1984. Research on babesiosis vaccines. *In* Malaria and Babesiosis: Research Findings and Control Measures. M. Ristic, P. Ambroise-Thomas & J.P. Kreier, Eds.: 103–122. Martinus Nijhoff. Boston, MA.
8. Shkap V. & E. Pipano. 2000. Culture-derived parasites in vaccination of cattle against tick-borne diseases. Ann. N.Y. Acad. Sci. **916:** 154–171.
9. Young, A.S. & S.P. Morzaria. 1986. Biology of *Babesia*. Parasitol. Today **2**(8): 211–219.
10. Mehlhorn, H. & E. Schein. 1984. The piroplasms: life cycle and sexual stages. *In* Advances in Parasitology. J.R. Baker & R. Muller, Eds. Vol. 23: 37–103. Academic Press. New York, NY.
11. Mosqueda, J. *et al.* 2002. *Babesia bovis* merozoite surface antigen 1 and rhoptry-associated protein 1 are expressed in sporozoites, and specific antibodies inhibit sporozoite attachment to erythrocytes. Infect. Immun. **70**(3): 1599–1603.
12. Mosqueda, J., T.F. McElwain & G.H. Palmer. 2002. *Babesia bovis* merozoite surface antigen 2 proteins are expressed on the merozoite and sporozoite surface, and specific antibodies inhibit attachment and invasion of erythrocytes. Infect Immun. **70:** 6448–6455.
13. Potgieter, F.T. & H.J. Els. 1977. Light and electron microscopic observations on the development of *Babesia bigemina* in larvae, nymphae and non-replete females of *Boophilus decoloratus*. Onderstepoort J. Vet. Res. **44:** 213–232.
14. Mehlhorn, H., E. Schein & W.P. Voigt. 1980. Light and electron microscopy study on developmental stages of *B. canis* within the gut of the tick *Dermacentor reticulatus*. J. Parasitol. **66:** 220–228.
15. Tardiuex, I. 1998. A *Plasmodium falciparum* novel gene encoding a coronin-like protein which associates with actin filaments. FEBS Lett. **441:** 251–256.
16. Morrissette, N.S. & L.D. Sibley. 2002. Cytoskeleton of apicomplexan parasites. Microbiol. Mol. Biol. Rev. **66:** 21–38.
17. Suarez, C. *et al.* 1991. Characterization of the gene encoding a 60-kilodalton *Babesia bovis* merozoite protein with conserved and surface exposed epitopes. Mol. Biochem. Parasitol. **46:** 45–52.
18. Sambrock, J. & D.W. Russell. 2001. Molecular Cloning: A Laboratory Manual. 3rd edit. Cold Spring Harbor Laboratory Press. Cold Spring Harbor, NY.
19. Sanger, F., S. Nicklen & A.R. Coulson. 1977. DNA sequencing with chain-terminating inhibitors. Proc. Natl. Acad. Sci. USA **74:** 5463–5467.

20. PALMER, D.A., G.M. BUENING & C.A. CARSON. 1982. Cryopreservation of *Babesia bovis* for in vitro cultivation. Parasitol **84:** 567–572.
21. VEGA, C.A. *et al.* 1985. In vitro cultivation of *Babesia bigemina.* Am. J. Vet. Res. **46:** 416–420.
22. GORENFLOT, A. *et al.* 1991. Cytological and immunological responses to *Babesia divergens* in different hosts: ox, gerbil, man. Parasitol. Res. **77:** 3–12.
23. FIGUEROA, J.V. & G.M. BUENING. 1991. In vitro inhibition of multiplication of *Babesia bigemina* by using monoclonal antibodies. J. Clin. Microbiol. **29:** 997–1003.
24. FIGUEROA, J.V., D.A. KINDEN & G.M. BUENING. 1998. Use of monoclonal antibodies for the identification of a common surface antigen. Ann. N.Y. Acad. Sci. **849:** 433–437.
25. LAEMMLI, U.K. 1972. Cleavage of structural proteins during the assembly of the head of bacteriophage T4. Nature **227:** 680–685.
26. TOWBIN, H., T. STAEHELIN & J. GORDON. 1979. Electrophoretic transfer of proteins from polyacrylamide gels to nitrocellulose sheets: procedures and some applications. Proc. Natl. Acad. Sci. USA **76:** 4350–4354.
27. SUZUKI, K. *et al.* 1995. Molecular cloning of a novel actin-binding protein, p57, with a WD repeat and a leucine zipper motif. FEBS Letters. **364:** 283–288.
28. ALTSCHUL, S.F. *et al.* 1997. Gapped BLAST and PSI-BLAST: a new generation of protein database search programs. Nucleic Acids Res. **25:** 3389–3402.
29. NEER, E.J. *et al.* 1994. The ancient regulatory-protein family of WD-repeat proteins. Nature **22:** 297–300.
30. DE HOSTOS, E.L. *et al.* 1991. Coronin, an actin binding protein of *Dictyostelium discoideum* localized to cell surface projections, has sequence similarities to G protein β subunits. EMBO J. **13:** 4097–4104.

# TaqMan-Based Detection of *Leishmania infantum* DNA Using Canine Samples

F. VITALE, S. REALE, M. VITALE, E. PETROTTA, A. TORINA, AND S. CARACAPPA

*Istituto Zooprofilattico Sperimentale of Sicily, Palermo, Italy*

**ABSTRACT: Leishmaniasis is a typical example of a worldwide diffused zoonosis. Geographic distribution depends on the presence of sand fly vectors and animal reservoirs. In Southern Europe, canines are considered the main reservoir of infection, and the phlebotomines are the vectors. In Sicily, as in all Mediterranean areas, sand flies are present almost all year around because the climate permits an uninterrupted lifecycle for the vectors. Visceral leishmaniasis is becoming a real public health concern especially in endemic areas; in fact, it is an opportunistic infection in immunocompromised patients and in HIV-positive subjects. In Italy, the visceral form of the disease is due exclusively to *Leishmania infantum* ZMON1, and its prevalence is growing. We have developed a highly accurate, reproducible, and sensible real-time polymerase chain reaction (PCR) assay. In a procedure that used a specific couple of primers, a 117-bp fragment was amplified from minicircle kinetoplast DNA (kDNA). The assay was able to detect even a single parasite (200 fg of DNA). In fact, a single parasite contains hundreds of kinetoplast minicircles for each class. We applied a rapid extraction method coupled with the real-time PCR assay. It was not only as sensitive as a conventional PCR assay for detection of *Leishmania* kDNA, but also more rapid. The assay is useful for the diagnosis of leishmaniasis in dogs and humans, and it facilitates the monitoring of parasite levels during pharmacological treatment.**

**KEYWORDS: real-time PCR; TaqMan; *Leishmania infantum*; ZMON1; amastigotes; zoonosis; DNA; minicircle kinetoplast DNA (kDNA); primers; probe**

## INTRODUCTION

Leishmaniasis is an infectious disease of people and wild/domestic animals in temperate, subtropical, and tropical regions of the world. The geographical distribution of leishmaniasis is limited by the distribution of the sand fly, the sand fly's susceptibility to cold climates, its tendency to take blood from humans or animals only, and its capacity to support the development of specific species of *Leishmania*. In the last 10 years, the regions that are *Leishmania*-endemic have expanded significantly, accompanied by a sharp increase in the number of recorded cases of the disease.

Address for correspondence: Dr. S. Reale, Istituto Zooprofilattico Sperimentale of Sicily, Palermo, Italy. Voice: +390916565314; fax: +390916565313.
sreale@pa.izs.it

**Ann. N.Y. Acad. Sci. 1026: 139–143 (2004). © 2004 New York Academy of Sciences.**
**doi: 10.1196/annals.1307.018**

Diagnosis is based on finding amastigotes free or in macrophages in giemsa-stained smears from lymph nodes or bone marrow. Serology (immunofluorescent assay; IFA) is used to verify the presence of antibody indicating a current infection. A qualitative polymerase chain reaction (PCR) assay is useful when it is necessary to have an immediate and sensitive diagnosis, no matter the kind of biological matrix, but especially in dubious cases requiring serological work. A novel method that incorporates a real-time PCR assay allows the quantification of nucleic acid template by analyzing the kinetics of PCR during the cycles.[1,2] We developed a microtitration method to identify the concentration of parasites in different type of dog samples. We plotted a standard curve on the basis of serial dilutions of DNA extracted from cultivated amastigotes. This was the starting point to perform the analysis on samples from ill animals.[3]

## MATERIALS AND METHODS

The parasitic culture loads of *Leishmania infantum* were determined using a real-time quantitative PCR test targeted at the kinetoplast DNA (kDNA) of the parasite. The exogenous DNA was used as control of the test in each tube. The reactions were performed in replicate for each sample to avoid the errors of the absolute measures.

### *Extraction*

The Prepman kit (Applied Biosystems, Foster City, CA) was employed to perform the quantitative extraction of total DNA from the samples. Different protocols were used for the various matrices. The blood samples were centrifuged, and the resulting pellet was lysed. Lymph node aspirates were resuspended directly in the lysis buffer. A small piece of skin was homogenized in the Prepman buffer. After incubation of the samples at 95°C for 15 min and centrifugation at 12,000 × g, the extracted DNA was employed for PCR test. A standard curve was obtained with the TaqMan *L. infantum* PCR test for an appropriate concentration range of DNA. The reference DNA was extracted from the growth of amastigotes in Tobie-modified medium. The cells from the cultures were collected by centrifugation and washed. After they were counted, the *Leishmania* cells were resuspended in 1× phosphate-buffered saline to have a starting charge of $1 \times 10^9$ cells/mL. The quantity of DNA in the examined samples was detected by comparison of the Ct (threshold cycle) values plotted on common log scale.

### *TaqMan PCR Test*

The PCR test was targeted on the constant region of the minicircle kDNA (NCBI accession number AF291093).[4,5] The primers and the FAM (6 carboxy-fluorescein)-labeled probe were chosen with the assistance of the computer program Primer express (Applied Biosystems) in accordance with technical parameters indicating a low level of penalty-coupling factor. The DNA target amplified was a 68-bp fragment. The exogenous internal control was supplied by the TaqMan exogenous internal positive control reagents kit (Applied Biosystems). It was labeled with VIC™ (Applied Biosystems) fluorochrome. The reactions were performed on 7700

ABIPrism sequence detection system (Applied Biosystems). Each amplification was performed in duplicate, in a 25-μL reaction mixture containing the TaqMan Universal Master Mix (Applied Biosystems), the specific primers, and the probe in the optimized concentration. The thermal cycling conditions comprised an initial incubation for 2 min at 50°C for uracyl-*N*-glycosylase activity. This step was followed by a 10-min denaturation at 95°C and 45 cycles of: 95°C for 15 sec and 60°C for 60 sec.

## RESULTS

The initial experiment was the determination of the optimal quantitative range for the TaqMan PCR test. We constructed two kinds of standard curves by making serial dilutions of *Leishmania* promastigotes and extracted DNA. The serial dilutions consisted of equivalents of DNA from $1 \times 10^6$ cells to 1 cell per amplified sample (FIG. 1). The difference for each point of the curve was one log factor. The *Leishmania* DNA copy number was normalized to the number of the exogenous DNA target control in order to indicate the amplification efficiency in each test. All the amplified dilution samples of the standard curve gave a measurable signal (FIG. 2). The amplification curves shift to the right as the input target quantity is reduced; this shift occurs because reactions with fewer target molecules require more amplification cycles to produce a detectable quantity of reporter molecules than do reactions with more target molecules. The reproducibility was tested by repeating the same experiment six times in independent runs with serial dilutions of DNA and cells. We were able to detect even a single parasite, which corresponded to 100 fg of DNA per re-

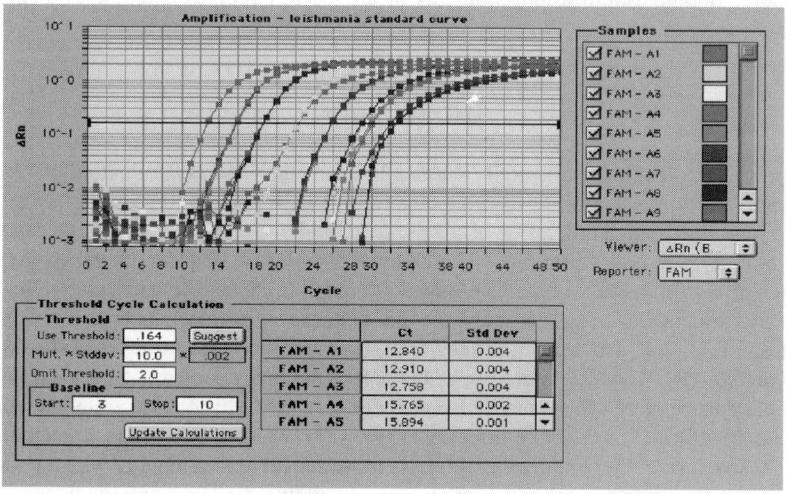


**FIGURE 1.** Plot registration obtained by amplification of serial dilution of *Leishmania* in a range from $10^6$ cells to 1 cell.

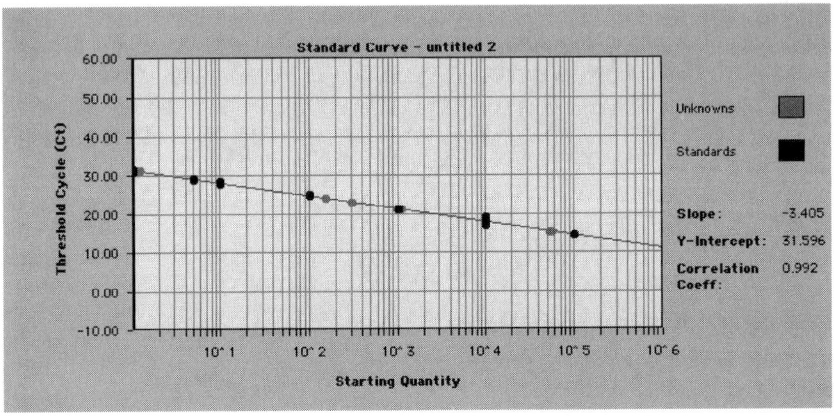


**FIGURE 2.** Standard curve obtained with the TaqMan *Leishmania infantum* PCR test. Serial 10-fold diution of *L. infantum* DNA from 1 equivalent to $10^6$ equivalents per reaction (5 to $5 \times 10^6$).

action. Because of the population of the minicircle kDNA per each cell, the curve permitted us to measure less than one cell's DNA equivalent. We decided to use 100 fg of DNA at a lower point of the curve. In a preliminary study, we used the curve to measure the DNA level in canine samples from animals with or without clinical signs. The amount of revealed DNA could be correlated to the physiological status of the animals in the previous phases of the infection and during the pharmacological treatment. The goal of the work could be to make a comparison through the data obtained in different moments of the therapy (copies/reaction) and canine samples.

## CONCLUSIONS

A new molecular real-time PCR assay for detection and quantification of *Leishmania infantum* was described. This quantitative PCR assay allows highly sensitive and reproducible quantification of the parasite burden over a wide range (at least six logs of parasite concentrations). The assay was based on TaqMan chemistry to detect, in real time, the molecules produced during the PCR. The technique relies on the fluorescence of a reporter molecule that increases as product accumulates during the amplification. The very high sensitivity is due to high copy number of the target minicircle kDNA, which is present at about 10,000 copies per parasite. The possibility of detecting DNA in a quantitative way could improve the diagnosis and the management of the *Leishmania* infection. Moreover, it results in a faster and less expensive test.[6,7] Quantitative PCR seems more sensitive if compared with the qualitative PCR, especially at the lower charges. Including the DNA extraction step, the assay can be performed within 4–5 h without risk of contamination. The development of a real-time PCR assay for qualitative and quantitative detection of *Leishma-*

*nia* could replace the classical PCR technology and improve the diagnosis based on serology technique.[8,9] In addition, our assay has applications in monitoring *Leishmania* infections for research experiments and to monitor pharmacological treatment.

## REFERENCES


1. Bretagne, S. *et al.* 2001. Real-time PCR as a new tool for quantifying *Leishmania infantum* in liver in infected mice. Clin. Diagn. Lab. Immunol. **8**(4): 828–831.
2. Buffet, P. *et al.* 1995. Culture microtitration: a sensitive method for quantifying *Leishmania infantum* in tissues of infected mice. Antimicrob. Agents Chemother. **39**(9): 2167–2168.
3. Lachaud, L. *et al.* 2000. Optimized PCR using patient blood samples for diagnosis and follow-up of visceral *Leishmaniasis,* with special reference to AIDS patients. J. Clin. Microbiol. **38**(1): 236–240.
4. Vitale, F. & S. Reale. Cloning and sequences of mini circle DNA kinetoplast variable region from Leishmania infantum *IPT1*. Unpublished.
5. Reale, S. *et al.* 1999. Detection of *Leishmania infantum* in dogs by PCR with lymph node aspirates and blood. J. Clin. Microbiol. **37**(9): 2931–2935.
6. Nicolas, L. *et al.* 2002. Real-time PCR for detection and quantitation of Leishmania in mouse tissues. J. Clin. Microbiol. **40**(5): 1666–1669.
7. Solano-Gallego, L. *et al.* 2001. Prevalence of *Leishmania infantum* infection in dogs living in an area of canine leishmaniasis endemicity using PCR on several tissues and serology. J. Clin. Microbiol. **39**(2): 560–563.
8. Lachaud, L. *et al.* 2002. Comparison of six PCR methods using peripheral blood for detection of canine visceral leishmaniasis. J. Clin. Microbiol. **40**(1): 210–215.
9. Lachaud, L. *et al.* 2001. Comparison of various sample preparation methods for PCR diagnosis of visceral leishmaniasis using peripheral blood. J. Clin. Microbiol. **39**(2): 613–617.

# Immune Response to *Babesia bigemina* Infection in Pregnant Cows

T.D. GARCÍA, M.J.V. FIGUEROA, A.J.A. RAMOS, M.C. ROJAS, A.G.J. CANTÓ, N.A. FALCÓN, AND M.J.A. ÁLVAREZ

*Centro Nacional de Investigaciòn Disciplinaria en Parasitologia Veterinaria, Instituto Nacional de Investigaciones Forestales, Agrícolas y Pecuarias, Civac, Morelos, Mexico*

**ABSTRACT: The objective of this study was to characterize the immune response of *Babesia bigemina*–infected cows during the second trimester of pregnancy. Twelve animals were divided into four groups (I, II, III, IV); groups I and II were pregnant cows, groups III and IV were non-pregnant cows. Groups I and III were infected with a virulent strain of *Babesia bigemina*, the doses utilized was $1 \times 10^7$ infected red blood cells IM. Groups II and IV were noninfected control groups. All the infected animals were severely affected; at days 5–7 post-inoculation (DPI) they showed clinical signs: fever (40–41.5°C), packed cell volume reduction, and parasitemia, and specific treatment was required. The immune response was monitored daily from 0–11 DPI. As shown by flow cytometry analysis, in infected animals the distribution in peripheral blood of the T-cells subpopulations ($CD4^+$, $CD8^+$, γδ T-cells) was not affected when compared to the control groups. By ELISA, IFN-γ production showed a trend to increase in plasma between 6–10 DPI; noninfected cows showed the lowest optical density values. By RT-PCR, a Th1 predominant response was observed, TNFα, INF-γ and iNOs were detected. In contrast IL-4 and IL-10 were weak or undetected. The results of this trial will be discussed.**

**KEYWORDS: *Babesia bigemina*; immune response; inflammatory cytokines**

## INTRODUCTION

Babesiosis is a tick-borne disease of cattle caused by *Babesia bigemina* and *Babesia bovis*, and transmitted by the tick vector *Boophilus microplus*. It is one of the most economically important infectious diseases in the tropical regions of the world.[1] In Mexico, it is a serious obstacle to improvement of the genetic background cattle in tropical areas.[2] Different studies have been performed *in vivo* and *in vitro* to understand the immune response. For example, *B. bovis*–infected erythrocytes and a membrane-enriched fraction of merozoites are able to stimulate inducible nitric oxide synthase (iNOs) transcription and NO production, but there is no report on the induction of inflammatory cytokines by *B. bigemina in vivo*.[3] In this study, we in-

Address for correspondence: M.J.A. Álvarez, Centro Nacional de Investigaciòn Disciplinaria en Parasitologia Veterinaria (CENID-PAVET), Instituto Nacional de Investigaciones Forestales, Agrícolas y Pecuarias (INIFAP), Apartado Postal 206, Civac, Morelos, Codigo Postal 62500, Mexico. Voice: 52-777-3260850; fax: 52-777-3192850, ext. 129.

alvarez.jesus@pavet.inifap.gob.mx

**Ann. N.Y. Acad. Sci. 1026: 144–148 (2004). © 2004 New York Academy of Sciences.**
**doi: 10.1196/annals.1307.019**

vestigated *B. bigemina* infection with regard to clinical infection, T cell distribution, and cytokine profile, during the acute phase of an experimental infection in cows in the second trimester of pregnancy.

## MATERIALS AND METHODS

### *Experimental Animals*

Twelve 24-month-old Holstein cows were estrus synchronized and divided at random into four groups (I, II, III, IV). Groups I and II were selected based on the diagnostic of pregnancy; groups III and IV were not pregnant. Groups I and III were infected with *B. bigemina*; groups II and IV were uninfected controls.

### *Parasites and Infection*

The *B. bigemina* strain was derived from a field outbreak, and specimens were kept frozen. The cows were injected with *B. bigemina*–infected red blood cells ($1 \times 10^7$), and the injections were administered intramuscularly. The control groups received uninfected red blood cells.

### *Clinical Monitoring and Sampling*

The cows were observed daily, from day 0 to day 11 postinfection (PI), and clinical signs were recorded. Rectal temperatures were registered, packed cell volume was measured by the microhematocrit method, and the presence of parasites was determined by microscopic examination of Giemsa-stained smears. Blood samples were taken daily, and vacuum tubes were used to separate sera and peripheral blood mononuclear cells (PBMCs).

### *Flow Cytometry Analysis*

PBMCs were isolated over Lymphoprep gradients (Nycomed Pharma As, Oslo, Norway), and T cell distribution was determined by one-color staining analysis (fluorescein isothiocyanate, FITC), as described previously.[4]

The monoclonal antibodies used were as follows:

- anti-bovine $\gamma\delta$ (mouse anti-bovine ILA29, IgG1).
- anti-bovine $CD4^+$ (CACT138A, IgG1).
- anti-bovine $CD8^+$ (BAQ111A1, IgM) (VMDR, Pullman, WA).

Secondary antibodies were rat anti-mouse IgG1-FITC and rat anti-mouse IgG2a-FITC (PharMingen, Becton Dickinson, San Diego, CA). The analysis was performed on a FACS Vantage (Becton Dickinson, Franklin Lakes, NJ), and data were analyzed using CellQuest software (BD, Franklin Lakes, NJ).

### *RNA Extraction*

Total RNA was extracted from $1 \times 10^7$ PBMC by using the RNeasy kit (Qiagen, Chatsworth, CA), according to the manufacturer's recommendations.

### *RT-PCR*

Equal amounts of RNA were used from PBMCs from each animal to analyze cytokine levels. Real-time polymerase chain reaction (RT-PCR) protocols were performed as previously described.[4] Both the cDNA synthesis and PCR were performed in a single tube using the SuperScript One-Step RT-PCR with platinum *Taq* (Invitrogen, Carsbad, California). Forward and reverse primers (20 μM each) were added to each reaction in individual tubes, and the 2× reaction mix yielded a final concentration of 1.2 μM $MgSO_4$, total RNA 0.15 μg/reaction, 200 μM each dNTP, RT/Platinum *Taq* mix 1 μL, in a final volume of 50 μL. The cDNA synthesis was performed at 45°C for 15 min and at 99°C for 5 min in a thermocycler model ICycler (Bio-Rad Hercules, CA). The PCR amplification was performed under the following conditions: 93°C for 1 min, 55°C for 1 min, and extension at 72°C for 2 min for 35 cycles, with a final extension at 72°C for 5 min. The PCR products (25 μL) were electrophoresed on a 1.8% agarose gel containing ethidium bromide. The primers used were as follows:

*$G_3PDH$*
Forward: 5′-GGA GAA ACC TGC CAA GTA TGA T-3′
Reverse: 5′-TCG CTG TTG AAG TCG CAG GAG AC-3′ (120-bp product)[4]

*TNF-α*
Forward: 5′-CCC AGA GGG AAG AGC AGT-3′
Reverse: 5′-CCC TGA AGA GGA CCT GTG-3′ (253-bp product)[5]

*IL-10*
Forward: 5′-TGT CTG ACA GCA GCT GTA TCC-3′
Reverse: 5′-CAC TCA TGG CTT TGT AGA CAC-3′ (405-bp product)[5]

*IL-4*
Forward: 5′-ACA TCC TCA CAA GCA GAA AG-3′
Reverse: 5′-GTC TTG GCT TCA TTC ACA GA-3′ (220-bp product)[5]

*iNOs*
Forward: 5′-TAG AGG AAC ATC TGG CCA GG-3′
Reverse: 5′-TGG CAG GGT CCC CTC TGA TG-3′ (372-bp product)[3]

*IL-12*
Forward: 5′-TGG TAT CCT GAT GCT CCT GGA G-3′
Reverse: 5′-TGC TCC AAG CTG ACC TTC TCT G-3′ (444-bp product)[3]

### *Determination of IFN-γ in Plasma*

Plasma was harvested for interferon-γ (IFN-γ) analysis with a commercial enzyme-linked immunosorbent assay (ELISA) kit (Bovigam, Biocor, Omaha, NE) according to the manufacturer's directions.

## RESULTS AND DISCUSSION

Infected pregnant and nonpregnant animals were severely affected on days 5 to 7 PI, showed fever up to 41.5°C, and packed cell volume decreased to 50%.The parasite was observed in groups I and II on day 6 PI (0.3%). Specific treatment was re-

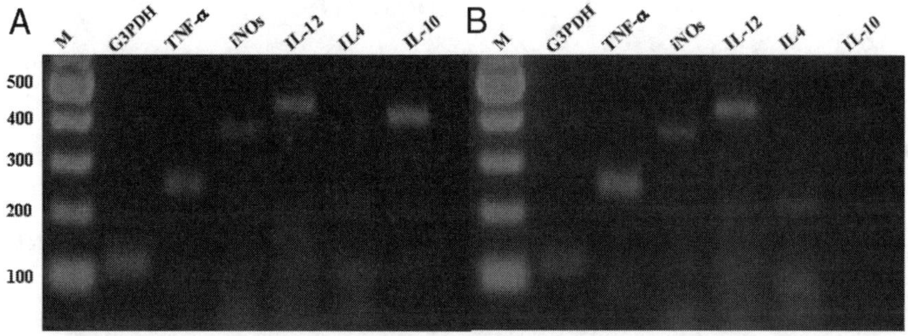


**FIGURE 1.** RT-PCR analysis of enhanced transcription of cytokine RNA in bovine PBMCs following *B. bigemina* infection. (**A**) Marker analysis of TNF-α (253 bp), iNOs (372 bp), IL-12 (466 bp), IL-4 (220 bp), and IL-10 (405 bp) at 1 DPI in one pregnant *B. bigemina*–infected cow. (**B**) Analysis of TNF-α, iNOs, IL-12, IL-4, and IL-10 at 5 DPI in one pregnant *B. bigemina*–infected cow.

quired to prevent death. Flow cytometry analysis determined the T cell subpopulations, and $CD4^+$, $CD8^+$, and γδ T cells showed no significant change in values at 6–9 DPI in peripheral blood when compared to the control groups (data not shown).

The induction of inflammatory cytokines was demonstrated at two different periods of time, 0–3 DPI and 5–7 DPI. Tumor necrosis factor-α (TNF-α) was induced at 1–3 DPI (FIG. 1A) with a marked presence at 5–6 DPI (FIG. 1B). TNF-α mRNA expression had been induced by addition of *Babesia bovis* merozoites plus IFN-γ in monocyte-derived macrophages taken *ex vivo*.[6] When *B. bovis*–infected erythrocytes induce TNF-α production, IFN-γ potentates the effect.[3] In this study, IFN-γ and interleukin-12 (IL-12) production were induced particularly during the clinical phase (5–7 DPI) (FIG. 1B). A similar condition was observed during an acute infection with WA1 *Babesia* in mice. These cytokines and the induction of macrophage-derived effector molecules such as NO are important elements of the response.[7,8] Another study showed that stimulation of NO was IFN-γ dependent in mononuclear phagocytes exposed to *B. bigemina* merozoites.[6] iNOs showed a weak presence at 1 DPI (FIG. 1A) and 5 DPI (FIG. 1B). Some Th cell lines have shown an *in vitro* ability to produce IFN-γ in response to antigen (*B. bovis* membrane) and antigen-presenting cells, which activate macrophages to produce NO.[9,10]

IL-12 production was better observed on 5–7 DPI. The *Plasmodium chabaudi* infection involves production of IL-12, IFN-γ, and TNF-α associated with an NO-dependent mechanism, as protective immunity.[7,11] IL-4 was not detected (FIG. 1) and IL-10 was detected at 1 DPI. The pregnancy condition had no effect on the susceptibility to the infection, or on the immune response. This study suggests the importance of the innate immunity against *B. bigemina* and suggests that the Th1 response is induced as a protective condition during the acute phase of the disease *in vivo*. The isolation of T cells and/or *in vivo* studies using new techniques such as RT-PCR and microarrays are required for a better understanding of the different elements involved in the immune response.

## ACKNOWLEDGMENTS

This research was supported in part by the Consejo Nacional de Ciencia y Tecnología, Mexico (Project 34477-B).

## REFERENCES

1. RISTIC, M. 1981. Babesiosis. *In* Diseases of Cattle in the Tropics. M. Ristic & I. McIntyre, Eds. Vol. 6: 443–468. Martinus Nijhoff. The Hague.
2. FIGUEROA, J.V. *et al.* 1993. Use of a multiplex polymerase chain reaction-based assay to conduct epidemiological studies on bovine hemoparasites in Mexico. Rev. Elev. Med. Vet. Pays Trop. 46, (1-2), 71-75.
3. SHODA, L.K.M. *et al.* 2000. *Babesia bovis*-stimulated macrophages express interleukin-1b, interleukin-12, tumor necrosis factor alpha, and nitric oxide and inhibit parasite replication in vitro. Infect. Immun. **68**(9): 5139–5145
4. SMITH, R. *et al.* 1999. Role of $CD8^+$ and $WC1^+$ $\gamma\delta$ T cells in resistance to *Mycobacterium bovis* infection in the SCID-bo mouse. J. Leuk. Biol. **65:** 28–34.
5. MWANGI, D.M. *et al.* 1998. Immunization of cattle by infection with *Cowdria ruminatum* elicits T lymphocytes that recognize autologous, infected endothelial cells and monocytes. Infect. Immun. **66:** 1855–1860
6. GOFF, W.L. *et al.* 2002. IL-4 and IL-10 inhibition of IFN-$\gamma$ and TNF-$\alpha$ dependent nitric oxide production from bovine mononuclear phagocytes exposes to *Babesia bovis* merozoites. Vet. Immunol. Immunopathol. **84:** 237–251.
7. AGUILAR-DELFÍN, I. *et al.* 2002. Resistance to acute babesiosis is associated with interleukin-12 and gamma interferon-mediated responses and requires macrophages and natural killer cells. Infect. Immun. **71**(4): 2002-2008.
8. BROWN, W.C. & G.H. PALMER. 1999. Designing blood-stage vaccines against *Babesia bovis* and *B. bigemina*. Parasitol. Today **15:** 275–280.
9. BROWN, W.C. *et al.* 1998. Immunodominant T-cell antigens and epitopes of *Babesia bovis* and *Babesia bigemina*. Ann. Trop. Med. Parasitol. **92**(4): 473–482.
10. STICH, R.W. *et al.* 1998. Stimulation of nitric oxide production in macrophages by *Babesia bovis*. Infect. Immun. **66:** 4130–4136.
11. NAHREVANIAN, H. & M.J. DASCOMBE. 2001. Nitric oxide and reactive nitrogen intermediates during lethal and nonlethal strains of murine malaria. Parasite Immunol. **23**(9): 491–501.

# Use of the Miniature Anion Exchange Centrifugation Technique to Isolate *Trypanosoma evansi* from Goats

CARLOS GUTIERREZ,[a] JUAN A. CORBERA,[a] FRANCISCO DORESTE,[a] AND PHILIPPE BÜSCHER[b]

[a]*University of Las Palmas, Canary Islands, Spain*

[b]*Department of Parasitology, Institute of Tropical Medicine, Antwerp, Belgium*

**ABSTRACT: DEAE (anion exchanger diethylaminoethyl)-cellulose and mini Anion Exchange Centrifugation Technique (mAECT) allow salivarian trypanosomes to be separated from the blood of affected animals. The purpose of this study was to assess the mAECT in goats infected with *T. evansi*. Five adult Canary goats were inoculated intravenously with at least 1 × 10$^5$ *T. evansi* isolated from a dromedary camel in the Canary Islands. The goats were monitored for specific antibodies and parasite detection. The inoculated goats became infected and the parasitemia remained very low but was persistent. For mAECT columns, the DEAE gel was equilibrated with phosphate-buffered saline glucose. *T. evansi* was detected by its mobility with a microscope at low magnification (10 × 10). The mAECT proved to be more sensitive than blood smear and buffy coat but less sensitive than mouse inoculation. We conclude that in cases of very low parasitemia in goats, mAECT can be used when other parasite-detection tests have failed.**

**KEYWORDS: *Trypanosoma evansi*; goat; experimental; miniature anion exchange centrifugation technique (mAECT)**

## INTRODUCTION

Salivarian trypanosomes can be separated from blood cells and platelets by passing blood from infected mammals through a column with the anion exchanger diethylaminoethyl cellulose (DEAE-cellulose).[1] The separation depends fundamentally on differences in surface charge; the DEAE-cellulose adsorbs the more negatively charged blood components while the less negatively charged flagellates are eluted.[2] An elaboration of the DEAE-cellulose technique, the miniature anion exchange centrifugation technique (mAECT), has been developed for use in humans in the field by Lumsden *et al.*[3] The mAECT allows for the elution of trypanosomes from venous blood and for their concentration at the bottom of a sealed glass tube by low-speed centrifugation (3000 rpm). The trypanosomes may then be detected via low-magni-

Address for correspondence: Carlos Gutierrez, Veterinary Faculty, University of Las Palmas, 35416, Arucas, Las Palmas, Canary Islands, Spain. Voice: 34 928451115; fax: 34 928451142. cgutierrez@dpat.ulpgc.es

**Ann. N.Y. Acad. Sci. 1026: 149–151 (2004). © 2004 New York Academy of Sciences. doi: 10.1196/annals.1307.020**

fication microscopy. Given that surface charges differ between species of salivarian trypanosomes, and that the negative charge on erythrocytes also varies with mammalian species,[4] the technique should be adapted to the species of salivarian trypanosome and to the mammalian host. With regard to goats, mAECT has been used previously to isolate *Trypanosoma vivax*, *T. congolense*, and *T. brucei brucei*.[5] The purpose of this study was to assess mAECT in goats infected with *T. evansi*.

## MATERIALS AND METHODS

### *Animals*

Five adult female Canary goats were selected for this study. The animals were purchased from a commercial dairy goat farm. The animals, which lacked any antecedents of trypanosomosis, had been recently vaccinated against *Salmonella* spp. and *Chlamydia* spp. and administered ivermectin against external and internal parasites. The animals were housed in an inoculation area of the University of Las Palmas. The animals tested negative in parasitological and serological tests for *T. evansi*. The goats were inoculated intravenously with at least $1 \times 10^5$ *T. evansi* isolated from a dromedary camel in the Canaries (Gran Canaria island). The animals were maintained for 8 months. They were examined daily for any clinical evidence of the disease, and they were monitored monthly for specific antibodies and parasite detection.

### *Parasite Detection Tests*

From each animal, blood samples were taken monthly from the jugular vein and collected in tubes containing EDTA and heparin as anticoagulants. The tests used included hematocrit centrifugation, wet film, stained thin smear, and mice inoculation (as described by the Office International des Ëpizooties[6]). The mAECT was also used. The test was provided by the Institute of Tropical Medicine, Antwerpen, Belgium. For goat blood, the DEAE gel was equilibrated with phosphate-buffered saline glucose ($Na_2HPO_4$ (anhydrous): 8,088 g/L; $Na_2PO_4{\cdot}2H_2O$: 0.468 g/L; NaCl: 2.55 g/L; glucose: 10 g/L). A volume of 300 µL of fresh heparinized blood was eluted on a 2.5-mL DEAE gel bed volume. After centrifugation of the eluate, the presence of *T. evansi* was evaluated by low-magnification microscopy ($10 \times 10$).

### *Indirect Hematological Parameters*

Packed cell volume (PCV) and total serum proteins were measured to detect effects on host health through the experiment.

## RESULTS AND DISCUSSION

The inoculated goats showed a particularly subclinical course of infection. The PCV dropped significantly ($P < 0.05$), and serum proteins also increased significantly ($P < 0.05$) because of globulin increases.

Parasitemias remained very low but were persistent through the experiment. The mAECT proved to be more sensitive than blood smear and buffy coat examination, but it was less sensitive than mouse inoculation. For mAECT, after centrifugation of the eluate, *T. evansi* was detected by its mobility under low magnification (10 × 10). The mAECT has been used in goats infected with *T. vivax*, *T. congolense*, and *T. brucei*, although the sensitivity obtained was lower.[5] However, similar sensitivities have been shown with *T. evansi* in antelopes[7] and water buffaloes.[8]

We conclude that in cases of very low parasitemia in goats, mAECT can be performed when other parasite detection tests have been ineffective.

## REFERENCES


1. LANHAM, S.M. 1968. Separation of trypanosomes from the blood of infected rats and mice by anion-exchangers. Nature **218:** 1273–1274.
2. LANHAM, S.M. & D.G. GODFREY. 1970. Isolation of salivarian trypanosomes from man and other mammals using DEAE-cellulose. Exp. Parasitol. **28:** 521–534.
3. LUMSDEN, W.H., C.D. KIMBER, D.A. EVANS & S.J. DOIG. 1979. *Trypanosoma brucei*: miniature anion-exchange centrifugation technique for detection of low parasitaemias: adaptation for field use. Trans. R. Soc. Trop. Med. Hyg. **73:** 312–317.
4. SEAMAN, C.V.F. & G. UNLENBRUCK. 1963. The surface structure of erythrocytes from some animal sources. Arch. Biochem. Biophys. **100:** 493–502.
5. KALU, A.U. & F.A.G. LAWANI. 1986. Caprine trypanosomiasis: comparative study of parasitological diagnostic techniques in mixed subpatent infections. Bull. Anim. Health Prod. Afr. **34:** 294–295.
6. OFFICE INTERNATIONAL DES EPIZOOTIES. 2000. Surra (*Trypanosoma evansi*). *In* Manual of Standards for Diagnostic Tests and Vaccines. 4th edit: chapter X.11. Office International des Epizooties. Paris.
7. SACHS, R. 1984. The superiority of the miniature anion-exchange centrifugation technique for detecting low grade trypanosome parasitaemias. Trans. R. Soc. Trop. Med. Hyg. **78:** 694–696.
8. HOLLAND, W.G., F. CLAES, L.N. MY, *et al.* 2001. A comparative evaluation of parasitological tests and a PCR for *Trypanosoma evansi* diagnosis in experimentally infected water buffaloes. Vet. Parasitol. **97:** 23–33.

# Performance of Serological Tests for *Trypanosoma evansi* in Experimentally Inoculated Goats

CARLOS GUTIERREZ,[a] JUAN A. CORBERA,[a] MANUEL MORALES,[a] AND PHILIPPE BÜSCHER[b]

[a]*Veterinary Faculty, University of Las Palmas, 35416, Las Palmas, Canary Islands, Spain*

[b]*Department of Parasitology, Institute of Tropical Medicine, Antwerp, Belgium*

**ABSTRACT: Natural *Trypanosoma evansi* infection in the Canary Islands has only been diagnosed in the camel population, but dissemination of the disease in other hosts has not been excluded. The objective of this work was to assess the performance of serological antibody tests in experimentally inoculated goats. Five Canarian goats were inoculated intravenously with at least $1 \times 10^5$ *T. evansi*. The animals were kept for 8 months and checked monthly for the presence of the parasite and specific antibodies. The serological tests investigated were the direct card agglutination test CATT/*T. evansi* and the indirect card agglutination test LATEX/*T. evansi*. All animals became positive in the CATT/*T. evansi* 1 month post-infection and remained positive with a minimum end-titer of 1/4. Similar results were obtained with the LATEX/*T. evansi*, although at lower end-titers (1/2). We conclude that CATT/*T. evansi* is adequate for assessing infection of Canarian goats by *T. evansi*.**

**KEYWORDS: *Trypanosoma evansi*; goat; serology; experimental**

## INTRODUCTION

*Trypanosoma evansi* was diagnosed for the first time in the Canary Islands (Spain) in 1997 in a dromedary camel presenting the chronic and terminal stage of the disease. The animal had been imported from the neighboring West African area, where *T. evansi* is highly prevalent.[1] Seroprevalences of 4.8% up to 9% were observed in camels on the Canary Islands between 1997 and 1999.[2,3] Affected animals were treated, but dissemination of the disease in other hosts is not excluded. Goats, in particular, could play an important role in the epidemiology of *T. evansi* in the Canary Islands. However, serological tests normally used in small ruminants for *T. evansi* antibody detection have not been validated for these species. Consequently, the objective of this work was to assess the performance of simple and rapid serological tests for antibody detection in experimentally inoculated goats.

Address for correspondence: Carlos Gutierrez, Veterinary Faculty, University of Las Palmas, 35416, Arucas, Las Palmas, Canary Islands, Spain. Voice: 34-92-8451115; fax: 34-92-8451142. cgutierrez@dpat.ulpgc.es

**Ann. N.Y. Acad. Sci. 1026: 152–153 (2004). © 2004 New York Academy of Sciences. doi: 10.1196/annals.1307.021**

## MATERIALS AND METHODS

Five adult Canary female goats were inoculated intravenously with at least $1 \times 10^5$ *T. evansi* isolated from a dromedary camel. The animals were kept for 8 months and checked monthly for the presence of the parasite and specific antibodies. The serological tests investigated were the direct card agglutination test CATT/*T. evansi* and the indirect card agglutination test LATEX/*T. evansi*, both developed at the Institute of Tropical Medicine, Antwerp, Belgium. For parasite detection, wet film, stained blood smears, buffy coat examination, and inoculation of mice were performed. Other indirect parameters of the disease such as packed cell volume (PCV) and serum total proteins were also determined.

## RESULTS AND DISCUSSION

The inoculated goats showed a subclinical course of the disease, and only a few episodes of fever (within the first weeks post-inoculation, PI) and arthritis (6 months PI) were evident. Parasitemias remained very low but were persistent. Drops in PCV (mean values: 29.5% before inoculation, BI; 20% at 4 months PI; 26% at 8 months PI), and total serum protein (mean values: 6.3 g/dL BI; 11.2 g/dL at 4 months PI; 8.6 g/dL at 8 months PI) were observed. All animals became positive in the CATT/*T. evansi* at 1 month PI and remained positive with a minimum end-titer of 1/4. Similar results were obtained with the LATEX/*T.evansi*, although at lower end-titers (1/2).

*T. evansi* infection has been reported in small ruminants, causing severe infection,[4] moderate infection,[5] or subclinical infection.[6] The low parasitemia found in this study could indicate that goats do not play an important role in the epidemiology of *T. evansi* in Canaries even though goats are receptive to experimental inoculation. Similar findings have also been reported by Jacquet *et al.*[1] in Mauritania and could indicate the prevalence of the subclinical form of the disease in goats.

We conclude that CATT/*T. evansi* is adequate for assessing infection of goats with *T. evansi*.

## REFERENCES

1. Jacquet, P., D. Cheikh, A. Thiam & M.L. Dia. 1993. La trypanosomose à *Trypanosoma evansi* (Stell 1885), Balbiani 1888 chez les petits ruminants de Mauritanie: résultats d'inoculation expérimentale et d'enquetes sur le terrain. Rev. Elev. Med. Vet. Pays Trop. **46:** 574–578.
2. Gutierrez, C., M.C. Juste, J.A. Corbera, *et al.* 2000. Camel trypanosomosis in the Canary Islands: assessment of seroprevalence and infection rates using the card agglutination test (CATT/*T. evansi*) and parasite detection tests. Vet. Parasitol. **90:** 155–159.
3. Molina, J.M., A. Ruiz, M.C. Juste, *et al.* 2000. Seroprevalence of *Trypanosoma evansi* in dromedaries (*Camelus dromedarius*) from the Canary Islands (Spain) using an antibody Ab-ELISA. Prev. Vet. Med. **47:** 53–59.
4. Ngeranwa, J.J., P.K. Gathumbi, E.R. Mutiga & G.J. Agumbah. 1993. Pathogenesis of *Trypanosoma (brucei) evansi* in small east African goats. Res. Vet. Sci. **54:** 283–289.
5. Hornby, H.E. 1952. Animal Trypanosomiasis in East Africa, 1949. Colonial Office London. Her Majesty's Stationary Office. Norwich, UK.
6. Rottcher, D., D. Schillinger & E. Zweygarth. 1987. Trypanosomiasis in the camel (*Camelus dromedarius*). Rev. Sci. Tech. Off. Int. Epiz. **6:** 463–470.

# Seroprevalence of *Leishmania infantum* in Northwestern Spain, an Area Traditionally Considered Free of Leishmaniasis

INMACULADA AMUSATEGUI,[a] ANGEL SAINZ,[a] ENARA AGUIRRE,[a] AND MIGUEL A. TESOURO[b]

[a]*College of Veterinary Medicine, Universidad Complutense de Madrid, Spain*

[b]*College of Veterinary Medicine, Universidad de León, Spain*

**ABSTRACT: Northwestern Spain has traditionally been considered to be free from leishmaniasis. The aim of this work was to determine the prevalence of canine leishmaniasis in this area and to assess the influence of several risk factors on the incidence of this disease. A total of 479 dogs attended at different veterinary clinics in northwestern Spain were tested for *L. infantum* with the immunofluorescent antibody (IFA) test. The seroprevalence of *L. infantum* in this area was 3.7%. Most of the seropositive dogs lived in two locations: Valdcorras (seroprevalence of 29.2%) and Ourense (seroprevalence of 7.5%). The detection of high antibody titers in most of the seropositive dogs (many of which presented clinical signs) coupled with the certainity that some of these dogs had never been outside their home areas indicates the presence of this zoonosis in these two sites. On the other hand, companion dogs were significantly less likely to acquire the disease than sheep dogs, hunting dogs, and those from kennels.**

**KEYWORDS: *Leishmania infantum*; canine leishmaniasis; immunofluorescent antibody test**

## INTRODUCTION

Canine leishmaniasis, caused by *Leishmania infantum*, is a chronic multisystemic disease found throughout the Mediterranean basin, including many areas of Spain.[1] Northern Spain, however, has traditionally been considered a leishmaniasis-free area, presumably because its cold and rainy climate limits the presence of sand flies.[2] Nevertheless, in the early 1990s, two cases of apparently autochthonous canine leishmaniasis were reported in Galicia (northwestern Spain).[3] The aim of the present work was to determine the prevalence of leishmaniasis in this area and to assess the influence of several risk factors on the incidence of the disease.

Address for correspondence: Inmaculada Amusategui, Departamento de Medicina y Cirugía Animal, Facultad de Veterinaria, Universidad Complutense de Madrid, Avda. Puerta de Hierro s/n. 28040, Madrid, Spain. Voice: 34-91-3943820; fax: 34-91-3943811.
angelehr@vet.ucm.es

**Ann. N.Y. Acad. Sci. 1026: 154–157 (2004). © 2004 New York Academy of Sciences.
doi: 10.1196/annals.1307.022**

## MATERIALS AND METHODS

A total of 479 dogs treated at different clinics in northwestern Spain were tested for *L. infantum* using the IFA (immunofluorescent antibody) test adapted by Tesouro in 1984; the cutoff was established at 1/100.[4] Dogs were classified according to the following qualitative variables: gender (213 males and 266 females); breed (119 mixed-breed dogs, 58 German shepherds, 45 greyhounds, 22 Irish setters, 19 boxers, 17 cocker spaniels, 17 rottweilers, and 180 dogs of 43 other breeds with less than 15 dogs each); age (215 young dogs ≤2 years old, 211 adults >2 but <8 years old, and 53 dogs ≥8 years old); habitat (372 lived in rural areas and 107 in urban areas); occupation (206 companion dogs, 150 hunting dogs, 12 kennel dogs, and 111 sheepdogs); and lifestyle (84 lived mainly indoors and 395 mainly outdoors).

Data were processed by the MedCalc (version 4.16g; MedCalc Software, Maria-Kerke, Belgium) computer program. The $\chi^2$ test was performed to compare the distribution of dogs within each qualitative variable considered as a risk factor. A statistically significant association was recognized when $P < 0.05$.

## RESULTS AND DISCUSSION

Our results show a seroprevalence of *L. infantum* in northwestern Spain of 3.7% (18/479). The great majority of the seropositive dogs lived in two locations of this area: Valdeorras, where seroprevalence was 29.2% (7/24), and Ourense, where seroprevalence was 7.5% (6/80). Sixteen of the 18 seropositive dogs displayed antibody titer above the cutoff (one had an antibody titer of 1/200, three displayed titers of 1/400, and the other 12 presented titers ≥1/800). Furthermore, 13 of these dogs displayed clinical signs compatible with leishmaniasis.

The presence of high antibody titers in the great majority of the seropositive cases (many of which presented clinical signs), together with the certainty that some of these dogs (three from Valdeorras and one from Ourense) had never been outside their home areas, indicates the presence of this zoonosis in these two sites. Moreover, the seroprevalence rates observed in these places are similar to those of endemic areas.[1,2,5,6] Nevertheless, the detection of three cases of apparently autochthonous canine leishmaniasis in other locations (two in Verín and one in Tui) leads us to suspect the existence of microfoci of infection throughout different areas of Galicia, traditionally considered a leishmaniasis-free region.

It is well known that sand flies require very specific environmental conditions (temperature, humidity, etc.) for their development and activity.[7] Such conditions may be found in limited districts with appropriate microclimates within larger areas whose overall conditions are unsuitable for their survival. As a result, in Galicia, this disease may propagate in the form of independent foci that offer favorable conditions for the survival of the vector involved in its transmission, rather than spreading out from one central site. This occurs even in endemic zones, in which areas with high densities of sand flies are found next to other areas in which these insects hardly exist.[5]

In agreement with other authors, we observed no significant differences between males (11/266 = 4.1%) and females (7/213 = 3.3%) ($P = 0.8009$).[6,8] Although significant differences were not reached with regard to age ($P = 0.0543$), *L. infantum* infection was much less frequent in young animals (5/215 = 2.3%) than in adults (13/211 = 6.2%), presumably because of a combination of a shorter period of exposure to the vector and the long leishmaniasis IFA seroconversion period.[9] As indicated in Materials and Methods, breeds with fewer than 15 dogs were included within a single category ("other breeds") to make statistical analysis possible. Among the 18 seropositive dogs, there were 8 mixed-breeds, two rottweilers, one German shepherd, one greyhound, and six representatives of other breeds. No breed predilection was found in the population included in this study ($P = 0.2819$); this observation agrees with those reported by others.[8]

It is clear that the prevalence of leishmaniasis was not related to habitat, because prevalence was statistically similar in dogs living in urban (6/107 = 5.6%) and rural (12/372 = 3.2%) areas ($P = 0.3945$). Canine leishmaniasis has traditionally been associated with rural and suburban areas, where sand flies appear to find the best environment to grow and reproduce.[7] Our results may reflect the geographic and urban characteristics typical of Galicia, with its small but numerous towns and townships sprinkled throughout the countryside; that is, our results may reflect that rural and urban environments in Galicia often practically overlap.

On the other hand, companion dogs (2/206) were significantly less prone to acquire the disease than sheepdogs (5/111), hunting dogs (9/150), and those from kennels (2/12) ($P = 0.0097$). Likewise, none of the 84 dogs living mainly indoors proved to be seropositive. The greater exposure to the vector of dogs that live outdoors may explain our findings with regard to occupation and lifestyle, both of which are intricately related.[8]

Our results demonstrate the presence of leishmaniasis in an area of Spain traditionally considered to be free of leishmaniasis. It is important to be aware of the existence of microclimates that may be suitable for the propagation of this kind of disease in areas that at first glance appear to be unfavorable. Furthermore, climatic and environmental changes may contribute to the spread of diseases that have been considered, until recently, restricted to tropical and subtropical areas.

## REFERENCES

1. Trotz-Williams, L.A. & A.J. Trees. 2003. Systematic review of the distribution of the major vector-borne parasitic infections in dogs and cats in Europe. Vet. Rec. **152:** 97–105.
2. Tesouro, M.A., F. Jiménez Mazzucchelli, C. Fragío, *et al.* 1989. Evolución del número de casos de leishmaniosis canina en Madrid y otras provincias: años 1981–1988. Vet. Rec. (Spanish edition) **2**(1): 6–7.
3. Puerta, J.L. 1993. Leishmaniosis. *In* Plan de Profilaxis y Lucha contra las Enfermedades del Perro. J.L. Puerta, Ed.: 62–64. Ed. Xunta de Galicia. Santiago de Compostela, Spain.
4. Tesouro Díez, M.A. 1984. Aspectos clínicos y laboratoriales del diagnóstico de la leishmaniosis canina: estudio epizootiológico en la provincia de Madrid. Ph.D. thesis. Facultad de Veterinaria. Universidad Complutense de Madrid, Spain.
5. Mancianti, F., L. Gradoni, M. Gramiccia, *et al.* 1986. Canine leishmaniasis in the isle of Elba, Italy. Trop. Med. Parasitol. **37:** 110–112.
6. Ozon, C., P. Marty, C. Veyssière, *et al.* 1995. Résultats d'une enquête sur la leishmaniose canine effectuée penddant une courte période chez les vétérinaires practiciens des Alpes-Maritimes. Prat. Méd. Chir. Anim. Comp. **30:** 199–201.

7. ALVAR, J., C. AMELA & R. MOLINA. 1995. El perro como reservorio de la leishmaniosis. Med. Vet. **12**(7,8): 431–439.
8. FERRER, L. 1992. Leishmaniasis. Kirk's Current Veterinary Therapy XI. Small Animal Practice. Saunders. Philadelphia, PA.
9. CARRERA, L., M.L. FERMÍN, M. TESOURO, *et al.* 1996. Antibody response in dogs experimentally infected with *Leishmania infantum*: infection course antigen markers. Exp. Parasitol. **82:** 139–146.

# Retrospective Study (1998–2001) on Canine Babesiosis in Belo Horizonte, Minas Gerais, Brazil

CAMILA DE VALGAS E BASTOS, SIMONE MAGELA MOREIRA, AND LYGIA MARIA FRICHE PASSOS

*Departamento de Medicina Veterinária Preventiva, Escola de Veterinária, Universidade Federal de Minas Gerais, Minas Gerais, Brazil*

**ABSTRACT: The present work describes a retrospective study of clinical cases of babesiosis in dogs examined at the Veterinary Hospital (Universidade Federal de Minas Gerais) from March 1998 to September 2001. From the clinical records of dogs with laboratory-confirmed *Babesia canis* infections, we analyzed: demography (age, breed, sex, time of year, and origin; clinical manifestations (concomitant infections, body temperature, presence of ticks, and clinical signs); and hematological alterations. From 194 records from animals with suspicious cases of hemoparasites, 145 were confirmed to be infected and among those 61 dogs (42%) were infected with *B. canis*. The results point to the importance of canine babesiosis in Brazil.**

**KEYWORDS: canine babeisosis; *Babesia canis*; hemoparasites; *Rhipicephalus sanguineus* (Latreille)**

## INTRODUCTION

Canine babesiosis is a disease caused by the intraerythrocytic protozoan *Babesia canis*, transmitted by ticks to domestic and wild canides.[1] In Brazil, *Rhipicephalus sanguineus* (Latreille, 1806) ticks have been considered the main vector of canine babesiosis. Because *R. sanguineus* is a three-host tick, very well adapted to urban areas, the disease assumes an endemic character, as reported in a previous study on the frequency of anti-*B. canis* antibodies in dogs.[2] The most common clinical signs are fever, lymph node enlargement, anorexia, jaundice, hemoglobinuria, and splenomegalia.[3] The great limitation of its diagnosis is related to the variety of clinical manifestations, making it difficult to have a definitive diagnosis solely on the basis of clinical examination. The laboratory diagnosis relies on the detection of infected erythrocytes in Giemsa-stained blood smears. However, the absence of parasites does not exclude possibility of infection. Recent studies have focused on the com-

Address for correspondence: Lygia Maria Friche Passos, Departamento de Medicina Veterinária Preventiva, Escola de Veterinária, Universidade Federal de Minas Gerais, Caixa Postal 567, Belo Horizonte, CEP 30123-970, Minas Gerais, Brazil.
lygia@dedalus.lcc.ufmg.br

**Ann. N.Y. Acad. Sci. 1026: 158–160 (2004). © 2004 New York Academy of Sciences.
doi: 10.1196/annals.1307.023**

prehension of immunological and pathophysiological aspects of the disease and on the search for more sensitive and specific methods of diagnosis.

Although the importance of canine babesiosis has increased in the state of Minas Gerais,[5] few epidemiologic studies have been carried out in the country, and little is known about its clinical and epidemiologic aspects in Brazil. Therefore, the present work had the objective of evaluating, through a retrospective study, the occurrence and some epidemiologic parameters of clinical cases of babesial infections in dogs examined at the Veterinary Hospital of the Veterinary School in Belo Horizonte, from March 1998 to September 2001.

## MATERIALS AND METHODS

The study reflected the spontaneous demand of the Veterinary Hospital and took into account clinical records of small animals with clinically suspicious hemoparasites, from which records with laboratory confirmation of *B. canis*, by direct examination of Giemsa-stained blood smears, were selected. For infected animals, the following information was analyzed: demographic parameters (age, breed, sex, time of year, and origin), clinical manifestations (concomitant infections, body temperature, presence of ticks, and clinical signals), and hematological alterations.

## RESULTS

From a total of 194 records from small animals with suspicious cases of hemoparasites, 145 were confirmed to be infected; among these infected animals, 61 dogs (42%) were infected with *B. canis*. As far as age is concerned, 19.2% of the infected dogs were from 13 to 24 months old and 21.2% were from 25 to 48 months old, with 73.1% of the cases being observed in dogs between 0 and 4 years of age.

Regarding breed, the higher frequencies of infection were detected in German sheepdogs (16.6%), poodles (13.3%), and rottweillers (11.6%), demonstrating that the disease occurs in dogs of different sizes, of both sexes. However, housing conditions appear to have an influence on frequency of disease because 78.8% of infected dogs were living in houses, while only 21.2% were living in apartments. All animals with history and/or presence of ticks were living in houses. Regarding precedence, the infected animals originated from a variety of urban and suburban regions.

The most frequent clinical signs were fever, apathy, anorexia,weight lost, dehydration, abdominal pain, and kidney sensitivity to palpation. Among the infected animals, 64.3% showed anemia, 19.6% had significant decrease in the packed cell volume (<15%), 62.5% had eosinopenia, 64.3% had an increase of neutrophils, and 89.3% presented anisocitosis.

## CONCLUSIONS

The results lead to the conclusion that there is a need for more attention from clinicians for *B. canis* infections in Brazil. These infections represent 42% of the diagnoses of hemoparasites. In addition, there is a higher risk for dogs living in houses.

## REFERENCES

1. Breitschwerdt, E. 1984. Babesiosis. *In* Clinical Microbiology and Infectious Diseases of Dog and Cat. C.E. Greene, Ed.: 796–805. Saunders. Philadelphia, PA.
2. Ribeiro, M.F.B., L.M.F. Passos, J.D. Lima, *et al.* 1990. Freqüência de anticorpos fluorescentes anti-*Babesia canis* em cães de Belo Horizonte, Minas Gerais. Arq. Bras. Med. Vet. Zootec. **42**(6): 511–517.
3. Spiewak, G. 1992. Aspectos epidemiológicos, clínicos e de diagnóstico da infecção por Babesia canis, em cães atendidos em clínicas veterinárias, em Belo Horizonte. Dissertação, 67pp. Escola de Veterinária, Universidade Federal de Minas Gerais.
4. Bicalho, K.A. 2001. O uso da citometria de fluxo para detecção de *Babesia canis*. Dissertação, 78pp. Escola de Veterinária, Universidade Federal de Minas Gerais.
5. Guimarães, A.M., T.M.F.S. Oliveira & I.C.A.S. Rosa. 2002. Babesiose canina: uma visão dos clínicos veterinários de Minas Gerais. Clin. Vet. **7**(41): 60–68.

# Identification of Antigenic Proteins of a *Theileria* Species Pathogenic for Small Ruminants in China Recognized by Antisera of Infected Animals

JOANA MIRANDA,[a,b,e] BARBARA STUMME,[c,e] DOREEN BEYER,[c] HELDER CRUZ,[a] ABEL GONZÁLEZ OLIVA,[a] MOHAMMED BAKHEIT,[c] DANIEL WICKLEIN,[c] HONG YIN,[d] JIANXUN LOU,[d] JABBAR S. AHMED,[c] AND ULRIKE SEITZER[c]

[a]*Instituto de Biologia Experimental e Tecnológica and* [b]*Instituto de Tecnologia Química e Biológica, Universidade Nova de Lisboa, Oeiras, Portugal*

[c]*Research Center Borstel, Borstel, Germany*

[d]*Lanzhou Veterinary Research Institute, Lanzhou, China*

**ABSTRACT: The antigenic proteins of the piroplasm stage of *Theileria* species (China), the causative agent of theilerosis of small ruminants in China, were analyzed by Western blot, revealing several specific immunoreactive proteins of different predicted molecular weights. Furthermore, sera from *Theileria* species (China)–infected animals were probed for reactivity with the TaSP protein of *T. annulata*, for which a homologue has been described in *Theileria* species (China). Affinity chromatography demonstrated the presence of TaSP-reactive antibodies, and the majority of the sera showed reactivity with this protein both in Western blots and in ELISA. The identified parasite antigens and TaSP will be assessed for their suitability for developing diagnostic methods as well as evaluated for their capacity to stimulated host immune competent cells.**

**KEYWORDS: *Theileria* species (China); piroplasm stage; TaSP cross-reactivity**

Several previously unidentified parasites have recently been described to be pathogenic for small ruminants. Among these is a parasite that causes a fatal disease of small ruminants in North China. This parasite, which is referred to as *Theileria* species (China),[1] is transmitted by *Haemaphysalis qinghaiensis* and often occurs in mixed infections with other parasites.[2,3] In a previous report, serum of infected animals was shown to specifically react with piroplasm homogenate of the parasite;[4] however, cross-reactivity with *Babesia ovis* could not be excluded. Therefore, a pre-

Address for correspondence: Professor Jabbar S. Ahmed, Veterinary Infectiology and Immunology, Research Center Borstel, Parkallee 22, 23845 Borstel, Germany. Voice: +49-(0)4537-188428; fax: +49-(0)4537-188627.
jahmed@fz-borstel.de

[e]Joana Miranda and Barbara Stumme contributed equally to this work.

**Ann. N.Y. Acad. Sci. 1026: 161–164 (2004).** **© 2004 New York Academy of Sciences.**
**doi: 10.1196/annals.1307.024**

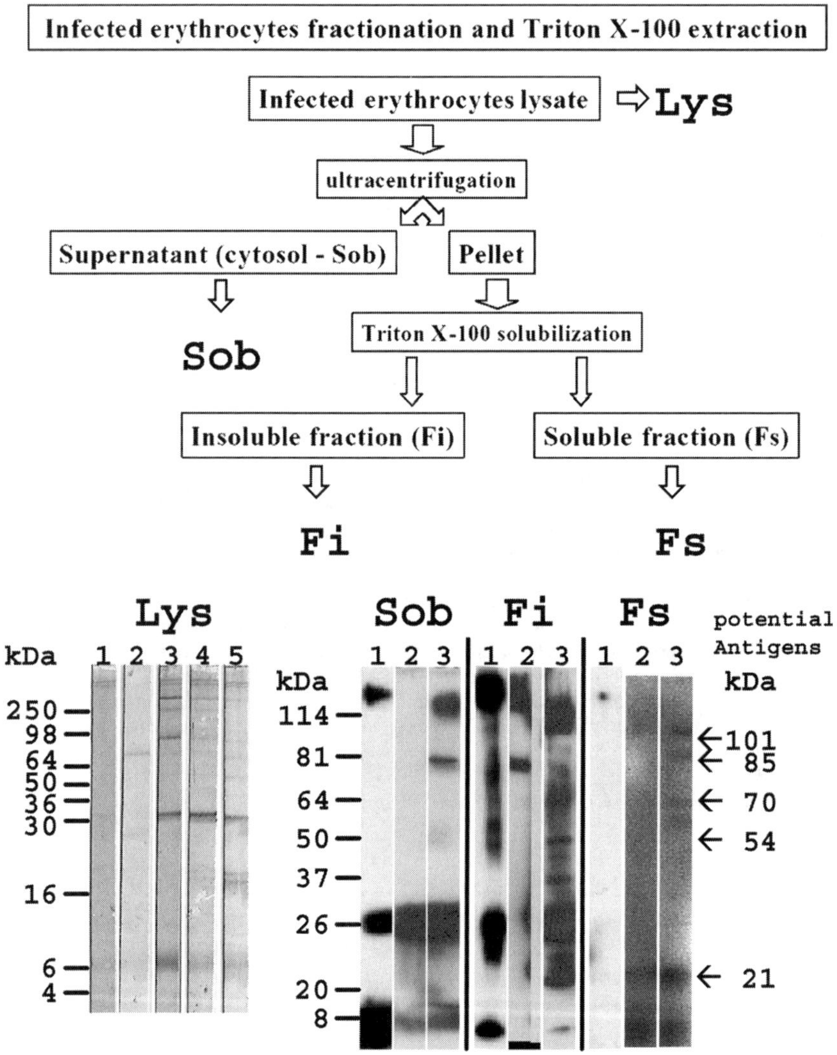


**FIGURE 1.** *Top panel:* Scheme of the fractionation procedure of *Theileria* species (China) piroplasm antigen and Western blot detection of antigenic proteins. *Bottom left panel:* Immunogenic reactivity pattern of the lysate fraction (Lys). *Lane 1*: secondary antibody controls; *lane 2*: antiserum from noninfected animals; *lanes 3–5*: antiserum from *Theileria* species (China)-infected animals. *Bottom right panel:* Western blot analysis revealing five different antigenic protein bands allocated to different fractions (supernatant: Sob; insoluble fraction: Fi; soluble fraction: Fs) of the fractionation procedure.

requisite for the evaluation of the humoral immune response is the use of parasite proteins specifically expressed in the *Theileria* species (China) parasite. To achieve this goal, lysed parasite material from the piroplasm stage was analyzed in this study to define immunogenic antigens.

## WESTERN BLOT ANALYSIS OF PIROPLASM IMMUNOGENIC ANTIGENS

Incubation of antiserum from infected animals with Western-blotted piroplasm antigen detected several antigenic protein bands from the 6-kDa to the greater than 250-kDa range. As shown in the lysate fraction in FIGURE 1 (Lys), the immunogenic reactivity pattern and intensity varied from different infected animals (lanes 3 to 5), although the majority appeared to be common. To resolve the relative complexity of the antigenic reactivity, the piroplasm antigen material was fractionated as depicted in the schematic of FIGURE 1. The resulting Western blot analysis detected the presence of five different specific antigenic protein bands allocated to different fractions (supernatant: Sob; insoluble fraction: Fi; soluble fraction: Fs) of the separation procedure after comparison with the appropriate secondary and noninfected antiserum controls. The antigenic signals below 40 kDa need to be resolved further to establish the specificity of the signals seen in the Lys blot, where several distinct and specific bands were also detectable.

The reactivity of the antisera with specific parasite antigens indicates that a humoral immune response is mounted in infected animals. Future experiments are directed at analysis by two-dimensional electrophoresis and subsequent sequencing of the reagenic protein bands. The identified parasite antigens will, on the one hand, be evaluated for their capacity to stimulate host immune competent cells; on the other hand, their suitability for developing diagnostic methods will be assessed. To this end, the immune response against these parasite proteins will be analyzed in correlation to the time course of infection. Attention will also be focused on the most potentially cross-reactive candidate protein of *Theileria* species (China) strains for the development of enzyme-linked immunosorbent assay (ELISA) and analysis of cell-mediated immunity.

## IMMUNOREACTIVITY OF SERA FROM *THEILERIA* (CHINA) SPECIES INFECTED SHEEP WITH *THEILERIA ANNULATA* SURFACE PROTEIN

Recently, a study reported on the identification of a gene encoding a macroschizont surface antigen from *T. annulata*, that is, *Theileria annulata* surface protein (TaSP).[5] This TaSP protein has been employed for the development of an ELISA for tropical theileriosis.[6] Because a gene with close identity to TaSP was also identified in *Theileria* species (China),[7] it was of great interest to see if there is any reactivity to this protein in sera from *Theileria* species (China)–infected animals. Interestingly, affinity chromatography demonstrated the presence of TaSP-reactive antibodies in *Theileria* species (China)–infected animals, and the sera showed specific reactivity with this protein in Western blots. When ELISA was used, 28 out of 33 animals (85%) showed a positive signal above the cutoff value calculated using negative serum, indicating that the majority of the sera tested had binding activity with the TaSP protein. These results suggest that, apparently, *Theileria* species (China)–infected animals may mount an immune response against the TaSP homologous protein of *Theileria* species (China). The diagnostic potential of the apparent TaSP-homologous protein of *Theileria* species (China) is under further investigation. A future improvement of this system would be the expression of the *Theileria* species (China) protein, which could lead to an increased sensitivity of the ELISA.

## ACKNOWLEDGMENTS

This work was supported in part by the European Union (ICA4-1999-30151) and the Portuguese Foundation for Science and Technology. Joana Miranda is the recipient of a scholarship (SFRH/BD/6494/2001).

## REFERENCES

1. SCHNITTGER, L., H. YIN, L. JIANXUN, *et al.* 2000. Ribosomal small-subunit RNA gene-sequence analysis of *Theileria lestoquardi* and a *Theileria* species highly pathogenic for small ruminants in China. Parasitol. Res. **86:** 352–358.
2. YIN, H., J. LUO, G. GUAN, *et al.* 2002. Transmission of an unidentified *Theileria* species to small ruminants by *Haemaphysalis qinghaiensis* ticks collected in the field. Parasitol. Res. **88:** S25–S27.
3. YIN, H., J. LUO, G. GUAN, *et al.* 2002. Experiments on transmission of an unidentified *Theileria* sp. to small ruminants with *Haemaphysalis qinghaiensis* and *Hyalomma anatolicum anatolicum*. Vet. Parasitol. **108:** 21–30.
4. GAO, Y.L., H. YIN, J. X. LUO, *et al.* 2002. Development of an enzyme-linked immunosorbent assay for the diagnosis of *Theileria* sp. infection in sheep. Parasitol. Res. **88:** S8–S10.
5. SCHNITTGER, L., F. KATZER, R. BIERMANN, *et al.* 2002. Characterization of a polymorphic *Theileria annulata* surface protein (TaSP) closely related to PIM of *Theileria parva*: implications for use in diagnostic tests and subunit vaccines. Mol. Biochem. Parasitol. **120:** 247–256.
6. BAKHEIT, M.A., L. SCHNITTGER, D.A. SALIH, *et al.* 2004. Application of the recombinant *Theileria* surface protein in an indirect ELISA for the diagnosis of tropical theilerosis. Parasitol. Res. **92:** 299–302.
7. SCHNITTGER, L., H. YIN, J. LUO, *et al.* 2002. Characterization of a polymorphic gene of *T. lestoquardi* and of a recently identified *Theileria* species pathogenic for small ruminants in China Parasitol. Res. **88:** 553–556.

# Use of a Monoclonal Antibody against *Babesia bovis* Merozoite Surface Antigen-2c for the Development of a Competitive ELISA Test

MARIANA DOMINGUEZ,[a] OSVALDO ZABAL,[a] SILVINA WILKOWSKY,[a] IGNACIO ECHAIDE,[b] SUSANA TORIONI DE ECHAIDE,[b] GUSTAVO ASENZO,[a] ANABEL RODRÍGUEZ,[a] PATRICIA ZAMORANO,[a] MARISA FARBER,[a] CARLOS SUAREZ,[c] AND MONICA FLORIN-CHRISTENSEN[a]

[a]*Instituto Nacional de Tecnología Agropecuaria , Centro de Investigación en Ciencias Veterinarias y Agronómicas, Castelar, Argentina*

[b]*Instituto Nacional de Tecnología Agropecuaria Rafaela, Estación Experimental Agropecuaria, Rafaela, Argentina*

[c]*Animal Disease Research Unit, Agricultural Research Service, United States Department of Agriculture, Pullman, Washington, USA*

**ABSTRACT: Bovine babesiosis caused by *Babesia bovis* is a disease that hampers the production of beef and dairy cattle in tropical and subtropical regions of the world. New diagnostic methods based on recombinant antigens constitute valuable biotechnological tools for the strategic control of this disease. We have developed a competitive enzyme-linked immunosorbent assay that includes a recombinant form of the merozoite surface antigen-2c and a novel monoclonal antibody against it. Preliminary results showed that this test is able to identify specific antibodies against *B. bovis* from experimentally and naturally infected cattle.**

**KEYWORDS: Bovine babesiosis; *Babesia bovis*; competitive enzyme-linked immunosorbent assay (competitive ELISA); merozoite surface antigen-2c (MSA-2c); monoclonal antibodies**

## INTRODUCTION

*Babesia bovis* is an intraerythrocytic parasite transmitted to bovines by *Boophilus* sp. tick vectors. The disease caused by this parasite is endemic in tropical and subtropical regions of the world and causes great economic losses because of the limitations it imposes on meat and milk production.[1] In an ideal scenario, equilibrium between bovines, ticks, and hemoparasites avoids the occurrence of outbreaks in en-

Address for correspondence: Monica Florin-Christensen, Instituto Nacional de Tecnología Agropecuaria Castelar, Centro de Investigación en Ciencias Veterinarias y Agronómicas, Los Reseros y Las Cabañas, 1712 Castelar, Argentina. Voice/fax: +54 114 4621-1743.
mflorin@cicv.inta.gov.ar

**Ann. N.Y. Acad. Sci. 1026: 165–170 (2004). © 2004 New York Academy of Sciences.
doi: 10.1196/annals.1307.025**

demic regions. Newborn calves from these zones are protected by innate mechanisms of defense and colostral antibodies. If these animals become naturally infected with babesias before 1 year of age, they remain persistently infected and are protected against babesiosis. However, this ideal equilibrium is broken when the tick population decreases in the region because of climatic changes or frequent use of acaricides. In these cases, a significant number of calves in a herd may not receive their natural immunization. Moreover, if the *Babesia*-infected tick population returns to its original high numbers, babesiosis outbreaks can occur with serious economic consequences. Therefore, diagnosis is necessary to periodically assess the epidemiological situation of babesiosis and to decide on the timely implementation of strategic control measures such as vaccination.

At present, detection of antibodies against *B. bovis* is mostly carried out by techniques that use crude antigen preparations, such as the immunofluorescence antibody test (IFAT) and the enzyme-linked immunosorbent assay (ELISA).[2] The main drawback of these techniques is the difficulty of purifying the *Babesia* antigens from the erythrocyte antigens, which are responsible for unspecific reactions and, in the case of IFAT, the subjectivity in the interpretation of results.

Diagnostic methods based on recombinant proteins are thus highly desirable. To this end, it is necessary to identify widely conserved, immunodominant antigens exposed to the bovine immune system. *B. bovis* rhoptry-associated protein 1 (RAP-1) meets these requirements, and an indirect ELISA test[3] and a competitive ELISA test[4] based on a recombinant form of this protein have been successfully developed.

Previous studies from our group have detected another valuable diagnostic candidate for *B. bovis* infections: merozoite surface antigen-2c (MSA-2c). This antigen has a surface localization and is highly conserved among geographically distant strains.[5,6] In addition, a variety of sera from experimentally and naturally *B. bovis*–infected bovines from Argentina recognized a recombinant form of MSA-2c (rMSA-2c) from the Argentine R1A-attenuated strain in Western blot analyses.[6] Thus, we produced monoclonal antibodies against rMSA-2c and explored their usefulness for the development of competitive ELISA tests for *B. bovis*. We here present the data obtained thus far, which indicate that rMSA-2c and a monoclonal antibody (MAb) against it can be useful diagnostic tools for bovine babesiosis.

## MATERIALS AND METHODS

### *Development of Monoclonal Antibodies*

Recombinant *B. bovis* MSA-2c was produced in the pBAD/thioTOPO (Invitrogen, Carlsbad, CA, USA) prokaryotic expression system using DNA from the Argentine R1A strain as starting material, and purified by affinity chromatography on Ni-agarose, as previously described.[6] Protein quantification was performed by comparison of band sizes with known amounts of bovine serum albumin after sodium dodecyl sulfate polyacrylamide gel electrophoresis (SDS-PAGE) and staining with Coomassie blue. Three mice were intradermally inoculated with 30 μg of rMSA-2c emulsified in a mineral oil adjuvant developed at the National Institute of Agricultural Technology, Argentina. This adjuvant was composed of Arlacel C (Farma International, Miami, FL, USA), Markol 52 (Univar, Richmond, BC, Canada) and

Tween 80 (Sigma Chemical Co., St. Louis, MO, USA). After the first inoculation, a second followed 15 days later, and a third 30 days later. Development of anti-MSA-2c antibody titers was confirmed by indirect ELISA using rMSA-2c as antigen, following conditions described before.[6] Mice were then euthanized; their spleens were excised and homogeneized; and the preparation of MAbs was carried out according to standard protocols. The screening of clones secreting anti-MSA-2c antibodies was performed by indirect ELISA. Parallel ELISA assays were performed using recombinant *Anaplasma marginale* MSP-5, produced in the same expression system, to discard those clones secreting antibodies against nonspecific B cell epitopes, including those in the co-expressed thioredoxin fusion protein. A clone identified in our research as H9P2C2 displayed the highest antibody titers against rMSA-2c and showed no reaction against rMSP-5. This clone was amplified in mice ascitis and further characterized. The MAb titer was determined in serial dilutions by indirect ELISA using rMSA-2c as antigen. Detection was carried out by incubation with peroxidase-conjugated anti-mouse immunoglobulin G followed by the orthophenylene diamine (OPD)-$H_2O_2$ chromogenic substrate and absorbance measurements at 490 nm. A titer of 4.4 was calculated for the ascitis preparations used in this study (20 mg/mL protein concentration) and corresponded to –log10 of the maximal dilution of MAb that duplicated the $A_{490}$ values obtained with the unrelated MAb Tryp1.

### *Serum Samples*

Serum samples from bovines experimentally infected with *B. bovis* R1A (attenuated) and S2P (pathogenic) strains were collected before infection as well as at different time points (T21, T35, and T65) after infection. In addition, sera from naturally infected bovines from a babesiosis-endemic region of Salta, a province in Northern Argentina, were also used in this study. Before the assay, serum samples were diluted 1:50 in phosphate-buffered saline/0.5% Tween 20 and incubated with 0.2 mg/mL of an *Esherichia coli* lysate for 1 hr at 37°C, to adsorb antibodies against *E. coli* antigens, as described previously.[5,6] After this period, samples were centrifuged (10 min, 14,000 × g, 4°C), and the supernatants were used in the assays.

### *Competitive ELISA*

Recombinant MSA-2c in carbonate/bicarbonate buffer, pH 9.6, was bound to Immulon II plates (Dynatech, Chantilly, VA) in a concentration of 8.5 ng/well. After blocking the plates with phosphate-buffered saline/0.5% Tween 20/0.5% gelatin, they were sequentially incubated with serial dilutions of test sera, an adequate dilution of MAb H9P2C2, peroxidase-conjugated anti-mouse immunoglobulin G (Kirkegaard and Perry Laboratories, Gaithersburg, MD), and the chromogenic substrate OPD-$H_2O_2$. Each incubation was followed by thorough washes with phosphate-buffered saline/0.5% Tween 20. Reactions were stopped with $H_2SO_4/H_2O$ (1:8, v/v), and absorbance was read at 490 nm. All determinations were carried out in triplicate.

### *Isotype Determination*

The immunoglobulin G isotype of MAb H9P2C2 was determined by indirect ELISA. rMSA-2c was bound to Immulon 2 HB plates, as above, and sequentially incubated with 1) MAb H9P2C2; 2) goat anti-mouse immunoglobulin $G_1$, immunoglobulin M, immunoglobulin $GG_2a$, immunoglobulin $GG_2b$, or immunoglobulin $GG_3$, conjugated to biotin (Caltag, Burlingame, CA); 3) streptavidin-alkaline phosphatase; and 4) *p*-nitrophenyl phosphate (Sigma Chemical, St. Louis, MO), followed by absorbance measurements at 405 nm. Recombinant forms of *Brucella abortus* bp26 and bovine herpes virus-1 gD and polyclonal sera against them were used as positive controls for the anti-isotypes.

## RESULTS AND DISCUSSION

In this work, a MAb against *B. bovis* MSA-2c, MAb H9P2C2, was developed. We first tested if this MAb was able to react with *B. bovis* antigens. Western blot analyses demonstrated that it recognized a 30-kDa protein band in *B. bovis* R1A-infected erythrocytes, while no reaction was observed towards non-infected erythrocytes. Reaction of MAb H9P2C2 with recombinant MSA-2c/thioredoxin fusion protein yielded the expected band of approximately 46 kDa, while no reactivity was observed towards *B. bovis* rRAP-1 or *A. marginale* rMSP-5 produced in the same expression system.

We also determined the immunoglobulin G isotype of this anti-MSA-2c MAb in indirect ELISA tests. Significant absorbance readings were only obtained with anti-immunoglobulin $GG_1$. All anti-isotypes reacted with the positive controls from *B. abortus* bp26 and bovine herpes virus-1 gD. Thus, our results clearly showed that MAb H9P2C2 is of type immunoglobulin $GG_1$.

A competitive ELISA test was then developed using rMSA-2c and monclonal antibody H9P2C2. This test is based on the reactivity of antibodies present in *B. bovis*-positive sera with rMSA-2c bound to ELISA plates. Upon further incubation with anti-MSA-2c MAb H9P2C2, the MAb binds to available MSA-2c molecules. Incubation with anti-mouse immunoglobulin G-peroxidase conjugate and a chromogenic substrate allows the detection of MSA-2c molecules not initially bound to bovine antibodies, and the amount of color developed is inversely related to anti-MSA-2c antibody titers.

This assay was first tested with bovine sera from *B. bovis* R1A experimentally infected bovines (65 dpi), which had been previously characterized as positive by IFAT. A clear decrease in $A_{490}$ measurements was observed with positive sera with respect to controls. We then determined which concentrations of reagents gave maximal sensitivity to the test using serial dilutions of *B. bovis*-positive and *B. bovis*-negative sera. Competitive ELISA tests were carried out with these sera against dilutions of rMSA-2c and a fixed MAb H9P2C2 concentration or a fixed amount of rMSA-2c and different dilutions of MAb. Our results showed that an antigen amount of 8.5 ng/well and a monclonal antibody dilution of 7500 gave maximal $A_{490}$ differences between positive and negative sera at all tested dilutions. These conditions were selected for the rest of the experiments.

Samples withdrawn at different time points from bovines experimentally infected with the *B. bovis* R1A ($N = 6$) or S2P ($N = 2$) strains were tested with this competitive ELISA. The results obtained are shown in FIGURE 1. It can be observed that anti-MSA-2c antibodies can be clearly detected at 35 days post-infection in animals infected with the R1A strain. In the case of bovines infected with the S2P strain, anti-

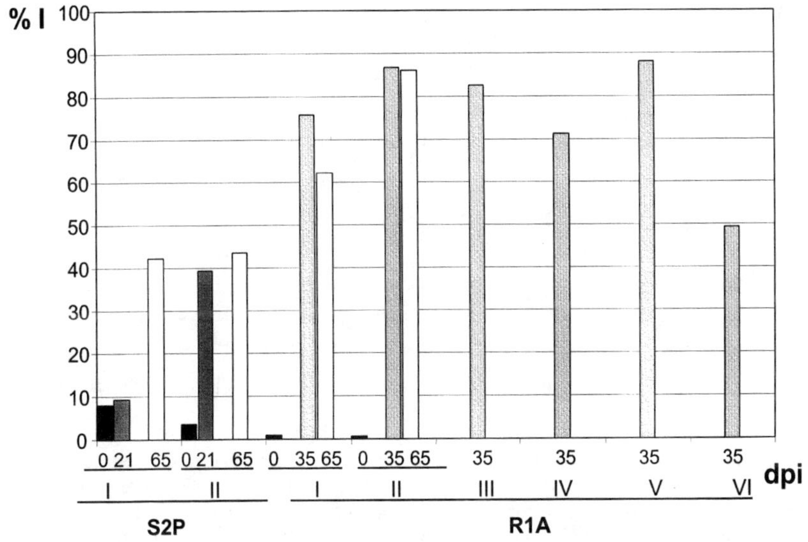


**FIGURE 1.** Detection by competitive ELISA of infections in bovines experimentally inoculated with the Argentine *B. bovis* S2P (pathogen) or R1A (attenuated) strains. Serum samples were withdrawn just before (T0) and at different time points after inoculation, and diagnosed by competitive ELISA using rMSA-2c and MAb H9P2C2 (serum dilution: 1:50). Roman numerals indicate different bovines. Results are expressed as percent inhibition (%I), calculated as 100 – [($A_{490}$ test serum × 100)/mean $A_{490}$ of negative sera], and represent the mean of triplicate determinations. Standard deviations between wells did not exceed 20%.

MSA-2c antibodies could be detected at 65 dpi in both cases In one of these animals (II), a high percent inhibition (%I) was also obtained at 21 dpi. Interestingly, this bovine was inoculated with $10^7$ parasites, while the inoculum used with bovine I was of $10^5$. Samples that showed significant %I according to this test were also positively diagnosed for *B. bovis* by IFAT.

In addition, we tested sera from bovines naturally infected with *B. bovis* field strains from a babesiosis endemic region in Salta province, Argentina. These samples were characterized as positive by IFAT. All tested sera ($N = 8$) gave a significant decrease in $A_{490}$ readings and therefore high %I values (FIG. 2).

Our results clearly show that a recombinant form of the *B. bovis* surface antigen MSA-2c and a MAb against it can be employed in the development of a competitive ELISA test for this pathogen. Trials using a larger amount of serum samples from different endemic regions of Argentina and neighboring countries are now underway in our laboratory. In addition, we will explore other competitive ELISA designs using the reagents we have developed to achieve maximal sensitivity and minimal background.

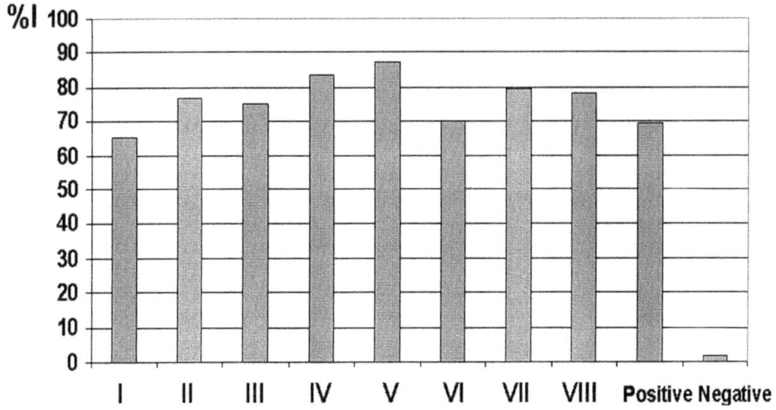


**FIGURE 2.** Detection of natural *B. bovis* infections from a babesiosis-endemic region of Argentina. *B. bovis*-infected sera, characterized as positive by IFAT, were diagnosed by competitive ELISA (serum dilution: 1:50). Roman numbers indicate different bovines. The negative and positive serum samples correspond to an R1A experimentally infected bovine (T0 and 65 dpi, respectively).

## ACKNOWLEDGMENTS

This work was supported by Grant PICT 98/08-3838 from Agencia Nacional Promocion Cientifica y Technologica and by Fundacion Antorchas, Argentina.

## REFERENCES

1. Uilenberg, G. 1995. International collaborative research: significance of tick-borne hemoparasitic diseases to world animal health. Vet. Parasitol. **57:** 19–41.
2. Bose, R., W.K. Jorgensen, R.J. Dalgliesh, *et al.* 1995. Current state and future trends in the diagnosis of babesiosis. Vet. Parasitol. **57:** 61–74.
3. Boonchit, S., X. Xuan, N. Yokoyama, *et al.* 2002. Evaluation of an enzyme-linked immunosorbent assay with recombinant rhoptry-associated protein 1 antigen against *Babesia bovis* for the detection of specific antibodies in cattle. J. Clin. Microbiol. **40:** 3771–3775.
4. Goff, W., T. McElwain, C. Suarez, *et al.* 2003. Competitive enzyme-linked immunosorbent assay based on a rhoptry-associated protein 1 epitope specifically identifies *Babesia bovis*-infected cattle. Clin. Diagn. Lab. Immunol. **10:** 38–43.
5. Florin-Christensen, M., C.E. Suarez, S.A. Hines, *et al.* 2002. The *Babesia bovis* merozoite surface antigen-2 locus contains four tandemly arranged and expressed genes encoding immunologically distinct proteins. Infect. Immun. **70:** 3566–3575.
6. Wilkowsky, S., M. Farber, I. Echaide, *et al.* 2003. *Babesia bovis* merozoite surface antigen-2c contains highly immunogenic, conserved B-cell epitopes that elicit neutralization-sensitive antibodies in cattle. Mol. Biochem. Parasitol. **127**(2): 133–141.

# Use of the Serial Analysis of Gene Expression (SAGE) Method in Veterinary Research

## A Concrete Application in the Study of the Bovine Trypanotolerance Genetic Control

JEAN-CHARLES MAILLARD,[a] DAVID BERTHIER,[b] SOPHIE THEVENON,[c] RONAN QUÉRÉ,[d] DAVID PIQUEMAL,[e] LAURENT MANCHON,[e] AND JACQUES MARTI[e]

[a]*Cirad-Prise, c/oNational Institute for Animal Husbandry (NIAH), Thuy Phuong, Tu Liem, Hanoï, Vietnam*

[b]*Cirad-Emvt Baillarguet, Animal Health Programme, Montpellier, France*

[c] *Centre International de Recherches Dévelopement en Zones Sub-humides (CIRDES), Urbio, Bobo Dioulasso, Burkina Faso*

[d]*Skuld Tech Cie, Montpellier, France*

[e]*Institut de Génétique Humaine/Centre National de Recherches Scientifiques (IGH/CNRS), Université Montpellier II, France*

**ABSTRACT: New postgenomic biotechnologies, such as transcriptome analyses, are now able to characterize the full complement of genes involved in the expression of specific biological functions. One of these is the Serial Analysis of Gene Expression (SAGE) technique, which consists of the construction of transcripts libraries for a quantitative analysis of the entire gene(s) expressed or inactivated at a particular step of cellular activation. Bioinformatic comparisons in the bovine genomic databases allow the identification of several up- and downregulated genes, expressed sequence tags, and unknown functional genes directly involved in the genetic control of the studied biological mechanism. We present and discuss the preliminary results in comparing the expressed genes in two total mRNA transcripts libraries obtained during an experimental *Trypanosoma congolense* infection in one trypanotolerant N'Dama animal cow. Knowing all the functional genes involved in the trypanotolerance control will permit validation of some results obtained with the quantitative trait locus approach, to set up specific microarrays sets for further metabolic and pharmacological studies, and to design field marker-assisted selection by introgression programs.**

**KEYWORDS: transcriptomics; SAGE; trypanotolerance; N'Dama cattle; *Trypanosoma congolense*; mRNA**

Address for correspondence: Dr. J.C. Maillard, Cirad-Prise, c/o NIAH, Thuy Phuong, Tu Liem, Hanoï, Vietnam. Voice: (84) 4 757 0521; fax: (84) 4 757 2177.
maillard@fpt.vn; maillard@cirad.fr

**Ann. N.Y. Acad. Sci. 1026: 171–182 (2004). © 2004 New York Academy of Sciences.
doi: 10.1196/annals.1307.026**

## INTRODUCTION

Several recent biotechnologies permit exhaustive functional analysis using transcriptomic approach, an efficient way to characterize the pool of genes involved in the expression of specific biological functions. The serial analysis of gene expression (SAGE) technique[1–3] enhances the power and the speed of such a transcriptome analysis. SAGE generates complete expression profiles of tissues or cell lines and results are quantitative and absolute. In the work reported here, we show the efficiency of the SAGE technique in comparing up- and downregulated genes involved in the control of a *Trypanosoma congolense* infection in one trypanotolerant N'Dama (*Bos taurus*) cow.

In central and sub-Saharan Africa, the most important constraint to livestock production is trypanosomosis. This tsetse fly–transmitted disease represents an important risk for about 60 million cattle on 7 million $km^2$ spread over 37 countries. Trypanosomosis affects productivity, such as milk and meat production, fertility, and strength. Trypanotolerance is the genetic ability of some breeds from several mammalian species (such as cattle, small ruminants, pig, wild buffalos, and antelopes) to live and remain productive in tsetse fly–infested areas.[4] This phenomenon was described in Africa as early as the beginning of the twentieth century.[5–10] Trypanotolerance results from various biological mechanisms under multigenic control and relate either to the control of trypanosome infection, as measured by parasitemia,[11–14] or the control of the pathogenic effects of the parasites, the most prominent of which is anemia.[15–18] Two different pools of genes are probably involved in determining the two characteristics and the various methodologies used so far did not succeed in identifying them. Zootechnical studies,[19–23] quantitative genetics approaches,[24] the electrophoretic analysis of targeted proteins,[25,26] MHC typing[27] have not made real progress in understanding trypanotolerance. Quantitative trait locus (QTL) studies, developed more recently (first in mice,[28–30] then in cattle[31]), produced more interesting results, but the studies were restricted to small parts of the cattle genome. Considering the limited number of experimental animals used, the confidence interval of bovine QTLs is too wide to be useful in a marker assisted selection program or in a positional candidate approach. The homologous comparison between murine and bovine genomes is limited and there is no proof that the same genes are involved in both species. Finally, the QTL approach could provide information on genes involved in the innate immunity, but not in those controlling the acquired immunity, as both gene types are involved in global trypanotolerance mechanisms. Furthermore the crossbreeding plan to study QTL segregation in the bovine model is expensive and would take a long time.

## METHODS AND MATERIALS

### *The SAGE Method*

This technique is based on the construction of total mRNA libraries for a quantitative analysis of the whole transcripts expressed or inactivated at particular steps of a cellular activation. It is based on three principles: (1) a short sequence tag (9–14 bp) obtained from a defined region within each mRNA transcript contains sufficient information to identify one specific transcript; (2) sequence tags can be linked together to form long DNA molecules (concatemers) that can be cloned and sequenced. Sequencing of the concatemer clones results in the quick identification of numerous individual tags; (3) the expression level of the transcript is quantified by the number of times a particular tag is observed.

We used the I-SAGE™ kit from Invitrogen (catalog number T5000-01) to construct from peripheral blood mononuclear cells of one N'Dama cow, two total mRNA transcripts libraries at Day 0 of a *Trypanosoma congolense* experimental infection (T0L) and at Day 10 postinfection, corresponding to the peak of parasitemia (MPL).

With bioinformatic comparisons[32] in several genomic databases (Unigen and Tigr, for example) we identified the different activated and inactivated tags [known genes, expressed sequence tags (ESTs), or unknown genes] and compared their respective frequencies in both T0L and MPL.

### *Experimental Animal and Design*

We used one cow of the N'Dama breed (*Bos taurus*), a Longhorn indigenous to West Africa, well known to be resistant to trypanosomosis infection. This animal was taken from a tsetse fly–infected area. A serological control verified the presence of specific *T. congolense* antibodies. Before the beginning of the experiment, this animal was treated for blood parasites (Veriben: diminazene aceturate, 7 mg/kg) and gastrointestinal parasites (Vermitan: albendazole, 7.5 mg/kg). After a few days of rest, a first blood sample was taken using a PAXgene Blood RNA tube (Quiagen, catalog number 762125), which contains a total RNA conservation medium. This first blood sample, taken at Day 0, was used to develop the first reference SAGE library

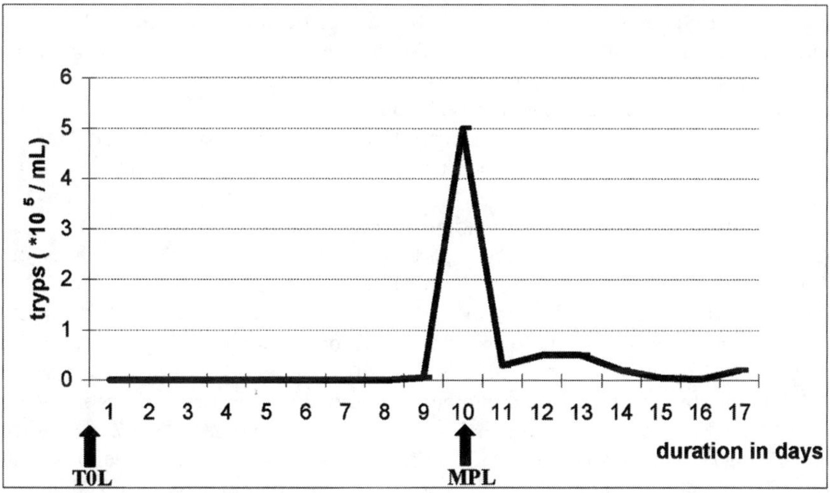


**FIGURE 1.** Curve of the parasitemia evolution in a N'Dama animal after a *T. congolense* experimental infection. The arrows (↑) indicate the blood samples to construct the T0L (Day 0) and MPL (Day 10) SAGE libraries.

**TABLE 1. Statistics of tag distribution**

| | Totally sequenced tags 4763<br>Different tags 2281 | Tags differentially expressed<br>187 ($P < 0.001$) |
|---|---|---|
| Genes | 386 | 92 |
| cDNAs/ESTs | 920 | 23 |
| No Match | 975 | 72 |

Definitions: Genes, tags matching well-identified genes; cDNAs/ESTs, tags matching anonymous described sequences; No Match, tags failing to match SAGEmap (rank 1 or 2) or UniGene sequences.

(T0L) from total white blood cells. Then, according to the experimental design, a *T. congolense* infection (Ser/71/STIB/212) was introduced using a unique syringe inoculation of $8{\times}10^5$ parasites.[33–36] Every couple of days, a blood control on the buffy coat monitored the appearance of the parasites and the kinetics of their development (FIG. 1). The second blood sample was taken at the peak of parasitemia, which appeared at Day 10, to develop the second reference SAGE library (MPL). So these two SAGE libraries (T0L and MPL) were used in a differential comparison of expressed genes in this N'Dama cow before and after a *T. congolense* infection.

## RESULTS

The analysis of the whole identified tags are summarized in the TABLE 1. From 4,763 sequenced tags, we identified 2,281 distinct transcripts, 187 of them showing significant frequency variations ($P < .001$). Rates of contamination by linker sequences were not significant. Repeated ditags (not taken into account for measurement of expression levels) represented 1.3% of the total ditag population, revealing a high complexity of the original mRNA population.

The tags showing the most significant frequency differences between T0L and MPL libraries are presented for the upregulated (TABLE 2) and the downregulated (TABLE 3) transcripts.

A different, interesting presentation of these results is a graphical scatterplot (FIG. 2), where each dot represents a particular tag. Some of these dots correspond to known genes (immunoglobulins, B and T cell receptors, interleukins, MHC Bola class I and II, metabolic and ribosomal proteins, etc.) or ESTs, but others correspond to unknown genes. These unknown genes could come from the N'Dama mRNAs but they could also come, to a small degree, from mRNAs of *T. congolense* parasites. To validate this hypothesis, we did another bioinformatic comparison of these whole "No Match" tags with the two existing *Trypanosoma* genome databases (*T. brucei* and *T. cruzi*). Not surprisingly, we identified five expressed genes actually coming from *T. congolense* genome. These are probably genes common to the *Trypanosoma* genus (TABLE 4). This result reveals a very interesting way to study the interactive mechanisms at the host-parasite interface in a parallel comparison of the parasite and the host SAGE libraries.

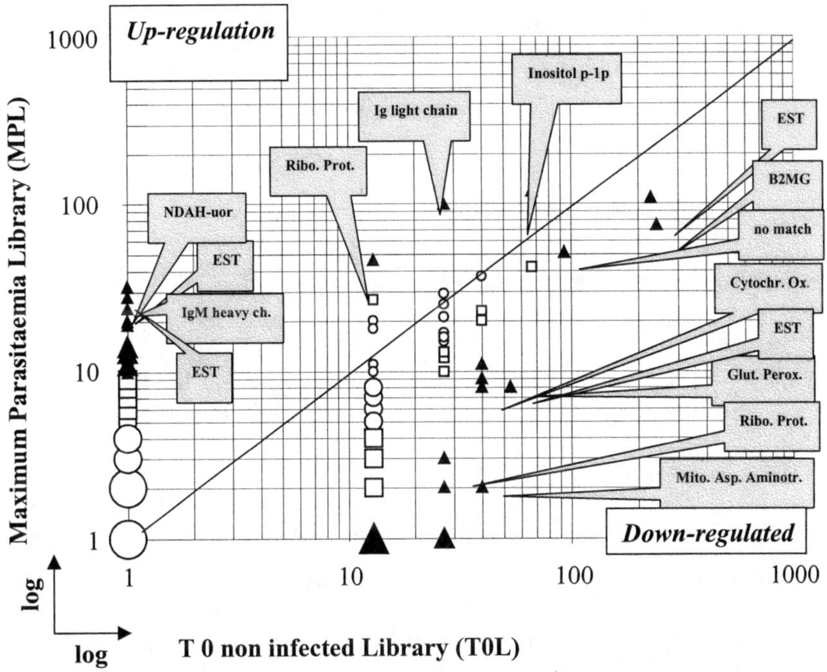


**FIGURE 2.** Comparative expression levels between the T0 non-infected library (T0L) and the maximum parasitemia library (MPL). The occurrence numbers of each tag were plotted on a logarithmic scale. The tags with no expression in one library were set to a value of one. Statistical significance (▲): $P < .001$; (□): $.001 < P < .01$; (○): $P > .01$.

## DISCUSSION

Among the 187 regulated tags in the pool of upregulated transcripts (TABLE 2), we found several known genes involved in the immune mechanisms, which confirmed several previous immunological results.[37] The most activated genes are those encoding different chains of immunoglobulin (IgG and IgM) molecules. This confirms their important role in the immune mechanisms involved in the control of trypanosome infections. Indeed, recent literature on this topic presents many corroborating results that indicate that, excepted for a primary-phase infection,[38] the ability of resistant animal to control the parasitemia is due to a more efficient specific antibody response. The T-independent responses producing IgM antibodies are sufficient to control the parasitemia and IgM is more efficient than IgG in neutralizing antibodies at the beginning of the infection.[39] The increase of the serological IgM level is simultaneous to the parasitemia appearance, whereas the IgG antibodies generally appear later. The IgM are mainly directed at the parasite surface antigens whereas the IgG are generally directed at internal antigens.[40] It has often been reported[13] that the trypanosomes are responsible of B and T cells polyclonal proliferation. Here we confirm the activation

**TABLE 2. Expression of upregulated transcripts**

| Tags | TOL | MPL | *P* value | Id. | Names |
|---|---|---|---|---|---|
| Immunity proteins | | | | | |
| CATGGACCCCTGAG | 27 | 99 | 2,64E-10 | Bt.100316 | Immunoglobulin light chain mRNA |
| CATGGAGCCCGCAG | 0 | 24 | 2,98E-08 | TC132989 | Ig M heavy chain constant region, membrane form |
| CATGAGTGCAGACT | 13 | 18 | 4,80E-02 | TC132990 | Ig M heavy chain constant region, secretory form |
| CATGGGCGTCTCTG | 0 | 3 | 6,25E-02 | TC124168 | Ig G3 heavy chain constant region |
| CATGCAGAAGTCCA | 0 | 2 | 1,26E-01 | Bt.101309 | Ig G2a = Ig C gamma heavy chain constant region |
| CATGGCCACTTAGT | 0 | 12 | 1,32E-04 | Bt100505 | B-cell antigen receptor mRNA |
| CATGTGAGGGTGCC | 0 | 4 | 3,13E-02 | TC133149 | T-cell receptor beta 1, constant region |
| CATGTGGGAAAAA | 0 | 2 | 1,26E-01 | Bt.100982 | T-cell receptor alpha chain, constant and 3′ UTR |
| CATGCATTGGAGAA | 0 | 1 | 2,50E-01 | Bt.100801 | Interleukin 1-beta (IL-1-beta) mRNA |
| CATGATGTGGCCAG | 0 | 1 | 2,50E-01 | TC135095 | Similar to interleukin 10 receptor beta (*Homo sapiens*) |
| CATGGATCTGGCTG | 0 | 3 | 6,25E-02 | Bt.506 | MHC class II BoLA-DBQ-chain mRNA |
| CATGATTATGAGTT | 0 | 4 | 3,19E-02 | Bt.99993 | MHC class II (BoLA-DQB)mRNA |
| Other genes | | | | | |
| CATGTGACACGTAT | 0 | 32 | 1,16E-10 | TC133213 | NADH–ubiquinone-oxidoreductase chain 1 |
| CATGGGCTGGGGGC | 0 | 20 | 4,77E-07 | TC132831 | PRO1_BOVIN Profilin I [Bovine] |
| CATGCTGGGAAATT | 0 | 13 | 6,10E-05 | TC132798 | ORF |
| CATGCATATTTGGG | 0 | 14 | 3,35E-05 | Bt100010 | Ferritin H subunit mRNA |
| CATGACAACACATA | 67 | 119 | 3,16E-04 | Bt.5174 | Inositol polyphosphate 1-phosphatase |
| CATGAGGAAAGCGG | 0 | 13 | 6,10E-05 | TC142407 | Homologue to CDH1-D (*Gallus gallus*) |
| CATGCAGCTCCGCG | 0 | 13 | 6,10E-05 | Bt.697 | Cdc42-associated tyrosine kinase ACK-2 mRNA |
| CATGTGAGAACATT | 0 | 12 | 1,22E-04 | TC132773 | Actin |
| CATGGACCCCTTTT | 0 | 11 | 2,44E-04 | TC132770 | Beta actin |
| CATGTTGTCTGTCT | 0 | 11 | 2,44E-04 | TC132625 | HSHU33histone H3.3 [validated] |

**TABLE 2. (*continued*)Expression of upregulated transcripts**

| Tags | T0L | MPL | *P* value | Id. | Names |
|---|---|---|---|---|---|
| CATGAGTCCAAGCC | 0 | 11 | 2,44E-04 | TC143154 | Similar to SDHL_HUMAN L-serine dehydratase [Human] |
| CATGAAGGTAATAA | 0 | 10 | 4,88E-04 | TC132804 | CChain C Crystal Structure Of Arp23 Complex |
| CATGTGTGTCTGTA | 0 | 9 | 9,77E-04 | TC123714 | TBA1_CRIGRTubulin alpha-1 chain. [Chinese hamster] |
| Ribosomal | | | | | |
| CATGTAAGGATCCA | 0 | 20 | 4,77E-07 | TC124043 | RS26_HUMAN40S ribosomal protein S26. [Rat] |
| CATGCTCACCAATA | 13 | 46 | 4,41E-06 | Bt.101746 | Ribosomal protein (QM) mRNA |
| CATGGGCTTCGGCT | 0 | 14 | 3,35E-05 | Bt101531 | Acidic ribosomal protein P2 mRNA |
| CATGCTGTTGGTGA | 0 | 9 | 9,77E-04 | TC132795 | RS23_HUMAN40S ribosomal protein S23. [Rat] |
| CATGAGGAAAGCGG | 0 | 13 | 6,10E-05 | TC141303 | Ribosomal protein L36 |
| ESTs | | | | | |
| CATGTTGCATTACC | 0 | 28 | 1,86E-09 | TC133588 | EST |
| CATGGAGGAGGAAG | 0 | 19 | 1,09E-06 | Bt.116903 | EST–211333 *Bos taurus* cDNA |
| CATGAAGCCCAGCG | 0 | 15 | 1,69E-05 | Bt.95263 | EST–AV662735 *Bos taurus* cDNA |
| CATGGGCTGGGGCT | 0 | 14 | 3,05E-05 | TC142538 | EST |
| CATGGCCACAGCCA | 0 | 10 | 5,21E-04 | Bt77418 | EST–AV603489 *Bos taurus* cDNA |

Definitions: T0L, T0 non-infected library; MPL, Maximum Parasitaemia Library; Id., Genbank accession number.

of the genes encoding the B and T cell receptors (TABLE 2). Furthermore, and very interestingly, the T cell receptor beta cluster is located in the bovine chromosome 4 (*Bta4*) in the 4q3.1 and 4q3.6 region (*IDVGA51-TGLA159/MGTG4B*) where Hanotte and colleagues[31] described a QTL strongly associated with the Fewer parasites trait (*PARMLn*) in N'Dama. We also found genes encoding cytokines, such as interleukins (*IL1* and *IL10R*), confirming their role in the induction of a cell polyclonal activation, particularly for IgM antibodies.[41] Finally, MHC class II *BoLA-DQB* genes seem to be activated (TABLE 2) while other MHC genes of class I (β2 microglobulin) and class II (*BoLA-DRA* and *BoLA-DMA*) seem to be downregulated (TABLE 3). Apart from molecules of the immune system, we identified several genes involved in different up- and downregulated metabolic pathways (such as the NADH-ubiquinone oxidoreductase chain 1, the bovine profilin, or the glutathione peroxidase pathways). Several ribosomal genes are also regulated. Within regulated ESTs, one upregulated

**TABLE 3. Expression of downregulated transcripts**

| Tags | T0L | MPL | *P* value | Id. | Names |
|---|---|---|---|---|---|
| Immunity proteins | | | | | |
| CATGGCTAAGCCTA | 241 | 74 | 2.83E-21 | TC142442 | BOVIN Beta-2-microglobulin precursor, Lactollin |
| CATGTTACCATAAA | 13 | 1 | 3.89E-04 | Bt.100131 | *B. taurus* mRNA for beta 2-microglobulin |
| CATGAGGAGTTGGG | 40 | 20 | 1.81E-03 | TC133042 | MHC class I heavy chain |
| CATGGCGCCCCTTC | 27 | 17 | 1.79E-02 | Bt.100356 | *B. taurus* mRNA for MHC class 1 (clone 6) |
| CATGGGCATCATTG | 27 | 16 | 1.38E-02 | Bt.100358 | *B. taurus* MHC class 1 protein molecule D18.1 |
| CATGTCAAGGCAAT | 13 | 1 | 4.27E-04 | TC144589 | MHC class II DM alpha-chain MHC class II antigen |
| CATGTAATGCCTTT | 13 | 8 | 4.64E-02 | Bt.101032 | Bovine BoLA-DRA mRNA for MHC class II BoLA-DR-alpha |
| CATGGAAGCAATAA | 13 | 7 | 3.69E-02 | TC133143 | MHC class II antigen [*B. taurus*] |
| Other genes | | | | | |
| CATGAATAAAGTGC | 54 | 8 | 3.67E-10 | TC143206 | Glutathione peroxidase (AA 1-204) |
| CATGTACGAGAAAG | 27 | 1 | 5.22E-08 | Bt.1316 | *B. taurus* mitochondrial aspartate aminotransferase |
| CATGCCTCGACGAT | 40 | 11 | 1.06E-05 | TC132812 | Cytochrome oxidase subunit [*Bos taurus*] |
| CATGCAAAGGAGAT | 13 | 1 | 3.89E-04 | Bt.100650 | Bovine ATP synthase inhibitor protein mRNA |
| CATGTAATAAAGCA | 13 | 1 | 4.27E-04 | TC125185 | Seryl-tRNA synthetase |
| CATGTTAATCCTAA | 13 | 1 | 4.27E-04 | Bt.5517 | Bovine mRNA for retinal 2′,3′-cyclic nucle.3′-phospha |
| CATGCAAATAAAAA | 13 | 1 | 3.89E-04 | Bt.100227 | Bovine mRNA for beta-crystallin subunit beta B1 |
| CATGCTAATTATAA | 13 | 1 | 4.27E-04 | TC124062 | 4L NADH dehydrogenase subunit 4L [*B. taurus*] |
| CATGCCGGCCCAGA | 13 | 1 | 4.27E-04 | TC142818 | Homologue to KCRB_CANFA creatine kinase B chain [dog] |
| CATGTGTACCTTTT | 13 | 1 | 4.27E-04 | TC143287 | COPA_BOVIN Coatomer alpha subunit (alpha-coat protein) |
| CATGGCCTGATGGG | 94 | 51 | 5.35E-05 | TC133123 | BAB22470.putative {*Mus musculus*} |
| CATGGGAGAAGGGT | 13 | 1 | 4.27E-04 | TC142585 | Similar to BAB28161. putative {*Mus musculus*} |
| Ribosomal | | | | | |
| CATGCACAAACAGT | 40 | 2 | 9.79E-11 | TC124028 | Homologue to ribosomal protein S27 cytosolic–human |

**TABLE 3. (*continued*) Expression of downregulated transcripts**

| Tags | T0L | MPL | *P* value | Id. | Names |
|---|---|---|---|---|---|
| CATGTGGTGTTGAG | 40 | 8 | 6.70E-07 | TC123562 | RS18_HUMAN40S ribosomal protein S18 (KE-3) |
| CATGGAACATATCC | 13 | 1 | 4.27E-04 | TC132800 | 60S ribosomal protein L19. [Mouse] {*Mus musculus*} |
| CATGGCAGAGTTCG | 13 | 1 | 4.27E-04 | TC132832 | RS6_HUMAN40S ribosomal protein S6, *Rattus norvegicus* |
| CATGTGAAAGATGC | 13 | 1 | 4.27E-04 | TC142426 | Ribosomal protein S4 |
| ESTs | | | | | |
| CATGTAGGTTGTCT | 228 | 107 | 9.784E-11 | BE236829 | EST |
| CATGCATTCTAGAG | 27 | 0 | 3.73E-09 | BF890336 | EST |
| CATGTGAAAAAAAA | 27 | 0 | 3.04E-09 | Bt.57839 | EST–170434 *B. taurus* cDNA |
| CATGTTAATAAAAA | 27 | 2 | 3.13E-07 | Bt.127590 | EST–*B. taurus* cDNA |
| CATGCGGTCAGCCA | 27 | 3 | 1.89E-06 | BI540074 | EST |
| CATGATTCTTTGGT | 40 | 9 | 1.45E-06 | Bt.209523 | EST–*B. taurus* cDNA |
| CATGAACAGAGGAG | 13 | 1 | 4.27E-04 | BM431775 | EST |
| CATGGAGAAATATC | 13 | 1 | 3.89E-04 | Bt.185285 | EST–*B. taurus* cDNA |
| CATGAGTTTGCCCT | 13 | 1 | 4.27E-04 | Bt.6126 | EST |
| CATGCAGCAGAAGC | 13 | 1 | 3.89E-04 | Bt.95810 | EST–*B. taurus* cDNA |
| CATGAACAGAGGAG | 13 | 1 | 4.27E-04 | AW463909 | EST |
| CATGCATAAAGGAA | 13 | 1 | 3.89E-04 | Bt.215670 | EST - *Bos taurus* cDNA |
| CATGAACAGAGGAG | 13 | 1 | 4.27E-04 | BM482666 | EST |
| CATGCCGACGGGCG | 13 | 1 | 4.27E-04 | TC123808 | EST |
| CATGGCTGGCCTGC | 13 | 1 | 4.27E-04 | TC136232 | EST |
| Unknown | | | | | |
| CATGGTACATAGAC | 27 | 0 | 3.73E-09 | | No Match |
| CATGTGCTTGTCGG | 13 | 1 | 4.27E-04 | | No Match |
| CATGGTGTGATGCT | 13 | 1 | 4.27E-04 | | No Match |
| CATGTGAGAAGTCG | 13 | 1 | 4.27E-04 | | No Match |
| CATGATGAACCCTG | 13 | 1 | 4.27E-04 | | No Match |
| CATGTTTGTCATCT | 13 | 1 | 4.27E-04 | | No Match |
| CATGCAGCAAGGAA | 13 | 1 | 4.27E-04 | | No Match |
| CATGTCGGCTTCTA | 13 | 1 | 4.27E-04 | | No Match |
| CATGGTATTTGCAA | 13 | 1 | 4.27E-04 | | No Match |

Definitions: T0L, T0 non-infected library; MPL, Maximum Parasitemia Library; Id., Genbank accession number; No Match; unknown gene.

(TC133588) and three downregulated (BE236829, BF890336, Bt. 57839) ESTs are of interest with highly significant statistics, but we have to wait for the bovine mapping progress to clearly identify these ESTs. Concerning the up- and downregulated unknown tags—even without present identification names they can be spotted on microarrays for further applications.

**TABLE 4. Expression of upregulated *T. congolense* transcripts**

| Tags | T0L | MPL | Names | Contig |
|---|---|---|---|---|
| CATGCCACACAAGC | 0 | 1 | TCJ3 Protein 241 3e-62 | CONTIG6688 |
| CATGTGTCACCCAC | 0 | 1 | Septation 119 8e-26 | CONTIG9150 |
| CATGGGACTTGGAC | 0 | 1 | Ribonucleoside-disphosphate r... 293 2e-78 | EM_NEW:AL473377 |
| CATGTGTGTCTGTG | 0 | 1 | Protein phosphatase-2C 203 2e-51 | EM_NEW:AL457897 |
| CATGTTGCTGTGTG | 0 | 2 | Possible amino acid transporter 164 7e-77 | CONTIG9490 |

Definitions: T0L, T0 non-infected library; MPL, Maximum Parasitemia Library.

In the future, to identify the genes implicated in the trypanotolerance mechanisms, we need to implement another differential analysis by conducting a similar experiment involving cattle of trypanotolerant breeds (*Bos taurus*), such as the Baoule, and cattle of a trypanosusceptible zebu breed (*Bos indicus*). Because trypanotolerance character is defined by two main criteria, one of which is the control of the anemia after a *Trypanosoma* infection, we need to follow the kinetics of the packed cell volume (PCV) to collect blood samples for SAGE libraries at the precise time when the PCV increases as a result of efficient mechanisms of anemia control. When comparing the results obtained with the different types of cattle, it will be possible to deduce that the genes commonly up- or downregulated in the N'Dama and Baoule trypanotolerant animals (and not in the control trypanosusceptible zebu) constitute the two global pools of genes involved in the trypanotolerance genetic character, either in the control of parasitemia and/or in the control of anemia.

The comparative SAGE libraries applied to the *T. congolense* parasite will also allow identification of parasite genes that are specifically up- and/or downregulated by the host defense mechanisms.

## CONCLUSION

The SAGE method, a very powerful and efficient technique of functional genomics, can compare changes in the status of cells or tissues in veterinary research. Identifying the genes involved in the trypanotolerance mechanisms will allow us to set up field marker–assisted selection and specific microarrays for more detailed metabolic and pharmacological studies. Finally, these results can be compared with those of the QTL approach for mutual validation and to identify positional candidate genes useful for future selection/introgression programs in different cattle breeds.

The progress expected in the comparative SAGE analysis in the host and parasite interface could be very interesting for drug development in animal and human trypanosomoses.

## ACKNOWLEDGMENTS

We acknowledge the scientific direction of Cirad for its special funding.

## REFERENCES

1. BERTELSEN, A.H. & V.E. VELCULESCU. 1998. High-throughput gene expression analysis using SAGE. Drug Discovery Today **3:** 152–159.
2. VELCULESCU, V.E., L. ZHANG, B. VOGELSTEIN & K.W. KINZLER. 1995. Serial analysis of gene expression. Science **270:** 484–487.
3. VELCULESCU, V.E., B. VOGELSTEIN & K.W. KINZLER. 2000. Analyzing uncharted transcriptomes with SAGE. Trends Genet. **16:** 423–425.
4. DWINGER, R.H., K. AGYEMANG, W.F. SNOW, *et al.* 1994. Productivity of cattle kept under traditional management conditions in the Gambia. Vet. Q. **16:** 81–86.
5. CHANDLER, R.L. 1952. Comparative tolerance of West African N'Dama cattle to trypanosomiasis. Ann. Trop. Med. Parasitol. **46:** 127–134.
6. DESOWITZ, R.S. 1959. Studies on immunity and host-parasite relationships. I. The immunological response of resistant and susceptible breeds of cattle to trypanosomal challenge. Ann. Trop. Med. Parasitol. **53:** 293–313.
7. DESROTOUR, J., P. FINELLE, P. MARTIN & E. SINODINOS. 1967. Les bovins trypanotolérants: leur élevage en République centrafricaine. Rev. Elev. Méd. Vét. Pays Trop. **20:** 589–594.
8. PIERRE, C. 1906. L'élevage dans l'Afrique Occidentale Française. Gouvernement Général de l'Afrique Occidentale Française. Inspection de l'Agriculture. Chalamel, Paris.
9. ROBERTS, C.J. & A.R. GRAY. 1973. Studies on trypanosome-resistant cattle. II. The effect of trypanosomiasis on N'Dama, Muturu and Zebu cattle. Trop. Anim. Health Prod. **5:** 220–233.
10. STEWART, J.L. 1951. The West African Shorthorn cattle. Their value to Africa as trypanosomiasis-resistant animals. Vet. Rec. **63:** 454–457.
11. DOLAN, R.B. 1987. Genetics and trypanotolerance. Parasitol. Today **3:** 137–143.
12. GIBSON, J.P. 2001. Towards an understanding of genetic control of trypanotolerance. Newsletter on Integrated Control of Pathogenic Trypanosomes and their Vectors **4:** 12–14.
13. MURRAY, M., W.I MORRISON & D.D. WHITELAW. 1982. Host susceptibility to African trypanosomiasis: trypanotolerance. Adv. Parasitol. **21:** 1–68.
14. ROELANTS, G.E. 1986. Natural resistance to African trypanosomiasis. Parasite Immunol. **8:** 1–10.
15. AGYEMANG, K., R.H. DWINGER, D.A LITTLE, *et al.* 1992. Interaction between physiological status in N'Dama cows and trypanosome infections and its effects on health and productivity of cattle in The Gambia. Acta Trop. **50:** 91–99.
16. TRAIL, J.C.M., G.D.M. D'IETEREN, A. FERON, *et al.* 1991. Effect of trypanosome infection, control of parasitaemia and control of anaemia development on productivity of N'Dama cattle. Acta Trop. **48:** 37–45.
17. TRAIL, J.C.M., G.D.M. D'IETEREN, J.C. MAILLE & G. YANGARI. 1991. Genetic aspects of control of anaemia development in trypanotolerant N'Dama cattle. Acta Trop. **48:** 285–291.
18. TRAIL, J.C.M., G.D.M. D'IETEREN, P. VIVIANI, *et al.* 1992. Relationships between trypanosome infection measured by antigen detection enzyme immunoassays, anaemia and growth in trypanotolerant N'Dama cattle. Vet. Parasitol. **42:** 213–223.
19. D'IETEREN, G.D.M., E. AUTHIÉ, N. WISSOCQ & M. MURRAY. 1998. Trypanotolerance, an option for sustainable livestock production in areas at risk from trypanosomosis. Rev. sci. techn. Off. Int. Epiz. **17:** 154–175.
20. D'IETEREN, G.D.M., E. AUTHIÉ, N. WISSOCQ & M. MURRAY. 1999. Exploitation of resistance to trypanosomes. *In* Breeding for Disease Resistance in Farm Animals, 2nd edit. R.F.E Axford, *et al.*, Eds.: 195–216. CABI Publishing, Wallingford, England.
21. HOSTE, C., E. CHALON, G.D.M. D'IETEREN & J.C.M. TRAIL. 1988 Le bétail trypanotolérant en Afrique occidentale et centrale. Vol. 3: Bilan d'une décennie. Etude FAO

Production et Santé animale, no. 2/3, Organisation des Nations Unies pour l'Alimentation et l'Agriculture. Rome.
22. HOSTE, C. 1992. Contribution du bétail trypanotolérant au développement des zones affectées par la trypanosomiase animale africaine. Rev. Mond. Zootech. **70-71:** 21–29.
23. MURRAY, M., M.J. STEAR, J.C.M. TRAIL, *et al.* 1991. Trypanosomiasis in cattle: prospects for control. *In* Breeding for Disease Resistance in Farm Animals, 1st edit. J.B. Owen & R.F.E. Axford, Eds.: 203–223. CABI Publishing. Wallingford, England.
24. TRAIL, J.C.M., N. WISSOCQ, G.D.M. D'IETEREN, *et al.* 1994. Quantitative phenotyping of N'Dama cattle for aspects of trypanotolerance under field tsetse challenge. Vet. Parasitol. **55:** 185–195.
25. QUEVAL, R. & J.-P. PETIT. 1982. Polymorphisme biochimique de l'hémoglobine de populations bovines trypanosensibles, trypanotolérantes et de leurs croisements dans l'Ouest africain. Rev. Elev. Méd. Vét. Pays Trop. **35:** 137–146.
26. QUEVAL, R. & L. BAMBARA. 1984. Le polymorphisme de l'albumine dans la race Baoulé et une population de zébus de type soudanien. Rev. Elev. Méd. Vét. Pays Trop. **37:** 288–296.
27. MAILLARD, J-C., S.J. KEMP, H. LEVEZIEL, *et al.* 1989. Système majeur d'histocompatibilité de bovins ouest-africains. Typage d'antigènes lymphocytaires (BoLA) de taurins Baoulé (*Bos taurus*) et de Zébus soudaniens (*Bos indicus*) du Burkina Faso (*Afrique occidentale*). Rev. Elev. Méd. Vét. Pays Trop. **42:** 275–281.
28. IRAQI, F., S. CLAPCOT, P. KUMAN, *et al.* 2000. Fine mapping of trypanosomiasis resistance QTLs in mice using advanced intercross lines. Mamm. Genome **11:** 645–648.
29. KEMP, S., F. IRAQI. A. DARVASI, *et al.* 1997. Localization of genes controlling resistance to trypanosomiasis in mice. Nat. Genet. **16:** 194–196.
30. KEMP, S.J. & A.J. TEALE. 1998. Genetic basis of trypanotolerance in cattle and mice. Parasitol. Today **14:** 450–454.
31. HANOTTE, O., Y. RONIN, M. AGABA, *et al.* 2003. Mapping of quantitative trait loci (QTL) controlling trypanotolerance in a cross of tolerant West African N'Dama cattle and susceptible East African Boran cattle. Proc. Natl. Acad. Sci. USA. **100** (13): 443–448.
32. PIQUEMAL, D., T. COMMES, L. MANCHON, *et al.* 2002. Transcriptome analysis of monocytic leukemia cell differentiation. Genomics **80:** 316–371.
33. DWINGER, R.H., D.J. CLIFFORD, K. AGYEMANG, *et al.* 1992. Comparative studies on N'Dama and Zebu cattle following repeated infections with *Trypanosoma congolense*. Res. Vet. Sci. **52:** 292–298.
34. NANTULYA, V.M., A.J MUSOKE, F.R. RURANGIRWA & S.K. MOLOO. 1984. Resistance of cattle to tsetse-transmitted challenge with *Trypanosoma brucei* or *Trypanosoma congolense* after spontaneous recovery form syringe passed infections. Infect. Immun. **43:** 735–738.
35. PALING, R.W., S.K. MOLOO, J.R. SCOTT, *et al.* 1991. Susceptibility of N'Dama and Boran cattle to tsetse-transmitted primary and rechallenge infections with a homologous serodeme of *Trypanosoma congolense*. Parasite Immunol. **13:** 413–425.
36. PALING, R.W., S.K. MOLOO, J.R. SCOTT, *et al.* 1991. Susceptibility of N'Dama and Boran cattle to sequential challenges with tsetse-transmitted clones of *Trypanosoma congolense*. Parasite Immunol. **13:** 427–445.
37. AUTHIÉ, E. 1993. Contribution à l'étude des mécanismes immunologiques impliqués dans la trypanotolérance des taurins d'Afrique. Thèse doctorat es sciences. Univ. Bordeaux II.
38. NAESSENS, J., A.J. TEALE & M. SILEGHEM. 2002. Identification of mechanisms of natural resistance to African trypanosomiasis in cattle. Vet. Immunol. Immunopathol. **87:** 187–194.
39. LUCKINS, A.G. 1974. The immune response of zebu cattle to infection with *Trypanosoma congolense* and *T. vivax*. Ann. Trop. Med. Parasitol. **70:** 133–145.
40. DE RAADT, P. 1974. Immunity and antigenic variation: clinical observations suggestive of immune phenomena in trypanosomiasis. *In* Trypanosomiasis and Leishmaniasis with special reference to Chaga's disease. CIBA Found. Symp. **20:** 199–211.
41. OKA, M. & I. YOSHIRITO. 1987. Polyclonal B-cell activating factors produced by spleen cells of mice stimulated with a cell homogenate of *Trypanosoma gambiense*. Infection and immunity **55:** 3162–3167.

# Feline Babesiosis in South Africa

## A Review

BANIE L. PENZHORN, TANYA SCHOEMAN, AND LINDA S. JACOBSON

*Departments of Veterinary Tropical Diseases and Companion Animal Clinical Studies, Faculty of Veterinary Science, University of Pretoria, Onderstepoort, 0110, South Africa*

**ABSTRACT: *Babesia felis*, originally identified in wild cats in the Sudan, was subsequently found to cause clinical disease in domestic cats. Although babesiosis in domestic cats has been reported sporadically from various countries, as a significant disease it appears to be a distinctly South African phenomenon. Apart from an inland focus, feline babesiosis is reported regularly only from coastal regions. The infection is assumed to be tick-borne, but the vector has not been identified. Feline babesiosis tends to be an afebrile, chronic, low-grade disease. The most frequently reported complaints by owners are anorexia and lethargy. The main clinical findings are anemia, depression, and occasionally icterus. Concurrent infections (e.g., *Mycoplasma haemofelis*, FeLV, FIV) may contribute to the clinical picture. Laboratory findings commonly include regenerative anemia, elevation of alanine transaminase (but not alkaline phosphatase) and total bilirubin concentrations, and a variety of electrolyte disturbances. Secondary immune-mediated hemolytic anemia can be seen occasionally. Drugs effective against other *Babesia* species give variable and questionable results. The drug of choice is primaquine phosphate, which effects a clinical cure but does not sterilize the infection. Repeated or chronic therapy may be required.**

**KEYWORDS: *Babesia felis*; feline babesiosis; cats; treatment; primaquine phosphate; regenerative anemia; immune-mediated hemolytic anemia**

Babesiosis of domestic cats has been reported sporadically from various countries, including France,[1,2] Germany,[3] Thailand,[4] India,[5] and Zimbabwe,[6] but does not seem to be a regularly occurring significant clinical disease in any country other than South Africa. Apart from an inland focus,[7] feline babesiosis is reported regularly only from South African coastal regions.[8] Incidence is highest along the southern Cape coast, with more sporadic occurrence in the eastern coastal areas and minimal disease reported from the west coast beyond Cape Town.[8] Overall incidence is highest in summer, but seasonality is less pronounced in nonseasonal and winter-rainfall areas.[8] Although feline babesiosis is assumed to be a tick-borne infection, the vector has not been identified.

Address for correspondence: Banie L. Penzhorn, Department of Veterinary Tropical Diseases, Faculty of Veterinary Science, University of Pretoria, Private Bag X04, Onderstepoort, 0110, South Africa. Voice: 27-12-529 8253; fax +27-12-529 8312.
banie.penzhorn@up.ac.za

**Ann. N.Y. Acad. Sci. 1026: 183–186 (2004). © 2004 New York Academy of Sciences.
doi: 10.1196/annals.1307.027**

*Babesia felis*, a small piroplasm first described in the Sudan from a wild-caught African wild cat *Felis sylvestris* (syn: *Felis ocreata*),[9] was found to be transmissible to domestic cats, but did not cause clinical illness.[10] The name was subsequently applied to the causative organism of a potentially fatal disease in domestic cats in South Africa.[11–13] A small piroplasm recently isolated from lions (*Panthera leo*) was found to be morphologically similar to, but serologically distinct from, *B. felis*.[14] This isolate as well as typical disease-causing *B. felis* and isolates from a caracal (*Felis caracal*) have recently been sequenced. The lion isolate was distinct from true *B. felis* and has subsequently been described as a separate species, while the two caracal isolates were very similar to *B. felis*.[15] This raises the question whether *B. felis* reported from domestic and wild felids is, in fact, a single species. Other piroplasms reported from domestic cats are the small *Babesia cati* from India[5] and a large, unnamed piroplasm found in cats in Zimbabwe.[6]

In contrast to babesiosis in other domestic animals, feline babesiosis is not associated with pyrexia.[16] Most affected cats seem to be young adults of less than three years of age.[17] No breed or sex predisposition is evident, but Siamese cats might be overrepresented among purebred cats. Owner complaints typically include anorexia, lethargy and weight loss. The most common clinical signs are anorexia, listlessness, and anemia, followed by icterus.[17] Less common signs are weakness, constipation, and pica.[17] Parasitemias are variable and range between very low and extremely high. The strong correlation between central and peripheral parasitemias indicates that sequestration is not a feature of the disease.[16,17]

Macrocytic, hypochromic, regenerative anemia is the most consistent hematological finding, although a large proportion of cats are not anemic.[17] Anemia, which can become severe in advanced cases,[17,18] is hemolytic, presumably resulting from both intravascular and extravascular erythrolysis. No characteristic changes are seen in total or differential leukocyte counts; when abnormal values are present they are often accompanied by concurrent illness or infection. Thrombocyte counts are variable and thrombocytopenia is an inconsistent finding. The in-saline agglutination test can be positive in a number of cases, indicating that secondary immune-mediated hemolytic anemia can also be a feature of this disease.[17]

The most remarkable clinico-pathological changes are elevation of hepatic cytosol enzyme activities and total bilirubin concentrations.[17,19] Serum alanine transaminase is significantly elevated in the majority of cases, whereas alkaline phosphatase and gamma-glutamyltransferase are generally within normal limits. This provides evidence of primary hepatocellular injury or inflammation in feline babesiosis. The hyperbilirubinemia is most likely a result of hemolysis, but secondary hepatocellular injury is probably an additional contributing factor. No characteristic changes in renal parameters are observed and serum urea and creatinine levels are mostly within normal limits, indicating that gross renal damage is not a consistent feature of the disease. No characteristic changes in serum electrolytes (sodium and potassium) are seen, although a variety of electrolyte disturbances can occur.[17] Serum protein values are mostly normal, but elevations are seen in some cases. Polyclonal gammopathies are observed in all cats with hyperglobulinemia; this can be ascribed to a combination of acute- and chronic-phase proteins produced in response to the *Babesia* antigens.

Concurrent infections with *Mycoplasma haemofelis*, feline leukemia virus and/or feline immunodeficiency virus have been identified in a number of cats and this seems to have profound effects on response to treatment and outcome of the disease.[17] Although response to therapy is usually good, and premunity is assumed to develop over time, the average mortality from feline babesiosis is estimated at 15%.[15]

Until 1980 tetracyclines, sometimes used in combination with trypan blue, and cephaloridine were recommended for treatment of feline babesiosis.[20–22] When efficacy of 10 drugs against *B. felis* infections was investigated experimentally, most drugs used against babesiosis in other domestic animals gave variable and questionable results.[23] Primaquine phosphate (Primaquine, Kyron Laboratories, Benrose, South Africa) at 0.5 mg/kg was highly effective, but frequently caused vomiting when administered orally and was lethal at dosages exceeding 1 mg/kg.[23] In a recent investigation, rifampicin and a sulfadiazine-trimethoprim combination appeared to have an anti-parasitic effect but were inferior to primaquine, which had a dramatic effect on parasitemia, but failed to sterilize *B. felis* infections.[24] Buparvaquone, enrofloxacin, and danofloxacin had no significant anti-babesial effect.[24] Despite its drawbacks, primaquine remains the drug of choice. Repeated or chronic therapy may be required. Doxycycline may add potential benefits in treatment of this disease.

## REFERENCES

1. BOURDEAU, P. 1996. Feline babesiosis. Point Vet. **27:** 947–953.
2. LEGER, N., H. FERTE, P. BERTHELOT, *et al.* 1992. A case of feline babesiosis in Haute-Saone, France. Sci. Vet. Med. Comp. **94:** 249–252.
3. MOIK, K. & R. GOTHE. 1997. *Babesia* infections of felids and a report on a case in a cat in Germany. Tierärztl. Prax. **25:** 532–535.
4. JITTAPALAPONG, S. & W. JANSAWAN. 1993. Preliminary survey on blood parasites of cats in Bangkhen District Area. Kasetsat. J. Nat. Sci. **27:** 330–335.
5. MUDALIAR, S.V., G.R. ACHARY & V.S. ALWAR. 1950. On a species of *Babesia* in an Indian wild cat (*Felis catus*). Ind. Vet. J. **26:** 391–395.
6. STEWART, C.G., K.J.W. HACKETT & M.G. COLLETT. 1980. An unidentified *Babesia* of the domestic cat (*Felis domesticus*). J. S. Afr. Vet. Assoc. **51:** 219–221.
7. PENZHORN, B.L., E. STYLIANIDES, M.A. COETZEE, et al. 1999. A focus of feline babesiosis at Kaapschehoop on the Mpumalanga escarpment. J. S. Afr. Vet. Assoc. **70:** 60.
8. JACOBSON, L.J., T. SCHOEMAN & R.G. LOBETTI. 2000. A survey of feline babesiosis in South Africa. J. S. Afr. Vet. Assoc. **71:** 222–228.
9. WILSON, D.E. & D.M. REEDER. 1993. Mammal Species of the World—A Taxonomic and Geographic Reference, 2nd edit. Smithsonian Institution. Washington DC.
10. DAVIS, L.J. 1929. On a piroplasm of the Sudanese wild cat (*Felis ocreata*). Trans. R. Soc. Trop. Med. **22:** 523–534.
11. JACKSON, C. & F.J. DUNNING. 1937. Biliary fever (nuttalliosis) of the cat: A case in the Stellenbosch District. J. S. Afr. Vet. Med. Assoc. **8:** 83–88.
12. MCNEIL. 1937. Piroplasmosis of the domestic cat. J. S. Afr. Vet. Med. Assoc. **8:** 88–90.
13. FUTTER, G.J. & P.C. BELONJE. 1980. Studies on feline babesiosis. 1. Historical review. J. S. Afr. Vet. Assoc. **51:** 105–106.
14. LÓPEZ-REBOLLAR, L.M., B.L. PENZHORN, D.T. DE WAAL & B.D. LEWIS.1999. A possible new piroplasm in lions from the Republic of South Africa. J. Wildl. Dis. **35:** 82–85.
15. PENZHORN, B.L., A.E. KJEMTRUP, L.M. LÓPEZ-REBOLLAR & P.A. CONRAD. 2001. *Babesia leo* n. sp. from lions in the Kruger National Park, South Africa, and its relation to other small piroplasms. J. Parasitol. **87:** 681–685.
16. FUTTER, G.J. & P.C. BELONJE.1980. Studies on feline babesiosis. 2. Clinical observations. J. S. Afr. Vet. Assoc. **51:** 143–146.
17. SCHOEMAN, T., R.G. LOBETTI, L.S. JACOBSON & B.L. PENZHORN. 2001. Feline babesiosis: signalment, clinical pathology and co-infections. J. S. Afr. Vet. Assoc. **72:** 4–11.

18. FUTTER, G.J., P.C. BELONJE & A. VAN DEN BERG. 1980. Studies on feline babesiosis. 3. Haematological findings. J. S. Afr. Vet. Assoc. **51:** 271–280.
19. FUTTER, G.J., P.C. BELONJE, A. VAN DEN BERG & A.W. VAN RIJSWIJK. 1981. Studies on feline babesiosis. 4. Chemical pathology, macroscopic and microscopic post mortem findings. J. S. Afr. Vet. Assoc. **52:** 5–14.
20. BROWNLIE, J.F. 1954. Aureomycin in the treatment of piroplasmosis in the cat. J. S. Afr. Vet. Med. Assoc. **25:** 654.
21. DORRINGTON, J.E. & W.J.C. DU BUY. 1966. Ceporan: efficacy against *Babesia felis*. J. S. Afr. Vet. Med. Assoc. **37:** 93.
22. ROBINSON, E.M. 1963. Biliary fever (nuttalliosis) in the cat. J. S. Afr. Vet. Med. Assoc. **34:** 45–47
23. POTGIETER, F.T. 1981. Chemotherapy of *Babesia felis* infection: efficacy of certain drugs. J. S. Afr. Vet. Assoc. **52:** 289–293.
24. PENZHORN, B.L., B.D. LEWIS, L.M. LÓPEZ-REBOLLAR & G.E. SWAN. 2000. Screening of five drugs for efficacy against *Babesia felis* in experimentally infected cats. J. S. Afr. Vet. Assoc. **71:** 53–57.

# Study of Gastrointestinal Nematodes in Sicilian Sheep and Goats

A. TORINA,[a] S. DARA,[a] A.M.F. MARINO,[a] O.A.E. SPARAGANO,[b] F. VITALE,[a] S. REALE,[a] AND S. CARACAPPA[a]

[a]*Istituto Zooprofillattico Sperimentale della Sicilia, Palermo, Italy*

[b]*School of Agriculture, Food and Rural Development, King George VI Building, Newcastle-upon-Tyne, United Kingdom*

**ABSTRACT: Parasitic gastroenteritis is one of the major causes of productivity loss in sheep and goats. This report records two studies of the helminth fauna from post-mortem examination. The first study, performed on the digestive tract of 72 sheep from a central part of Sicily in a high hill village (1,360 meters above sea level), between April 1996 and March 1997, showed an infection rate of 78%. The second study targeted goats from the western part of Sicily and showed an infection rate of 90%. For sheep, a total of 23 species of helminths were identified belonging to the family of Trichostrongyloidea, with the genera *Haemonchus*, *Ostertagia* (*Teladorsagia*), *Trichostrongylus*, *Cooperia,* and *Nematodirus*; Strongilolidea with the genera *Oesophagostomum* and *Chabertia*: Ancylostomidea with *Bunostomum;* and Tricuridea with *Tricuris*. *Teladorsagia circumcincta* was the most common in the sheep abomasum, *Bunostomum trigonocephalum* and *Trichostrongylus* spp. in the small intestine, and *Chabertia ovina* and *Trichuris ovis* in the large intestine. For goats, a total of 12 species were isolated in the abomasum with *Teladorsagia circumcincta* and *Trichostrongylus axei* the most common species. In the small intestine, five species were isolated and *Trichostrongylus capricola* was the dominant species. *T. ovis* and *O. venulosum* were dominant in large intestine and in the cecum. We also found species belonging to other ruminants such as *O. ostertagi* (in cattle) and *S. kolchida* and *O. leptospicularis* (in wild ruminants).**

**KEYWORDS: nematodes; sheep; goats; Sicily; *Teladorsagia*; *Haemonchus*; *Trichostrongylus*; *Oesophagostomum*; *Chabertia*; *Bunostomum*; *Trichuris***

## INTRODUCTION

Infections from gastrointestinal nematodes in livestock in countries with well-developed agricultural systems have always been an important production issue. Economic losses are estimated to be in the millions of dollars every year. Low production of meat, wool, and milk as well as the costs of anthelminthic treatment[1] are the major causes of production losses in animal production in the World[2] and appear

Address for correspondence: Dr. Alessandra Torina, Institute Zooprophilactic Experimental of Sicily (IZS), Via Marinuzzi n.3, 90129 Palermo, Sicily, Italy. Voice: 091-6565360; fax: 091-6565361.
torina@pa.izs.it

**Ann. N.Y. Acad. Sci. 1026: 187–194 (2004). © 2004 New York Academy of Sciences.
doi: 10.1196/annals.1307.028**

to be a major constraint to efficient sheep production. In Italy, parasites are responsible for 6% of the agriculture economic loss at a national level. It is estimated that for sheep and goats, parasites are responsible for 80% of the observed pathologies.[3] Parasite diseases due to gastrointestinal nematodes are widespread in goats and have been well documented. In France,[4] we showed that even with subclinical infections (with low nematode eggs excreted per grams of feces) there was even at that stage a low levels of pepsinogen and phosphorus and low hematocrit values.

Survival and development of the nematode larval populations require appropriate environmental conditions. This influences the clinical manifestations and transmission rates and differs between regions.[5] The distribution of parasites has an important implication for economical losses in the animal farming system.[6]

Studies on the helminths in Sicily had not been extensively studied.[7–9] We decided to study the gastrointestinal nematode infections in sheep in the central part of Sicily. This research is based on sheep samples from a pilot farm. This paper reports the epidemiological results on different sheep tissues collected in a slaughterhouse in the Ganci district, Palermo province. In the past, only a few studies were carried out in Sicily.[8,9] However, nowadays we need to understand the distribution of such diseases to develop appropriate control methods.

Goats are more susceptible than sheep to gastrointestinal nematode infections with a higher production of worm eggs and a higher parasitemia.[10,11] Anthelminthic treatments are less effective in goats[12] and could contribute to the nematode resistance.[13] These factors show that goats can be responsible for the dissemination of worm eggs from resistant nematodes in mixed populations of small ruminants.[14]

## MATERIALS AND METHODS

### *Location of the Study*

Sheep samples were collected in the Ganci district, Palermo province, situated at 1,360 meters in altitude in the Monte Zimmara region. The sampled area (around 60 hectares) is mainly made of natural pastures with an argilo-calcarous soil. Goat samples were collected from the western part of Sicily.

### *Parasite Collection*

Samples were collected between April 1996 and March 1997. We examined digestive tracts from 72 sheep and 30 goats post-mortem. To collect the samples, we used the double fastening technique.[15] Each sample was opened longitudinally and then the mucosa was washed and the material was collected by filtration using filters of different diameters (212, 75, and 38 μm). Using pressurized water we were able to isolate 10% of the material from each sample for further examination. Parasites were kept in lacto-phenol and identified using the taxonomic keys from several authors.[16–18] Samples were then stored in 70% alcohol.

### *Statistical Analysis*

For each tissue, we calculated the percentage of infected animals, the parasitemia level for the whole group and for the infected animal sub-group. Parasites were iso-

**TABLE 1. Percentages of infection with parasitosis in tissues from sheep and goat**

| Samples | Infection (%) | Parasitosis/All Examined Animals (Average) | Parasitosis/Infected Animals Only(Average) |
|---|---|---|---|
| Sheep | | | |
| Abomasum | 72 | 1,439 | 1,992 |
| Small intestine | 48 | 555 | 1,158 |
| Large intestine | 15 | 2 | 12 |
| Cecum | 20 | 1 | 9 |
| Goat | | | |
| Abomasum | 83 | 1,781 | 2,226.3 |
| Small intestine | 55 | 496.1 | 930.2 |
| Large intestine | 31 | 8.4 | 28.0 |
| Cecum | 34 | 9.5 | 28.5 |

lated and grouped according to a method[19] previously described based on the value of the Importance Index (I):

$$I_j = (A_j \times B_j / A_i \times B_i) \times 100,$$

where $A_j$ = number of parasite for species $j$; $B_j$ = Number of host infected with species $j$; $A_i$ and $B_i$ represent the total number of parasite observed and the total number of infected hosts with any parasite species, respectively. Species with an Importance Index higher than 1 were considered dominant, species with an Importance Index between 0.01 and 1 were considered co-dominant, and species with an Importance Index less than 0.01 were considered subordinate.

For each nematode species, we considered the intensity of the infection (total number of nematodes observed from the same species divided by the number of infected samples with this species), the abundance (total number of nematodes from the same species divided by the total number of samples) and its prevalence as previously described.[20]

## RESULTS

Of 72 samples from sheep, 56 showed a nematode infection (78%). In TABLE 1 we show the prevalence for each examined tissue with the abomasum, the small intestine, and the cecum being the most infected tissues (72%, 48%, and 20%, respectively). The parasite numbers also followed the same ranking. Twenty-three species were identified representing nine different genera: *Bunostomum*, *Chabertia*, *Cooperia*, *Haemonchus*, *Nematodirus*, *Oesophagostomum*, *Ostertagia* (*Skrjabinagia*, *Teladorsagia*), *Trichostrongylus*, and *Trichuris*. TABLE 2 shows the species identification for each tissue, the prevalence rate, the Importance Index, intensity, and abundance.

Of 30 goat samples, we found 27 (90%) infected with gastrointestinal nematodes. The prevalence of infection for different tissues after post-mortem examination was

**TABLE 2. Parasites in the abomasum, small intestine, large intestine, and cecum of sheep**

| | Positive Samples | % | Importance Index | Intensity | Abundance |
|---|---|---|---|---|---|
| Abomasum–Dominating Species | | | | | |
| *Teladorsagia circumcincta* | 48 | 94 | 68.7 | 1,483.6 | 989.1 |
| *Teladorsagia pinnata* | 18 | 35 | 4.2 | 241.0 | 60.2 |
| *Teladorsagia trifurcata* | 17 | 33 | 3.3 | 198.3 | 46.8 |
| *Trichostrongylus axei* | 25 | 49 | 18.1 | 747.7 | 259.6 |
| *Trichostrongylus vitrinus* | 14 | 27 | 3.3 | 245.2 | 47.7 |
| *Trichostrongylus colubriformis* | 6 | 12 | 1.6 | 274.5 | 22.9 |
| Abomasum–Co-dominating Species | | | | | |
| *Haemonchus contortus* | 7 | 14 | 0.48 | 70.4 | 6.8 |
| *Ostertagia leptospicularis* | 2 | 4 | 0.19 | 100.7 | 2.8 |
| *Skrjabinagia kolchida* | 2 | 4 | 0.05 | 26.2 | 0.7 |
| *Trichostrongylus capricola* | 1 | 2 | 0.02 | 21.1 | 0.3 |
| *Ostertagia ostertagi* | 1 | 2 | 0.02 | 23.0 | 0.3 |
| Small Intestine–Dominating Species | | | | | |
| *Bunostomum trigonocephalum* | 14 | 40 | 2.8 | 80.9 | 15.7 |
| *Trichostrongylus capricola* | 6 | 17 | 11.8 | 801.6 | 66.8 |
| *Trichostrongylus colubriformis* | 16 | 46 | 18.9 | 478.6 | 106.3 |
| *Trichostrongylus vitrinus* | 16 | 46 | 34.5 | 875.1 | 194.5 |
| *Nematodirus abnormalis* | 11 | 31 | 4.15 | 152.9 | 23.3 |
| *Nematodirus filicollis* | 12 | 34 | 8.43 | 284.8 | 47.5 |
| *Nematodirus spathiger* | 14 | 40 | 6.46 | 187.0 | 36.4 |
| *Nematodirus battus* | 4 | 11 | 7.65 | 775.6 | 43.1 |
| Small Intestine–Co-dominating Species | | | | | |
| *Cooperia curticei* | 1 | 3 | 1.00 | 411.4 | 5.7 |
| *Cooperia pectinata* | 2 | 6 | 0.29 | 58.0 | 1.6 |
| *Cooperia oncophora* | 2 | 6 | 0.33 | 66.0 | 1.8 |
| *Cooperia punctata* | 1 | 3 | 0.10 | 35.9 | 0.5 |
| Large Intestine–Dominating Species | | | | | |
| *Chabertia ovina* | 5 | 45 | 59.6 | 16.2 | 1.1 |
| *Trichuris ovis* | 6 | 54 | 36.2 | 8.2 | 0.7 |
| *Oesophagostomum venulosum* | 1 | 9 | 4.2 | 5.7 | 0.1 |
| Cecum–Dominating Species | | | | | |
| *Trichuris ovis* | 12 | 100 | 100 | 8.9 | 1.5 |

as follows: abomasum (83%), small intestine (53%), large intestine (27%), and cecum (37%). Prevalence data, Importance Index, intensity, and abundance are presented in TABLE 3. In the abomasum (TABLE 3) we found two dominating species, *T. circumcincta* and *T. axei,* with an Importance Index higher than 1; all the other species being co-dominant. The small intestine (TABLE 3) showed very low numbers of nematode species, with *T. capricola* the dominant species and four others consid-

**TABLE 3. Prevalence rates and Importance Indices (I) of gastrointestinal nematode species in the abomasum, small intestine, large intestine, and cecum of goats**

| Dominating Species | Positive Samples | % | I | Intensity | Abundance |
|---|---|---|---|---|---|
| Abomasum–Dominating Species | | | | | |
| *Teladorsagia circumcincta* | 24 | 100 | 2.39 | 549.5 | 439.6 |
| *Trichostrongylus axei* | 14 | 58 | 1.58 | 548.0 | 256 |
| Abomasum–Co-dominating species | | | | | |
| *Teladorsagia pinnata* | 10 | 42 | 0.19 | 71.3 | 23.8 |
| *Teladorsagia trifurcata* | 4 | 17 | 0.14 | 47.4 | 6.3 |
| *Haemonchus contortus* | 8 | 33 | 0.27 | 103.0 | 27.5 |
| *Trichostrongylus vitrinus* | 10 | 42 | 0.22 | 105.0 | 34.9 |
| *Trichostrongylus colubriformis* | 5 | 21 | 0.06 | 18.2 | 3.0 |
| *Ostertagia leptospicularis* | 4 | 17 | 0.26 | 74.7 | 9.9 |
| *Skrjabinagia kolchida* | 4 | 17 | 0.07 | 29.3 | 3.9 |
| *Trichostrongylus capricola* | 9 | 37 | 0.78 | 218.0 | 65.4 |
| *Marshallagia marshalla* | 1 | 4 | 0.14 | 73.0 | 2.4 |
| *Ostertagia ostertagi* | 1 | 4 | 0.55 | 1,899.2 | 63.3 |
| Small Intestine–Dominating species | | | | | |
| *Trichostrongylus capricola* | 11 | 69 | 8.14 | 788.0 | 288.9 |
| Small Intestine–Co-dominating species | | | | | |
| *Trichostrongylus colubriformis* | 1 | 6 | 0.83 | 124.2 | 4.14 |
| *Trichostrongylus vitrinus* | 8 | 50 | 0.73 | 93.4 | 24.9 |
| *Bunostomum trigonocephalum* | 2 | 13 | 0.10 | 7.3 | 0.5 |
| *Nematodirus abnormalis* | 8 | 50 | 0.49 | 39.6 | 10.6 |
| Large Intestine–Dominating Species | | | | | |
| *Oesophagostomum venulosum* | 6 | 67 | 14.9 | 25.7 | 5.1 |
| *Trichuris ovis* | 1 | 11 | 1.76 | 4.4 | 0.1 |
| Large Intestine–Co-dominating species | | | | | |
| *Chabertia ovina* | 1 | 11 | 0.79 | 2.0 | 0.1 |
| Cecum–Dominating species | | | | | |
| *Oesophagostomum venulosum* | 8 | 80 | 10.6 | 17.2 | 4.6 |
| *Trichuris ovis* | 3 | 30 | 5.15 | 13.2 | 1.3 |

ered co-dominant. These results are in contrast with those found in France.[21] There *T. capricola* was not found, whereas *T. colubriformis* was the dominating species (which was found in one sample only in our study). *Nematodirus* spp. and *Bunostomum* spp. were not found in the French study whereas *N. abnormalis* and *O. venulosum* were present in 50% and 13% of our samples, respectively. In the large intestine, *O. venulosum* was the dominant species, whereas it represented only 9% in a previous study.[9] In this study, *C. ovina* was observed in one sample only. *O. venulosum* was found in 80% of the cecum samples, while *T. ovis* was the only sheep parasite found there at a prevalence of 30% (TABLE 3).[9]

## DISCUSSION AND CONCLUSIONS

The results presented in this paper clearly show the importance of nematode parasites in the studied region—more than 78% and 90% of the samples were infected for sheep and goats, respectively.

The results for the different species found in sheep and goats (TABLES 2 and 3) characterize the high diversity of nematode species present in Sicily: 24 different species were identified (11 species found in both sheep and goats).

Six different dominant species in sheep abomasums was found: *Teladorsagia circumcincta* with two other species from the same genus (*T. pinnata* and *T. trifurcata*) and the species *Trycostrongylus axei*, *T. colubriformis*, and *T. vitrinus*. The co-dominant species brought some interesting findings as we observed the presence in sheep of nematodes usually associated with other ruminant species. Different ruminant species are usually grazing together on the same pastures, demonstrating some parasite adaptation to different hosts. *Ostertagia leptospicularis* and a very close species *Skrjabinagia kolchida* are nematode species of wild animals such as chamois and fallow deer,[22] *Trichostrongylus capricola* being more a goat nematode while *Ostertagia ostertagi* is mainly found in cattle.

In the sheep small intestine, the dominating species belonged to the following genera: *Trichostrongylus*, *Nematodirus*, and *Bunostomum*, the latter species belongs to the Ancylostomatidi and shows a lower parasitosis compared to Trichostrongili and has a completely different biological cycle and transmission method. It is interesting to point out that for the dominating species we still identified *Trichostrongylus capricola* and *Nematodirus battus*, nematodes usually found in goats and cattle, respectively in contradiction with the abomasum nematodes where "adapted" species were co-dominating species. This latter co-dominant group consisted of only *Cooperia* genus with low prevalence rates.

In the sheep large intestine and cecum the helminths were represented by a low number of species (all dominating species) but with a low parasitosis while in the cecum we found only one species, *Trichuris ovis*.

We found a different species situation in goat abomasum samples, where only two species were dominating (*T. circumcincta* and *T. axei*), all the other ten species are co-dominant. Results for goats (not treated with anthelminthic drugs in this study) demonstrate that each single helminth species can create its own ecologic niche in contradiction with what is observed for sheep, where anthelminthic products influence the nematode populations; six species belonging to the *Teladorsagia* and *Trichostrongylus* genera being then dominant showing that the other species are probably more susceptible to the treatment. However, in goats all the species showed a similar importance and none were found as subordinate species. In TABLE 3, we could also observe the adaptation of some species to other hosts with *O. ostertagi* usually found in cattle and *O. leptospicularis* and *S. kolchida* usually found in wild ruminants.[9]

## ACKNOWLEDGMENTS

We would like to thank Mrs. Rosalia D'Agostino, Mr. Gaspare Lo Bue, and Mr. Sig Salvatore Scimeca for their valuable technical support.

## REFERENCES

1. BARGER, I.A. 1982. Helminth parasites and animal production. *In* Biology and Control of Endoparasites. L.E.A. Symons, A.D. Donald & J.K. Dineen, Eds.: 133–155. Academic Press. Sydney, Australia.
2. AMERICAN ASSOCIATION OF VETERINARY PARASITOLOGISTS. 1983. Research needs and priorities for ruminant internal parasites in the United States. Am. J. Vet. Res. **44:** 1836–1847.
3. ARRU, E. 1985. Principali Nematodi e Trematodi degli animali domestici—Elmintiasi: problemi veterinari e zootecnici. Atti Pfizer **9:** 5–12.
4. CHARTIER, C. & H. HOSTE. 1996. Impact des strongyloses gastro-intestinales sur la physiologie digestive et sur la production laitière chez les caprins. Bull. GTV **109:** 85–93.
5. ZAJAC, A.M. & G.A MOORE. 1993. Trattamento e controllo dei nematodi gastroenterici degli ovini. *In* The Compendium on Continuing Education for the Practicing Veterinarian. **7:** 999–1010.
6. BARGER, I.A. 1985. The statistical distribution of trichostrongilid nematode parasites in grazing lambs. Int. J. Parasitol. **15:** 645–649.
7. GALLO, C. 1960. I Tricostrongilidi degli ovini nella Sicilia sud-orientale. Parassitologia **2:** 179–180.
8. CARACAPPA, S., S. RIILI, F. PRATO, *et al.* 1994. Indagine sulla fauna elmintica abomasale in ovini e caprini siciliani. Atti Congr. SIPAOC. p. 287–290.
9. CARACAPPA, S., V. DI MARCO, S. DARA, *et al.* 1996. Indagine sulla diffusione dei nematodi intestinali in allevamenti di ovini e caprini della Sicilia. Atti Congr. SIPAOC. p. 385–388.
10. LE JAMBRE, L.F. & V.M. ROYAL. 1976. A comparison of worm burdens in grazing Merino sheep and Angora goats. Aust. Vet. J. **52:** 181–183.
11. POMROY, W.E., M.G. LAMBERT & K. BETTERIDGE. 1986. Comparison of faecal strongilyte egg counts of goats and sheep on the same pasture. N. Z. Vet. J. **34:** 36–37.
12. SANGSTER, N.C., J.M. RICHARD, D.R. HENNESSY, *et al.* 1991. Disposition of oxfendazole in goats and efficacy compared with sheep. Res. Vet. Sci. **51:** 258–263.
13. CHARLES, T.P., J. POMPEU & D.B. MIRANDA. 1989. Efficacy of three broad-spectrum anthelmintics against gastrointestinal nematode infections of goats. Vet. Parasitol. **34:** 71–75.
14. VARADY, M., J. PRASLICKA & J. CORBA. 1994. Treatment of multiple resistant field strain of *Ostertagia spp.* in cashmere and Angora goats. Intern. J. Parasitol. **24:** 335–340.
15. EUZEBY, J. 1958. Diagnostic expérimental des Helminthoses animales. Vigot Fréres Ed. Paris, pp. 172–185.
16. SKRJABIN, K.I., N.P. SHIKHOBALOVA & R.S. SCHULZ. 1961. Key to parasitic nematodes. Strongylata. Vol. III. Israel Program for Scientific Translation. Jerusalem.
17. DURETTE-DESSET, M.C. 1983. Key to genera of the Superfamily Trichostrongyloidea. Ediz. CAB.

18. CABARET, J., N. GASNIER & P. JACQUIET. 1998. Faecal egg counts are representative of digestive tract strongyle worm burdens in sheep and goats. Parasite **5:** 137–142.
19. BUSH, A.O. 1973. An ecological analysis of the helminth parasites of the white ibis in Florida. M.S. Thesis, University of Florida. Gainesville, FL.
20. MARGOLIS, L., G.W. ESCH, J.C. HOLMES, *et al.* 1982. The use of ecological terms in parasitology report of an ad-hoc committee of the American Society of Parasitologists. J. Parasitol. **68:** 131–133.
21. CHARTIER, C. & B. RECHE. 1992. Gastrointestinal helminths and lungworms of French dairy goats: prevalence and geographical distribution in Poitou-Charente. Vet. Res. Commun. **16:** 327–335.
22. GENCHI, C., A. BOSSI & M.T. MANFREDI. 1985. Gastrointestinal nematode infections in wild ruminants *Rupicapra rupicapra* and *Dama dama*: influence of density and cohabitation with domestic ruminants. Parassitologia **27:** 211–223.

# Identification of an Expressed Gene in *Dipylidium caninum*

RODRIGO R. C. MIRANDA, LIVIO M. COSTA-JÚNIOR, ARTUR K. CAMPOS, HUDSON A. SANTOS, AND ÉLIDA M. L. RABELO

*Departamento de Parasitologia, Universidade Federal de Minas Gerais, Belo Horizonte, Brazil*

**ABSTRACT: Recombinant DNA studies have been focused on developing vaccines to different cestodes. But few studies involving *Dipylidium caninum* molecular biology and genes have been done. Only partial sequences of mitochondrial DNA and ribosomal RNA gene are available in databases. Any molecular work with this parasite, including epidemiology, study of drug-resistant strains, and vaccine development, is hampered by the lack of knowledge of its genome. Thus, the knowledge of specific genes of different developmental stages of *D. caninum* is crucial to locate potential targets to be used as candidates to develop a vaccine and/or new drugs against this parasite. Here we report, for the first time, the sequencing of a fragment of a *D. caninum* expressed gene.**

**KEYWORDS: *Dipylidium caninum;* expressed gene; NADH-dehydrogenase**

## INTRODUCTION

The dog tapeworm *Dipylidium caninum* is a cosmopolitan parasite of dogs and cats and occasionally causes human infection. Morphologically, this parasite is divided in three regions—scolex, neck, and strobila. The strobila is formed by proglottids, and these proglottids are further divided according to the reproductive developmental stage (immature, mature, and gravid) and which are specialized in egg-packet formation. Recombinant DNA studies have been focused on developing vaccines to different cestodes.[1–3] However, few studies involving *D. caninum* molecular biology and consequently genes isolation have been done. Up to now only partial sequences of mitochondrial DNA and ribosomal RNA gene are available in databases.[4,5] Therefore, any molecular work with this parasite, including the epidemiology, study of drug-resistant strains, and vaccine development, is hampered by the lack of knowledge of its genome. Thus, the knowledge of specific genes of different developmental stages of *D. caninum* is crucial to locate potential targets to be used as candidates to develop a vaccine and/or new drugs against this parasite. Here we report, for the first time, the sequencing of a fragment of a *D. caninum* expressed gene.

Address for correspondence: Élida M.L. Rabelo, Departamento de Parasitologia, ICB, Universidade Federal de Minas Gerais, Avenida Antônio Carlos 6627 CEP 31270-901, Belo Horizonte, MG-Brasil. Voice: 55 31 4992851; fax: 55 31 4992970.
rabelo@icb.ufmg.br

**Ann. N.Y. Acad. Sci. 1026: 195–198 (2004). © 2004 New York Academy of Sciences.
doi: 10.1196/annals.1307.029**

## MATERIALS AND METHODS

Using the Trizol reagent (GIBCO-BRL) mRNA from immature, mature and gravid individual proglottids from adults *D. caninum* were obtained. cDNA was prepared using reverse transcriptase and oligo(dT). A control cDNA without the reverse transcriptase enzyme was done to assure that the products obtained in the differential display[6] were due to the cDNA and not to genomic DNA. One microliter of each cDNA sample was submitted to the PCR reaction using 10 different arbitrary primers. The reaction conditions were: 200 μM dNTP; 2.6 μM primer; 1 unit Taq DNA polymerase enzyme, 1.5 mM $MgCl_2$; 1 × Taq buffer in 25 μL final volume. The PCR conditions were: two cycles at 37°C annealing temperature followed by 30 cycles with the temperature specific to each primer. PCR products were resolved on a 6% polyacrylamide gel. Specific bands were recovered from the gel and eluted in 100 μL of 1× PCR buffer at 95°C for 20 minutes. The bands were then re-amplified using the same conditions for the differential display PCR, except for the low annealing step that was omitted. The amplified bands were cloned in the vector PUC 18 using the Sure Clone kit (Pharmacia). The new plasmids were introduced in *Escherichia coli* DH 5-α competent cells and the plasmids extracted and sequenced using the DYEnamic™ ET dye terminator kit (MegaBACE™, Amersham). The sequence was submitted to homology search with sequences deposited in gene databanks using the Basic Local Alignment Search Tool program for nucleotide (BLAST N) or peptide (BLAST X).[7]

## RESULTS AND DISCUSSION

Ten different arbitrary primers were used in combination with the cDNAs produced. The profiles obtained with most of these primers were quite similar and not very informative (data not shown). FIGURE 1 shows an example of a differential display using the primer RB1.1 (5′ GCAGGTGTGTGAGCATGGGC 3′), where the three cDNAs were compared. The arrows shows two specific bands–one of about 230 bp from the gravid proglottid and the other of 390 bp of the immature proglottid. However, only the band of 230 bp was successfully reamplified, cloned, and sequenced. A sequence of 198 bp was obtained after the primer sequence has been taken out. A partial homology to a NADH dehydrogenase subunit 2 *Ostrinia nubilalis* was found (TABLE 1). Whether this gene is developmental stage–specific or not, re-

**TABLE 1. Homology search results**

| Clone | GenBank Access Number | Fragment Size | Stage | Homology with Accession Number |
|---|---|---|---|---|
| DcGdd2 | CD664187 | 198 | Gravid proglottid | NADH dehydrogenase subunit 2 *Ostrinia nubilalis* gi18314291 |

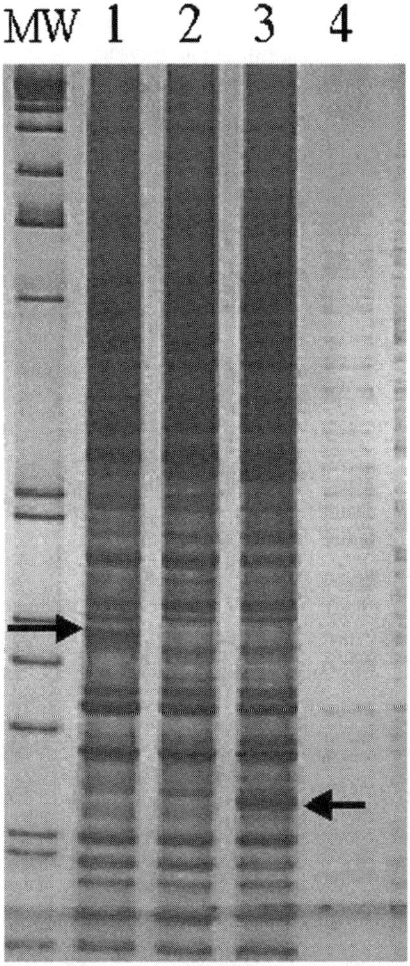


**FIGURE 1.** Differential display among different developmental stages of proglottids of *D. caninum.* The PCR products were resolved on a 6% polyacrylamide gel and silver stained. MW: 1 Kb DNA ladder (GIBCO-BRL). *Lane 1*, immature proglottids; *Lane 2*, mature proglottids; *Lane 3*, gravid proglottids; and *Lane 4*, no cDNA.

mains to be confirmed by others techniques. However due to the homology found, it isn't likely to be a developmental-specific gene. Even so, the first identification of a fragment to an expressed gene in *D. caninum* opens new avenues towards genome knowledge and consequent application of molecular biology tools to the study of this important parasite.

## ACKNOWLEDGMENTS

This work was supported by Coordenadoria de Aperfeiçomento de Pessoal de Ensino Superior (CAPES) and Conselho Nacional de Desenvolvimento Cientifico e Technologico (CNPq).

## REFERENCES

1. Carpio, A. 2002. Neurocysticercosis: an update. Lancet Infect. Dis. **2:** 751–762.
2. Lightowlers, M.W. & C.G. Gauci. 2001. Vaccines against cysticercosis and hydatidosis. Vet. Parasitol. **101:** 337–352.
3. Gauci, C., M. Merli, V. Muller, *et al.* 2002. Molecular cloning of a vaccine antigen against infection with the larval stage of *Echinococcus multilocularis*. Infect. Immun. **70:** 3969–3972.
4. Litvaitis, M.K. & K. Rohde. 1999. A molecular test of platyhelminth phylogeny: inferences from partial 28S rDNA sequences. Invertebr. Biol. **118:** 42–56.
5. Von Nickisch-Rosenegk, M., R. Lucius & B. Loss-Frank. 1999. Contributions to the phylogeny of the Cyclophyllidea (*Cestoda*) inferred from mitochondrial 12S rDNA. J. Mol. Evol. **48:** 586–596.
6. Liang, P. & A.B. Pardee. 1992. Differential display of eukaryotic messenger RNA by means of polymerase chain reaction. Science **257:** 967–971.
7. Altschul, S.F., W. Gish, W. Miller, *et al.* 1990. Basic local alignments search tool. J. Molec. Biol. **215:** 403–410.

# Identification of Specific Male and Female Genes in Adult *Ancylostoma caninum*

RODRIGO R.C. MIRANDA, LIVIO M. COSTA-JÚNIOR, ARTUR K. CAMPOS, HUDSON A. SANTOS, ÉLIDA M. L. RABELO

*Departamento de Parasitologia, Universidade Federal de Minas Gerais, Belo Horizonte, MG-Brasil*

**ABSTRACT: The hookworm *Ancylostoma canium* represents a serious health problem, not only for animals but also for humans. These blood-feeding parasites produce various proteolytic enzymes in order to digest the host hemoglobin. The female worm ingests more blood than does the male. It is not known whether this difference is accompanied by expression of sex-specific proteinases. The identification of new genes related either to the developmental process of maturation of each sex or to the proteinases secreted by these worms could provide researchers with new tools to be used in control programs for this important parasite. The differential-display technique was used to compare the gene expression patterns of adult male and female worms in order to find specific genes that could be used as new targets in the control strategies for this parasite.**

**KEYWORDS: *Ancyclostoma canium*; hookworms; sex-specific genes; EST; excretory-secretory proteins; differential display**

## INTRODUCTION

The hookworm *Ancylostoma caninum* (ERCOLANI, 1859) parasitizes dogs and can develop in the human gut inducing human eosinophilic enteritis. Excretory-secretory (ES) proteins are claimed to be the main cause of eosinophilic enteritis.[1] Various secreted proteins, which are proteases or proteases inhibitors, have been ascribed to *A. caninum*. Among these molecules are cathepsin B proteinases,[2] inhibitors of coagulation factor Xa,[3] and Kunitz-type proteinase inhibitor.[4] It was shown that the AceAP1 an *Ancylostoma ceylanicum* anticoagulant peptide 1 (AceAP1) homologue to the *A. caninum* AcAP5 (*A. caninum* anticoagulant peptide 5) is substantially less potent than the AcAP5.[5] These values correlate to previously reported differences in blood-feeding capabilities between these two species of hookworm,[6] suggesting that factor Xa inhibitory activity is predictive of hookworm blood-feeding capability *in vivo*.[5] Following this same line of reasoning, one would assume that the difference between the blood volume sucked by each sex— females from *A.*

Address for correspondence: Élida M.L. Rabelo, Departamento de Parasitologia, ICB, Universidade Federal de Minas Gerais, Avenida Antônio Carlos 6627 CEP: 31270-901, Belo Horizonte, MG-Brasil. Voice: 55 31 4992851; fax: 55 31 4992970.
rabelo@icb.ufmg.br

**Ann. N.Y. Acad. Sci. 1026: 199–202 (2004). © 2004 New York Academy of Sciences.
doi: 10.1196/annals.1307.030**

*caninum* suck more blood than males do[6]—would also be accompanied by a higher expression of hemoglobinase genes in the female, justifying a search for increased, and/or specific transcripts, which would account for this difference in blood ingestion. Here, we describe results comparing the gene expression pattern between males and females of *A. caninum* aiming at finding specific genes which would be used as new targets to control this parasite.

## MATERIALS AND METHODS

Trizol reagent (GIBCO-BRL) was used to extract mRNA from male and female adults of *A. caninum* parasites. The cDNA were prepared by using reverse transcriptase and oligo dt. A control cDNA without the reverse transcriptase enzyme was prepared in order to assure that the products obtained in the differential display[7] were due to the cDNA and not to genomic DNA. One μL of each cDNA sample was submitted to the PCR reaction using the following arbitrary primers (MAGO2U: 5′ATGGATTAACAAATGACAAG 3′; αPR:5′AAGTGGATATT-TGGAGCGTT 3′; DimpYU: 5′ ATGGCTAGTGTTTTGTTGC 3′; RB1.1: 5′ CAGGTGTGTGAG-CATGGGC 3′; DimpYL: 5′ TTATGAAACACGCTTATGAT 3′). The reaction conditions were: 200 μM dNTP; 2.6 μM primer; 1 unit Taq DNA polymerase enzyme, 1.5 mM MgCl2; 1 × Taq buffer in 25 μL final volume. The PCR conditions were: 2

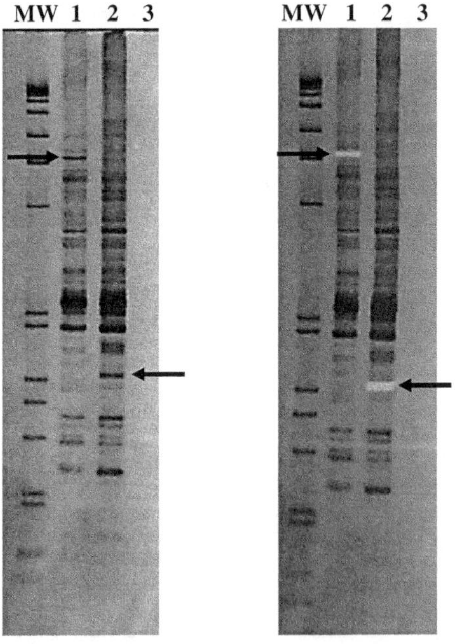


**FIGURE 1.** The PCR products were resolved on a 6% polyacrylamide gel and silver stained. MW: 1Kb DNA ladder (GIBCO-BRL). *Lane 1:* male cDNA; *lane 2:* female cDNA; *lane 3:* no cDNA. *Right panel* shows the gel after specific bands had been recovered.

**TABLE 1. Homology search results**

| Clone | GenBank access number | Fragment size | Sex | Homology with access number |
|---|---|---|---|---|
| Acmdd11 | CD664190 | 145 | male | fattyacid and retinol-binding protein 1 *Ancylostoma caninum* gi 22164324 |
| Acmdd19 | CD664188 | 263 | male | none |
| Acfdd21 | CD664195 | 396 | female | NADH dehydrogenase, *Caenorhabditis elegans* gi 17568379 |
| Acfdd24 | CD664191 | 137 | female | none |
| Acfdd25 | CD664192 | 168 | female | female-specific meiotic progression factor deleted in azoospermia *Caenorhabditis elegans* gi 25146806 |
| Acfdd26 | CD664196 | 113 | female | none |
| Acfdd39 | CD664199 | 154 | female | chromobox protein (CHCB1) *Gallus gallus* gi 3649783 |
| Acfdd40 | CD664200 | 122 | female | none |
| Acfdd48 | CD664197 | 446 | female | none |
| Acfdd51 | CD664198 | 126 | female | microsomal signal peptidase subunit *Caenorhabditis elegans* gi 17555658 |
| Acfdd52 | CD664193 | 196 | female | none |
| Acfdd53 | CD664194 | 200 | female | none |
| Acmdd57 | CD664189 | 151 | male | none |

cycles at 37°C annealing temperature followed by 30 cycles with the temperature specific to each primer. PCR products were resolved on a 6% polyacrylamide gel. Sex-specific bands were recovered from the gel, reamplified and cloned in the vector PUC 18 using the Sure Clone kit (Pharmacia). The new plasmids were introduced in *Escherichia coli* DH 5-α –competent cells, the plasmids extracted and at least three different colonies to each fragment were sequenced using the DYEnamic™ ET dye terminator kit (MegaBACE™- Amersham). The sequences were submitted to homology search with sequences deposited in gene databanks using the Basic Local Alignment Search Tool program for nucleotide (BLAST N) or peptide (BLAST X).[8]

## RESULTS AND DISCUSSION

FIGURE 1 shows an example of a differential display where male and female cDNAs were compared. Various different profiles were obtained when different arbitrary primers were used. Thirteen different bands, ten from female and three from male, were reamplified, cloned and sequenced, producing 13 new *A. caninum* ESTs (TABLE 1). Four sequences presented homology to genes from other organisms and one sequence (clone Acmdd11) presented a partial homology to a gene from *A. caninum*, indicating the existence of a family for this gene. No homology was found to eight sequences. The absence of significant homology to these ESTs suggests that they might be *A. caninum*–specific genes or it also could be due to the fact that the region sequenced represents 5′ or 3′ non-translated regions that are poorly conserved between homologous genes from different species. It is noteworthy the homology found to the clone Acfdd25 which was identified as a female-specific band, and presented partial homology to a *Caenorhabditis elegans* female-specific protein. These data validate the usefulness of this technique to find molecules specific to the female worms. Another important homology found was the one to the clone Acfdd51, also female specific, that presented homology to a microsomal signal peptidase. These peptidases catalyze the removal of peptide signal of secreted proteins. Sex differential behavior has already been detected during a vaccination trial where the number of female parasites located at the colon was much higher than the number of males.[9] This fact indicates that the females present different mechanisms of interaction with the host, which is probably reflected by different molecules that they produce. The knowledge of new molecules synthesized by parasites can also open new avenues to new areas of medicine as is the case with the *Ancylostoma caninum* anticoagulant peptide which was used to block metastasis *in vivo*.[10]

## ACKNOWLEDGMENTS

This work was supported by CAPES and CNPq /Brazil.

## REFERENCES

1. CROESE, J., A. LOUKAS, J. OPDEBEECK, *et al.* 1994. Human enteric infection with canine hookworms: an emerging problem in developed communities. Ann. Intern. Med. **120:** 369–374.
2. HARROP, S.A., N. SAWANGJAROEN, P. PROCIV, *et al.* 1995. Characterization and localization of cathepsin B proteinases expressed by adult *Ancylostoma caninum* hookworms. Mol. Biochem. Parasitol. **71:** 163–171.
3. CAPPELLO M, J.M. HAWDON, B.F. JONES, *et al.* 1996. *Ancylostoma caninum* anticoagulant peptide: cloning by PCR and expression of soluble, active protein in *E. coli.* Mol. Biochem. Parasitol. **80:**113–117.
4. HAWDON, J.M., B. DATU., & M. CROWELL. 2003. Molecular cloning of a novel multidomain Kunitz-type proteinase inhibitor from the hookworm *Ancylostoma caninum.* J. Parasitol. **89:** 402–407.
5. HARRISON, L.M., A. NERLINGER, D.B. RICHARD, *et al.* 2002. Molecular characterization of *Ancylostoma* inhibitors of coagulation factor Xa. J. Biol. Chem. **227:** 6223–6229.
6. WANG, Z.Y., X.Z. WHANG, Y.F. PENG, *et al.* 1983. Blood sucking activities of hookworms: routes and quantities of blood loss in *Ancylostoma caninum, A. duodenale* and *A. ceylanicum* infections. Chin. Med. J. **96:** 281–286.
7. LIANG, P. & A.B. PARDEE. 1992. Differential display of eukaryotic messenger RNA by means of polymerase chain reaction. Science **257:** 967–971.
8. ALTSCHUL, S.F., W. GISH, W. MILLER, *et al.* 1990. Basic local alignments search tool. J. Molec. Biol. **215:** 403–410.
9. HOTEZ, P.J., J. ASHCOM, Z. BIN, *et al.* 2002. Effect of vaccinations with recombinant fusion proteins on *Ancylostoma caninum* habitat selection in the canine intestine. J. Parasitol. **88:** 684–690.
10. DONNELLY, K.M., M.E. BROMBERG, A. MILSTONE, *et al.* 1998. *Ancylostoma caninum* anticoagulant peptide blocks metastasis in vivo and inhibits factor Xa binding to melanoma cells in vitro. Thromb. Haemost. **79:**1041–1047.

# Climatic Conditions and Gastrointestinal Nematode Egg Production

## Observations in Breeding Sheep and Goats

A. TORINA,[a] V. FERRANTELLI,[a] O.A.E. SPARAGANO,[b] S. REALE,[a] F. VITALE,[a] AND S. CARACAPPA[a]

[a]*Istituto Zooprofillattico Sperimentale della Sicilia, Palermo, Italy*

[b]*School of Agriculture, Food and Rural Development, King George VI Building, Newcastle upon Tyne, United Kingdom*

**ABSTRACT: Parasitic egg production was studied in sheep and goats affected by parasitic gastroenteritis. The herds studied were located at different altitudes and in different climatic conditions. Samples were taken every month and the number of eggs per grams of feces was calculated. Observation of preliminary data shows that the maximum peak of egg production was during the winter period, whereas in other countries winter is a period of hypobiosis. This study shows that understanding peak time of infection related to different climatic and environmental conditions will help improve anthelminthic treatments and animal health strategies.**

**KEYWORDS: gastrointestinal nematodes; sheep, goats; EFG; altitude; Sicily**

## INTRODUCTION

Economic losses from gastrointestinal nematode infestation are estimated to be in the millions of dollars every year worldwide. The low production of meat, wool, and milk and the costs of anthelminthic treatments,[1] are the major causes of losses in livestock production in the world[2] and appear to be a major constraint to efficient sheep production.

In sheep and goats, the more common gastrointestinal nematodes belong to Trichostrongylidae, genera *Haemonchus*, *Ostertagia* (*Teladorsagia*), *Trichostrongylus*, *Cooperia*, and *Nematodirus*, to Strongilidae, genera *Oesophagostomum* and *Chabertia*, to Ancylostomatidae genera *Bunostomum* and to Tricuridae.

Survival and development of the larval populations require appropriate environmental conditions that influence the clinical status of livestock and transmission of nematodes and these conditions differ from region to region

Address for correspondence: Dr. Alessandra Torina, Institute of Experimental Zooprophylaxy of Sicily (IZS), Via Marinuzzi n.3, 90129 Palermo, Sicily, Italy. Voice: 39-091-6565360; fax: 39-091-6565361.

torina@pa.izs.it

**Ann. N.Y. Acad. Sci. 1026: 203–209 (2004). © 2004 New York Academy of Sciences.**
**doi: 10.1196/annals.1307.031**

In this study, the number of nematode eggs excreted per gram of feces were measured in three herds of sheep and three herds of goats every month from May 1996 to April 1997.

Farms were chosen in areas with different environmental conditions to compare the influence of different climatic conditions present in Sicily. Goats were chosen for this study because they are more susceptible to gastrointestinal nematode infections than sheep and also have with a higher production of worm eggs and a more severe parasitosis than sheep.[4,5] Anthelminthic treatments are less effective for goats[6] and may contribute to the nematode resistance.[7] These factors show that goats can be responsible for the dissemination of worm eggs from resistant nematodes in mixed populations of small ruminants.[8] Also, because they are parasitized by the same gastrointestinal nematode species as sheep, they have an important role in the infestations of breeding sheep in Sicily.

## MATERIALS AND METHODS

The study was made from May 1996 to April 1997 using three herds of sheep (Sh1, Sh2, and Sh3) and three herds goats (Go1, Go2, and Go3) affected by parasitic gastroenteritis. The sheep and goat herds were located in different areas and there were no contact between herds of sheep.

Sheep herds were positioned at different altitudes and in different climatic conditions:

Sh1 was situated at 1,360 meters in altitude, in the Ganci district, Palermo province, in the Monte Zimmara region. The sampled area (around 60 hectares) is mainly made of natural pastures with an argilo-calcarous soil. The herd consisted of 160 free-breeding sheep, which had anthelminthic treatment in December and delivered lambs in September.

Sh2 was situated at 700 meters of altitude, in the Misita region, district Santo Stefano di Quisquinia, Agrigento province. The sampled area was 200 hectares of natural pastures. The herd consisted of 390 sheep, which were treated with anthelminthics in June and at the end of August with Ivermectine and delivered lambs in September.

Sh3 was situated at 200 meters of altitude, in Batia region, district St. Mauro Castelverde, Palermo province. The sampled area was 70 hectares of natural pastures and some scrub zone. The herd consisted of 100 sheep, which were treated with Ivermectin in May and delivered lambs in September.

Goat herds were all kept in the mountains, so they were exposed to limited differences in altitude. Generally, goats are kept in marginal areas that cannot be used for other animals:

Go1 was situated at 1,000 meters of altitude in the Canalicchio region, district St Mauro Castelverde, Palermo province, on 300 hectares of natural pastures and some scrub zone. The herd consisted of 320 free-breeding goats that received no anthelminthic treatments and delivered lambs in November.

Go2 was situated at 700 meters of altitude in the Vranca region, district St Mauro Castelverde, Palermo provinces, on 60 hectares of natural pastures and some scrub zone. The herd consisted of 200 free-breeding goats that received anthelminthic treatment in May (Fenbendazole) and delivered lambs in September.

Go3 was situated at 600 meters of altitude in the Cerrito Lorito region, district St Mauro Castelverde, Palermo province, on 600 hectares of natural pastures and some scrub zone with an argilous soil (pH around 6.5). The herd consisted of 400 free-breeding goats that received no anthelminthic treatments and delivered lambs in November.

Samples were taken every month from 10% of the animals and the number of eggs per gram of feces (EFG) was calculated using the modified McMaster method:[9] 5 g of feces were treated with an iodine-mercury solution of potassium (specific weight 1,450). Each egg found was equivalent to 30 eggs per gram of feces.[10] Samples were collected directly from the rectum or from the soil, if fresh. In some cases it was not possible to take the feces directly as the rectal ampule was empty.

## RESULTS

We have analyzed a total of 1,535 samples (682 from sheep and 853 from goats). EFG was estimated on a monthly average.

FIGURES 1–3 are related to sheep herds and show climatic conditions (temperature, humidity, and rainfall) associated with EFG counts.

For the three goat herds, the same sensor for the climatic conditions was used and results are therefore presented in a single figure (FIG. 4).

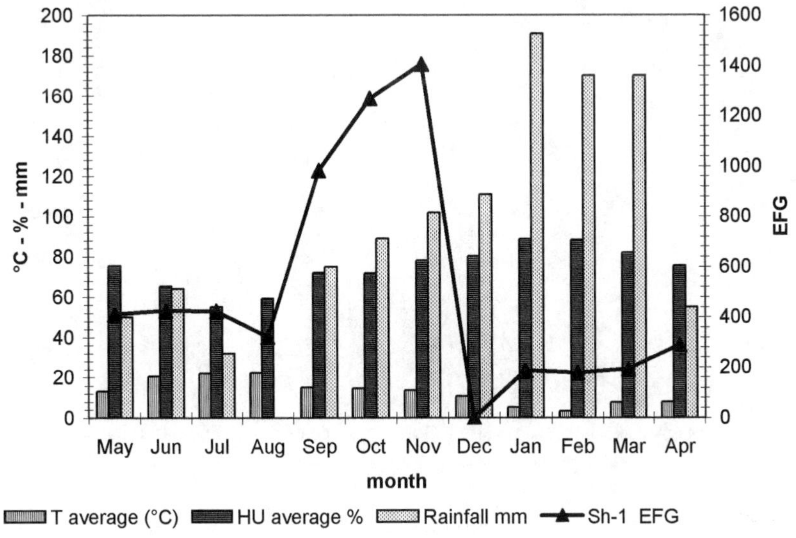


**FIGURE 1.** EFG and climatic conditions for sheep herd Sh1.

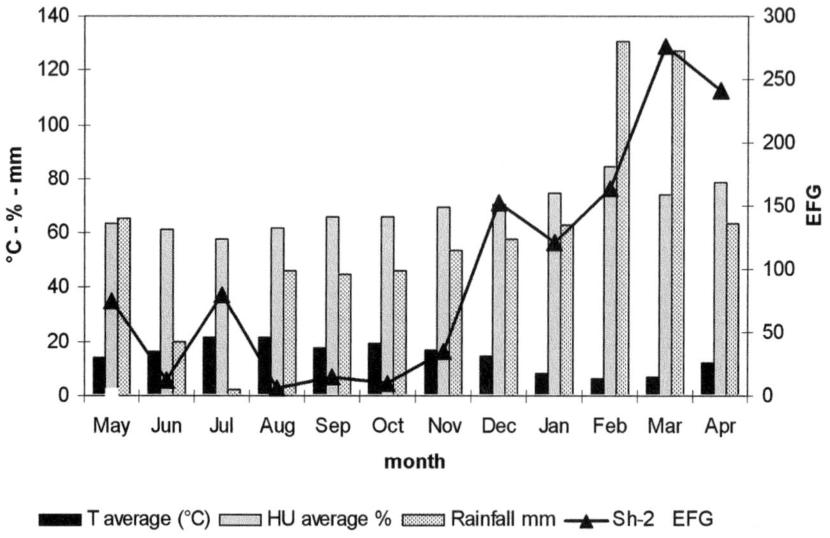


**FIGURE 2.** EFG and climatic conditions for sheep herd Sh2.

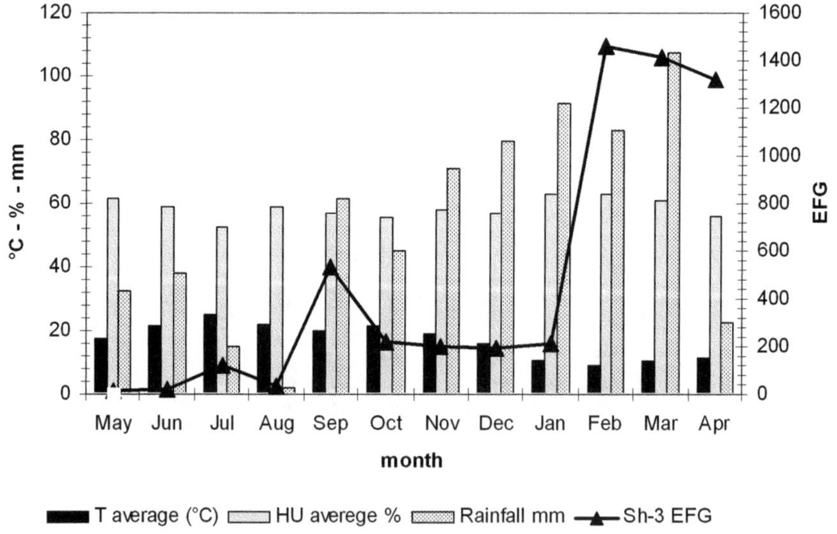


**FIGURE 3.** EFG and climatic conditions for sheep herd Sh3.

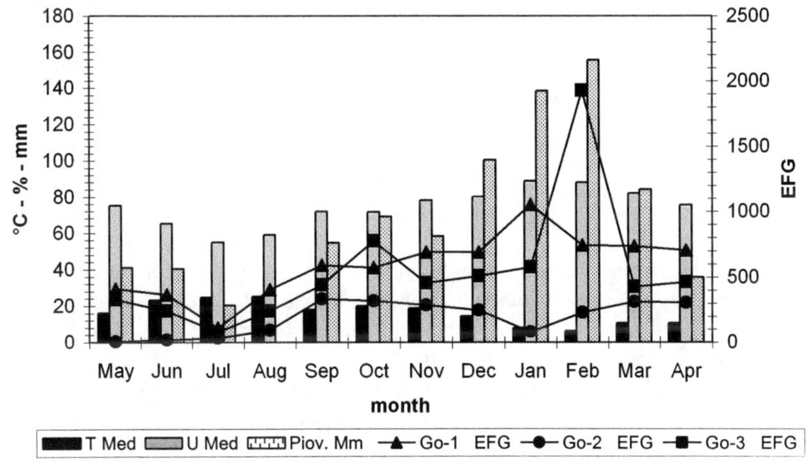


**FIGURE 4.** EFG and climatic conditions for goat herds GO1, GO2, and GO3.

## DISCUSSION AND CONCLUSIONS

The principal aspect of this study is that we considered six herds from the field without any experimental interaction and therefore the study results represent a regional situation.

In Sicily, as the data showed, anthelminthic treatments are given to sheep that are planned to give birth in September, while goats, usually not receiving treatment, are planned for November (except for herd GO2 that was treated in the same manner as the sheep).

Considering well-documented events, such as "spring rise" and "lactation rise," we were expecting to see EFG peaks—one in spring (April-May) and one later around October-November—which should be stopped by treating the animals accordingly with anthelminthic drugs.

Herd Sh1 (FIG. 1) shows an increase of EFG in September, during the period where birth was programmed and the subsequent "lactation rise," with the major peak in November and the lowest in December, after anthelminthic treatment. In the followings months, starting from January, there were a progressive increase from 200 to 400 EFG, but we can considered the treatment efficient when compared to the high level of grazing infestations (about 1,500 EFG in November). This course is very interesting while Sh1 was at 1,360 meters of altitude and, excepting December, there were a high level of EFG (more than 200) during the entire year, demonstrating the persistence of gastrointestinal nematodes at with coldest temperatures.

Egg excretion did not drop during the dry season (August rainfall: 0 mm) showing EFG counts doubling after just one month and going up to 1,500 EFG in November. Thanks to these observations it seems that a treatment in August would better maintain low EFG levels for the rest of the year and help to avoid high grazing infestations.

In FIGURE 2, Sh2, grazing at 700 m, shows lower EFG counts compared to Sh-1 (with a maximum EFG value of only 280 in March) and a different nematode kinetic, obviously due to the two anthelminthic treatment received in June and at the end of August.

The June treatment does not seem to have controlled the nematode population in contradiction with the results observed. The treatment at the end of August is efficacious because it limits egg production during the "lactation rise" but, once again, an EFG increase is observed during the winter.

For the third sheep herd, grazing at 200 m, the average temperature was higher that for the other herds, the relative humidity is always high, and there is no evident effect of salinity. The treatment given in May stopped the nematode proliferation for a few months with a first peak in September (EFG around 600) succeeded by a drop for a few months and a dramatic increase in February. The treatment maintained the proliferation for a few months, but one more treatment in August could have reduced the infection wave observed in winter (with an EFG peak of 1,500).

Herds GO1 and GO3, characteristic of the Sicilian situation, did not receive anthelminthic treatment and showed higher EFG values than GO2 (showing only a maximum peak of 300) whereas GO1 and GO3 EFG peaks reached 1,000 and 2,000, respectively.

Results show that without treatment herds are exposed to a small EFG increase in September-October and a more important one in January-February. The lowest peak for all the three herds is in July, probably reflecting the summer hypobiosis due to high temperature and low relative humidity and rainfall.

In temperate regions with wet and hot climates, like Sicily, it is impossible to organize anthelminthic treatment based on results from other countries, as the winter hypobiosis was not observed in our region (with winter average temperature around 7–9°C).

We even observed the highest EFG peaks during these months. Maximum EFG values were observed for herds grazing below 1,000 m; herd Sh1 grazing at 1,360 m was not exposed to a winter EFG peak either due to the December treatment or lower registered temperatures.[3–5] In our study, EFG increased during months with high rainfall.

The well-known "spring rise" was not evident in our study, whereas the "lactation rise" could explain the EFG increase in October-November for sheep herds and in January-February for the goats.

Therefore, these observations suggest that the study of the period of maximum infestation, taking into account the climate and environmental conditions, is useful to characterize the best period for anthelminthic treatments to have maximum benefit with lowest cost for the farmer and for the ecosystems involved in the biological cycles of parasites.

Considering the above results it seems that two treatments for sheep and goats (in August and in December) should reduce the worm burdens year round in Sicily. Treatments in May have only a limited effect on gastrointestinal infestations.

## ACKNOWLEDGMENTS

We would like to thank Mrs. Angela Alongi, Mr. Gaspare Lo Bue, and Mr. Salvatore Scimeca for their valuable technical support.


## REFERENCES


1. BARGER, I.A. 1982. Helminth parasites and animal production. *In* Biology and Control of Endoparasites. L.E.A. Symons *et al.*, Eds.: 133–155. Academic Press. Sydney, Australia.
2. AMERICAN ASSOCIATION OF VETERINARY PARASITOLOGISTS. 1983. Research needs and priorities for ruminant internal parasites in the United States. Am. J. Vet. Res. **44:** 1836–1847.
3. ZAJAC, A.M. & G.A MOORE. 1993. Trattamento e controllo dei nematodi gastroenterici degli ovini. The compendium on continuing education for the Practicing Veterinarian **15:** 999–1010.
4. LE JAMBRE, L.F. & V.M. ROYAL. 1976. A comparison of worm burdens in grazing Merino sheep and Angora goats. Aust. Vet. J. **52:** 181–183.
5. POMROY, W.E., M.G. LAMBERT & K. BETTERIDGE. 1986. Comparison of faecal strongilyte egg counts of goats and sheep on the same pasture. N. Z. Vet. J. **34:** 36–37.
6. SANGSTER, N.C., J.M. RICHARD, D.R. HENNESSY, *et al.* 1991. Disposition of oxfendazole in goats and efficacy compared with sheep. Res. Vet. Sci. **51:** 258–263.
7. CHARLES, T.P., J. POMPEU & D.B. MIRANDA. 1989. Efficacy of three broad-spectrum anthelmintics against gastrointestinal nematode infections of goats. Vet. Par. **34:** 71–75.
8. VARADY, M., J. PRASLICKA & J. CORBA. 1994. Treatment of multiple resistant field strain of *Ostertagia* spp. in Cashmere and Angora goats. Int. J. Par. **24:** 335–340.
9. RAYNAUD, J.P. 1970. Etude de l'efficacité d'une technique de coproscopie quantitative pour la diagnostic de routine et le contrôle des infestations parasitaires des bovins, ovins, équins, et porcins. Ann. Par. Hum. Comp. **45:** 321.
10. RAYNAUD, J.P., J.C. LEROY, N. VIRAT & J.A. NICOLAS. 1979. Une technique de coproscopie quantitative polyvalente par dilution et sédimentation en eau, flottaison en solution dense (d.s.f.) et numération en lame de Mc Master. Revue Méd. Vét. **130:** 377.

# Characterization of Excretory/Secretory Antigen from *Toxocara vitulorum* Larvae

WILMA A. STARKE-BUZETTI AND FABIANO P. FERREIRA

*Departamento de Biologia e Zootecnia, Universidade Estadual Paulista-Campus de Ilha Solteira, Ilha Solteira15385-000, SP, Brazil*

**ABSTRACT: *Toxocara vitulorum* is a nematode parasite of the small intestine of cattle and water buffalo, particularly buffalo calves between one and three months of age, causing high morbidity and mortality. The purpose of this research was to characterize the excretory/secretory (ES) antigens of *T. vitulorum* larvae by SDS-polyacrylamide gel electrophoresis (PAGE) and Western blot (WB), using immune sera and colostrum of buffalo naturally infected by *T. vitulorum*. The parasitological status of the buffalo calves was also evaluated using sequential fecal examinations. The results showed that the ES antigen revealed eight (190, 150, 110, 90, 64, 56, 48, and 19 kDa) protein bands by SDS-PAGE. The majority of these bands were recognized in the sera and colostrum of infected buffalo with *T. vitulorum* when analyzed by WB. However, particularly fractions of high molecular weight (190, 150, 110, and 90 kDa) were represented in more prominent bands and persisted in the groups of buffalo calves at the peak of egg output, as well as during the period of rejection of *T. vitulorum* by the feces of the calves. During the period of post-rejection of the worms (between the day 118 and 210 of age) the serum antibodies did not react with any protein bands. On the other hand, sera from buffalo calves at one day of age (after suckling the colostrum and at the beginning of infection) reacted with the same bands detected in the serum and colostrum of the buffalo cows.**

**KEYWORDS: *Toxocara vitulorum*; water buffalo; ES antigen; immunity**

## INTRODUCTION

*Toxocara vitulorum* is a parasite of the small intestine of ruminants, particularly buffalo calves one to three months of age. This parasite is acquired by calves when they suckle colostrum/milk contaminated with infective larvae from infected cows.[1–4]

Antibodies against larval excretory/secretory antigen (ES)[5] and larval soluble extract (Ex)[6] of *T. vitulorum* have been detected in sera of buffalo cows and calves naturally infected with *T. vitulorum*, indicating that *T. vitulorum* infection can stimulate the immune system of the buffalo. Similarly, the highest levels of anti-ES antibodies of *T. vitulorum* in buffalo cow sera were detected by enzyme-linked immunosorbent assay (ELISA)[7] during the perinatal period and were maintained at high levels

Address for correspondence: Wilma A. Starke-Buzetti, Departamento de Biologia e Zootecnia, UNESP-Campus de Ilha Solteira, 56 Av. Brasil, Ilha Solteira 15385-000, SP, Brazil. Voice: 55-18-3743-1152; fax: 55-18-3743-1186.
starke@bio.feis.unesp.br

**Ann. N.Y. Acad. Sci. 1026: 210–218 (2004). © 2004 New York Academy of Sciences.
doi: 10.1196/annals.1307.032**

through 300 days after parturition. Colostrum antibody concentration was highest on the first day post-parturition, but decreased sharply during the first 15 days. Antibodies passively acquired by the calves were in the highest concentrations 24 hours after birth and remained at high levels until 45 days—coincidentally the peak of *T. vitulorum* infection—but declined during the period of rejection of the worms,[7] suggesting a role of the antibody-mediated immunity in this process. On the other hand, the immunization of mice with ES antigen of *T. vitulorum* has induced protection superior to 92%[8] and up to 58%[9] against larval migration in their tissues, suggesting a potential protective antigen.

Using SDS-PAGE and Western blot (WB), the ES antigen was detected by antibodies from sera and colostrum of buffalo cows and sera of buffalo calves in different periods of *T. vitulorum* infection (during the beginning and the peak of the infection, during and after the rejection of the worm). In addition, the parasitic status of the buffalo calves naturally infected with *T. vitulorum* was also evaluated.

## MATERIALS AND METHODS

### *Buffalo Housing*

Water buffalo naturally infected with *T. vitulorum* were kept for about 12 months on a 12-hectare pasture of *Brachiaria decumbens* grass with a pond as the source of water. The cows were not milked and the calves were grazed together with the dams in this area.

### *Fecal, Serum, and Colostrum/Milk Samples from Buffalo Calves and Cows*

Rectal fecal samples were collected from the buffalo calves ($N = 10$) according to the following schedule: weekly (from birth to 90 days) and fortnightly (from day 91 until attaining two consecutive weeks with an absence of *T. vitulorum* eggs). Fecal examinations were performed according to Whitlock[10] and results expressed as eggs per gram of feces (EPG). Parasitic status of the buffalo calves for *T. vitulorum* infection was represented by a curve of four periods: Period 1, beginning of the infection, between 01 and 39 days after birth; Period 2, peak of egg out put, between 40 and 47 days; Period 3, rejection of the parasite, between 48 and 117 days; and Period 4, post-rejection or after a one month absence of eggs in the feces (FIG. 1).

Buffalo calf sera ($N = 5$) were previously and individually assayed by ELISA and sampled and pooled as follows: (1) at one day of age before suckling the colostrum (negative reference serum); (2) at one day after suckling the colostrum; (3) at beginning of *T. vitulorum* infection (Period 1); (4) at the peak of infection (Period 2); (5) at parasite-rejection period (Period 3); and (6) after parasite rejection period (Period 4). Colostrum and serum samples of buffalo cows were collected one day after parturition and pooled ($N = 10$). The serum of buffalo cow was considered Positive Reference Serum.

The samples of colostrum were centrifuged at 4°C in a refrigerated centrifuge at 460*g* for 15 minutes. After removal of solidified fat, the samples were left in an incubator at 37° C for one hour for casein precipitation with 1% rennin. Then the colostrum/milk serum was separated by centrifugation for 15 minutes at $460 \times g$ at 4°C.

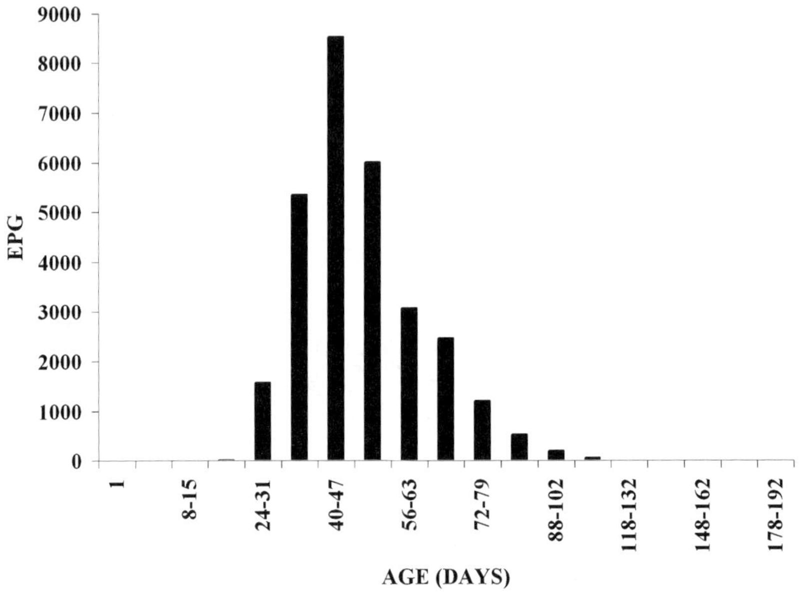


**FIGURE 1.** EPG counts of buffalo calves ($N = 10$).

Serum and colostrum/milk samples were separated, placed into aliquots and stored at –70°C.

### T. vitulorum *ES Antigen Preparation*

*T. vitulorum* adults were recovered by expulsion of this parasite through the feces of naturally infected water buffalo calves by administration of 100 mg/kg of piperazine. Mature females were dissected and the uteri and eggs removed. The eggs were incubated in phosphate-buffered saline (PBS, 0.1 M, 7.5 pH) with several drops of commercial sodium hypochlorite solution (1% available chlorine) in Petri dishes for 20–45 days at room temperature. The dishes with the egg suspension were gently stirred for daily aeration while the development of eggs was observed daily with an optical microscopic until the infective third-stage larvae developed. After that, the egg suspension was transferred to tubes (15 mL) and washed in distilled water by centrifugation. Sediment from eggs was collected, then combined with an equal volume of sodium hypochlorite solution (14% available chlorine), and incubated for 20 minutes at room temperature until the eggs were completely decorticated. PBS was added to this solution to increase the volume to 15 mL after which the mixture was centrifuged 10 times at $460 \times g$ for 2 minutes or until chlorine odor could not be detected. The decorticated eggs were suspended in PBS and placed in a water bath at 37°C for one hour while air was bubbled with a Pasteur pipette through the suspension until the $L_3$ were hatched (about 5 minutes). The larval suspension was recovered by centrifugation and suspended in a RPMI-1640 culture medium containing antibiotics/antimycotic in glass tissue-culture flats in a 5% $CO_2$ incubator according to Rajapakse *et al.*[11] Every week the culture medium without any larvae was removed and centrifuged at 460 × *g* for 5 minutes. The supernatant was filtered through a membrane with a 0.2-μm pore size to which protease inhibitor was added (Protease inhibitor cocktail, Sigma, P 2714 containing: 4-92-aminoethyl bensene-sulfonyl fluoride, AEBSF; *trans*-epoxysccinyl-L-leucyl-amido guanidino butane, E-64; sodium EDTA; bestatin; leupeptin; and aprotinin). This medium was then dialyzed in 25,000 Spectra/Por DispoDialysers, (Spectrum®) for 24 h at 4°C in 2 liters of PBS stirred constantly at 10 rpm. The dialyzed material was filtered through a membrane with a 0.2-μm pore size and then dehydrated in a vacuum centrifuge and stored at –70°C.

Protein concentration of each antigen was measured using a Protein Assay Kit (Sigma®,P-5656) using Lowry's reagent. The concentration of ES was 500 μg/mL.

### *Polyacrylamide Gel Electrophoresis*

One-dimensional polyacrylamide gel electrophoresis (PAGE) was carried out in 12% gels with the acrylamide/bis ratio of 36.5:1 in the presence of 10% sodium dodecyl sulfate (SDS) supplemented with Temed (Sigma, T-9281) and ammonium persulfate in TRIS-HCl buffer pH 8.8, according to Laemmli.[12] Molecular weight standard mixtures (M.W. 15,000–150,000, Sigma M-0671) were used for calibrating the gel. The antigen diluted in a Tris (pH 6.8) sample buffer (0.1 M Tris-HCl, 2% SDS, 10% glycerol, 0.2 M 2-mercaptoethanol, and 0.1% bromophenol blue) was loaded in the gel with 20 μg/lane. The electrophoresis was monitored using 0.1% bromophenol blue and the current was set at 30 mA. The protein fractions were visualized by staining with 0.1% Coomassie Brilliant Blue R 250 (Sigma, B-0149).

The relative molecular weights were calculated using prestained protein molecular weight standard according to the relative electrophoretical mobility (RM), which is the distance of the protein migration divided by the distance of bromophenol blue migration. The RM values (ordinate) were related to known molecular weights of the standard proteins (abscissa) in a semilogarithmic graph giving a base for interpolating the data of the ES-antigen proteins (FIG. 2).

### *Immunoblotting Assays*

Gels with *T. vitulorum* ES antigen were electrophoretically transferred to nitrocellulose sheets (0.22 μm) for immunoblotting according to the procedure described by Towbin and colleagues.[13] The transfer was performed in a Mini Trans-Blot Electrophoretic Transfer Cell (Bio-Rad apparatus) for 12 hours in a constant current of 35 V in a transfer buffer (Tris-Glycine-Methanol). The nitrocellulose papers were blocked in a blocking solution with 5% non-fat dried milk in TBS-Tween (0.01 M Tris, 0.15 M NaCl, 0.05% Tween-20) and incubated for 90 minutes with primary antibodies (pool of serum of buffalo cows and calves) diluted 1:50 in the blocking solution and 5% normal rabbit serum in a rotating homogenizer. After that, the nitrocellulose was washed three times (15 min each) in TBS-Tween and milk solution. Specifically bound antibodies in all filters were detected with anti–bovine alkaline phosphatase conjugate (Sigma, A-7914) diluted 1:30,000 in the blocking solution for 90 minutes. After rinsing three times in TBS-Tween and milk solution,

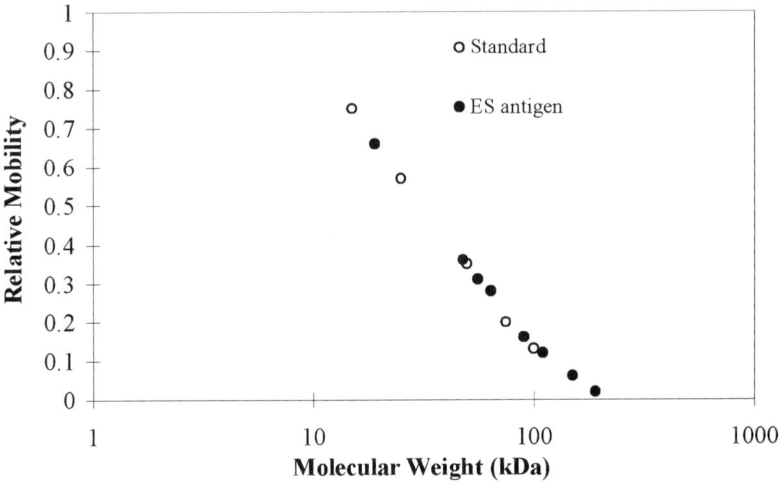


**FIGURE 2.** Relative electrophoretical mobility (RM) of *T. vitulorum* ES antigen in relation to the standard molecular weight (kDa) (Sigma, M-0671). Data obtained from the minigel of 12% SDS-PAGE stained with Coomassie Blue.

the blots were incubated at room temperature for about 10 minutes in enzymatic substrate [5-bromo-4-chloro-3-indolyl phosphate/nitroblue tetrazolium (BCIP/NBT), Calbiochem, 203790] until color developed. All incubations were done at room temperature (about 28°C) and under rotating homogenization.

## RESULTS

The first detection of *T. vitulorum* eggs in buffalo-calf feces occurred on day 16 and the peak maximum EPG counts was days 40–47. Immediately after the peak, the EPG counts started to decline until the day 117, then eggs were no longer seen in the feces of the calves (FIG. 1).

As shown in FIGURES 2 and 3, the SDS-PAGE pattern of larval *T. vitulorum* ES antigen indicated the presence of polypeptide bands of 190, 150, 110, 90, 64, 56, 48, and 19 kDa. The most prominent bands were at approximately 190 to 110 kDa. By WB, using buffalo cow sera and colostrum and buffalo calf sera collected from different periods of *T. vitulorum* natural infection, it was possible to identify all polypeptide bands ranging from 190 to 19 kDa. The most prominent antigenic bands were observed between 190 and 110 kDa. The antibodies present in serum and colostrum of buffalo cows at the parturition day and in the serum of calves one day after suckling the colostrum reacted with all antigenic bands revealed by electrophoresis (TABLE 1 and FIG. 4). However, at the peak of egg output (Period 2), between days 40 and 47, and during the period of rejection of the parasite (Period 3), between days 48 and 117, the antibodies present in the serum of the calves detected only the fol-

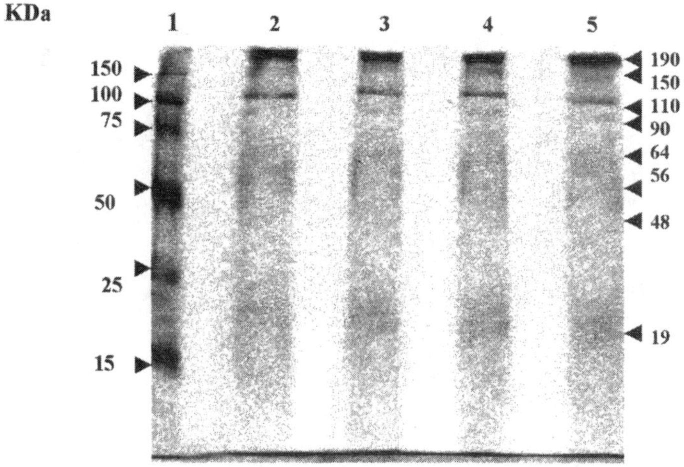


**FIGURE 3.** SDS-PAGE pattern of larval *T. vitulorum* ES antigen revealing eight bands: 19, 48, 56, 64, 90, 110, 150, and 190 kDa stained by Coomassie Blue. Column 1 = molecular size standards; columns 2–5 = ES run in quadruplicate.

**TABLE 1. Characterization of *T. vitulorum* ES antigen**

| | Molecular weight (kDa) of *T. vitulorum* ES antigen bands detected by antibodies present in the serum and colostrum of buffalo | | | | | | | |
|---|---|---|---|---|---|---|---|---|
| | | | | Sera of buffalo calves naturally infected with *T. vitulorum* | | | | |
| SDS-PAGE (kDa) | Positive Serum[a] | Negative Serum[b] | Colostrum[c] | One day of age[d] | Period 1[e] | Period 2[f] | Period 3[g] | Period 4[h] |
| 190 | 190 | – | 190 | 190 | 190 | 190 | | – |
| 150 | 150 | – | 150 | 150 | 150 | 150 | 150 | – |
| 110 | 110 | – | 110 | 110 | 110 | 110 | 110 | – |
| 90 | 90 | – | 90 | 90 | 90 | 90 | 90 | – |
| 64 | 64 | – | 64 | 64 | 64 | – | – | – |
| 56 | 56 | – | 56 | 56 | 56 | 56 | – | – |
| 48 | 48 | – | 48 | 48 | 48 | – | – | – |
| 19 | 19 | – | 19 | 19 | 19 | 19 | 19 | – |

Symbols: –, Without band (negative); [a]Buffalo cow serum on the parturition day; [b]Buffalo-calf serum before suckling the colostrum; [c]Colostrum on the parturition day; [d]Buffalo-calf serum one day after suckling the colostrum; [e]Beginning of infection; [f]Peak of infection; [g]Rejection of the parasite; [h]Post-rejection period (7 months of age).

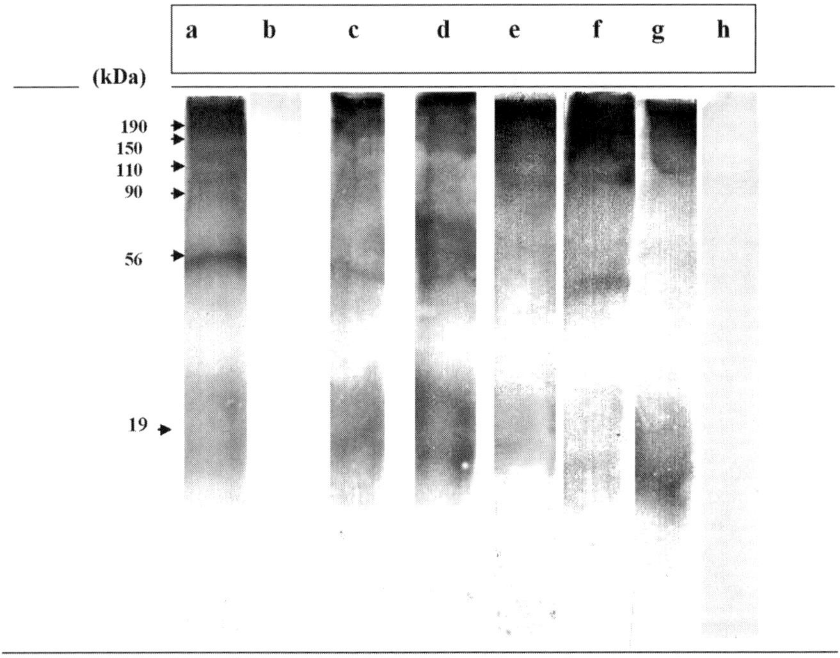


**FIGURE 4.** Characterization of *T. vitulorum* ES antigen by WB with pool of buffalo cow sera ($N = 10$) and calf sera ($N = 5$) following SDS-PAGE electrophoresis: positive reference serum (**a**); negative reference serum (**b**); colostrum (**c**); buffalo calves at one day of age after suckling the colostrum (**d**); buffalo calves at beginning of infection (**e**); buffalo calves at the peak of infection (**f**); buffalo calves during the period of rejection of the worm (**g**), and buffalo calves during the period post rejection of the worm (**h**).

lowing antigenic bands: 190, 150, 110, 90, 56, and 19 kDa (during the peak) and 190, 150, 110, 90, and 19 kDa (during the rejection). Moreover, during the post-rejection (Period 3), from 118 to 210 days post-birth, the WB was negative (TABLE 1 and FIG. 4).

## DISCUSSION

The first detection of *T. vitulorum* eggs in buffalo-calf feces occurring on day 16 confirmed the very short prepatent period also reported by other authors.[14,15] The precocity of this infection was due particularly to the infection acquired by calves through the ingestion of infective larvae in the colostrum/milk from buffalo cows during the first 10 days after birth.[2–4]

Serological and colostral antibodies anti–*T. vitulorum* from buffalo cows at one day of parturition as well as from buffalo calves at one day of age after suckling the colostrum and at the beginning of infection (Period 1) reacted with all eight bands revealed by electrophoresis (190, 150, 110, 90, 64, 56, 48, and 19 kDa), suggesting that IgG antibodies from serum of the cows were transferred to calves through the colostrum and then passed to the calf sera within 24 hours after birth. In fact, the calves that did not suckle the colostrum had no detectable IgG antibodies in their sera. Similarly, by ELISA, high levels of antibodies against *T. vitulorum* ES antigen were reported in the serum of 100% buffalo cows and calves that had suckled the colostrum on the first day of calving.[7] In the colostrum, on the other hand, the antibody concentration against this antigen was the highest on the day of parturition, but declined rapidly after the seventh day to reach a very low concentration on day 15.[7] Similar results for development of the antibody-mediated immune response in buffalo cows and calves against *T. vitulorum* infection was reported by other authors.[5,6]

At the peak of calf egg output (Period 2) and during the period of parasite rejection (Period 3), the antibodies reacted particularly with antigenic fractions having higher molecular weights. Antibodies present during the peak and rejection periods in the calf serum might be passively acquired by colostrum. By ELISA, the concentration of serological antibodies from buffalo calves decreases after the peak of *T. vitulorum* egg output, remains in very low concentration during the period of rejection of the worms, starts to increase slightly after 150 days of calf age, but remains below the cut-off point after the rejection of the worms.[7] The WB in the present work was negative for serum collected from calves during the post-rejection period (Period 3). Even though the calves had no adult worms present in the intestines, some migratory larvae could be present or migrating from the intestinal mucosa to other tissues and could stimulate the immune system. However, our results of WB in addition to very low levels of antibodies by ELISA[7] during the post-rejection period, indicate a very weak immunological stimulation during this period.

Based on the results that the buffalo cow antibodies (from serum and colostrum) recognized the ES antigenic bands (eight protein bands) of *T. vitulorum* larva; that all these protein bands were also detected in the sera of buffalo calves at one day after suckling the colostrum and at the beginning of the infection; and that only four bands of higher molecular weight (190 to 90 kDa) and one of lower molecular weight (19 kDa) remained present during the period of rejection, and no band was detected after the rejection of the worms, it is possible to conclude that the antibodies acquired by calves through the cow colostrum, particularly those against antigens with higher molecular weight, may have a role in the process of rejection of *T. vitulorum* by the calves, but may not participate in the inhibition of larval migration to their tissues.

## REFERENCES


1. CHAUHAN, P.P.S., B.B BHATIA & B.P. PANDE. 1974. Incidence of gastro-intestinal nematodes in buffalo and cow at State livestock farms in Uttar Pradesh. Indian J. Anim. Sci. **43:** 216–219.
2. MIA, S., M.L. DEWAN, M. UDDIN & M.V.A. CHOWDHURY. 1975. The route of infection of buffalo calves by *Toxocara* (*Neoascaris*) *vitulorum.* Trop. Anim. Health Prod. **77:** 153–156.
3. ROBERTS, J.A., S.T. FERNANDO & S. SIVANATHAN. 1990. *Toxocara vitulorum* in the milk of buffalo (*Bubalus bubalis*) cows. Res. Vet. Sci. **49:** 289–291.

4. STARKE, W.A., R.Z. MACHADO & M.C. ZOCOLLER. 1992. Transmammary passage of gastrointestinal nematode larvae to buffalo calves. II. *Toxocara vitulorum* larvae. Arq. Bras.Med. Vet. Zoot. **44:** 97–103.
5. RAJAPAKSE, R.P.V.J., S. LLOYD & S.T. FERNANDO. 1994. *Toxocara vitulorum*: maternal transfer of antibodies from buffalo cows *(Bubalus bubalis)* to calves and levels of infection with *T. vitulorum* in the calves. Res. Vet. Sci. **57:** 81–87.
6. STARKE-BUZETTI, W.A., R.Z. MACHADO & M.C. ZOCOLLER-SENO. 2001. An enzyme-linked immunosorbent assay (ELISA) for detection of antibodies against *Toxocara vitulorum* in water buffaloes. Vet. Parasitol. **97:** 55–64.
7. SOUZA, E.M. 2001. Immunological responses of buffaloes (*Bubalus bubalis*) naturally infected with *Toxocara vitulorum*. Master's Thesis. Faculdade de Engenharia de Ilha Solteira, Universidade Estadual Paulista. Ilha Solteira.
8. AMERASINGHE, P.H., R.P. RAJAPAKSE, S. LLOYD & S.T. FERNANDO. 1992. Antigen-induced protection against infection with *Toxocara vitulorum* larvae in mice. Parasitol. Res. **78:** 643–647.
9. PAULA, S.H.S. 2003. Mice immunization against *Toxocara vitulorum*. Master's Thesis. Faculdade de Engenharia de Ilha Solteira, Universidade Estadual Paulista, Ilha Solteira.
10. WHITLOCK, H.V. 1948. Some modifications of the McMaster helminth egg-counting technique and apparatus. J. Count. Sci. Ind. Res. Australian **21:** 177–180.
11. RAJAPAKSE, R.P.V.J., V.W. VASANTHATHILAKE, S. LLOYD & S.T. FERNANDO. 1992. Collection of eggs and hatching and culturing second-stage larvae of *Toxocara vitulorum in vitro*. J. Parasitol. **78:** 1090–1092.
12. LAEMMLI, U.K. 1970. Cleavage of structural proteins during the assembly of the head of Bacteriophage T4. Nature **227:** 680–685.
13. TOWBIN, H., T. STAEHELIN & J. GORDON. 1979. Electrophoretic transfer of proteins from polyacrylamide gels to nitrocellulose sheets: procedure and some applications. Proc. Natl. Acad. Sci. USA **76:** 4350–4354.
14. STARKE, W.A., R.Z. MACHADO, M.Z. HONER & M.C. ZOCOLLER. 1983. Natural course of gastrointestinal helminthic infections in buffaloes in Andradina County (SP). Arq. Bras. Med. Vet. Zoot.. **35:** 651–664.
15. ROBERTS, J.A. 1990. The egg production of *Toxocara vitulorum* in Asian buffalo (*Bubalus bubalis*). Vet. Parasitol. **37:** 113–120.

# *Stomoxys calcitrans* Parasitism Associated with Cattle Diseases in Espírito Santo do Pinhal, São Paulo, Brazil

AVELINO J. BITTENCOURT AND BRUNO G. DE CASTRO

*Universidade Federal Rural do Rio de Janeiro, Instituto de Veterinária, Departamento de Medicina e Cirurgia Veterinária, Seropédica, RJ-Brasil 23890-000*

**ABSTRACT: The stable fly has been of great significance to livestock production in the county of Espirito Santo do Pinhal; it has a painful bite, sucks blood, and carries many diseases. The aim of this study was to establish a relationship between the parasitism of *Stomoxys calcitrans*, manure management, cattle diseases, and technical support. According to the farmers the stable fly reaches its highest level in the rainy season, the same period in which diseases were detected. Most of the farmers said that they did not receive technical assistance. The association of inappropriate manure management, verified in this survey, with the low frequency of technical visits, resulted in a low level of technology utilization. Better technological assistance could moderate the stable fly infestation and help manage serious cattle diseases.**

**KEYWORDS: stable fly; *Stomoxys calcitrans*; parasitism; cattle diseases; Brazil**

## INTRODUCTION

The stable fly has been incriminated as a cause of great economic losses.[1] In addition to blood feeding, this fly transmits several pathogenic agents.[2] The results presented are part of a larger study that evaluated the infestations and the parasitism of this fly. This study used observations and a survey of farmers. The objective of this study was to verify the farmers' perception about the role of *S. calcitrans* in transmitting cattle diseases, about manure management, and veterinary assistance.

## MATERIALS AND METHODS

The municipality of Espírito Santo do Pinhal, São Paulo State, Brazil, is located in the Mesoregion of "Mantiqueira Paulista." The rainy season starts in September/October, reaching the highest precipitation levels in January/February, and decreasing levels between March and April. For this study, 52 dairy farms, corresponding to 23.74% of the farms of the studied area, were visited. Data were analyzed using the

Address for correspondence: Avelino J. Bittencourt, UFRRJ, IV, DMCV Rod. BR 465, Km 7, Seropédica, RJ- Brasil / 23890-000.
bittenc@ufrrj.br

**Ann. N.Y. Acad. Sci. 1026: 219–221 (2004). © 2004 New York Academy of Sciences.
doi: 10.1196/annals.1307.033**

IMPS 3.0 program of the US Bureau of the Census and analyzed the frequency of farmers' answers to questions about the presence of diseases and manure management (according to farmers' perception), as they related to the fly population increase and veterinary assistance.

## RESULTS AND DISCUSSION

All 52 interviewed farmers knew about the stable fly. The majority of them (61.6%) answered that the month of highest infestations is January, 26.9% of them answered December, 3.8% October, and 5.8% November. One farmer did not know the answer. Peaks of populations, with more than 100 flies per animal, had been observed in December 1993 and December 1994.[3]

Regarding frequency of foot diseases, eight farmers (15.4%) believed that the incidence is higher during the rainy months and four (7.7%) believed that the highest incidence occurs during the dry season. However, 40 farmers (76.9%) did not know or did not answer this question, revealing the farmers limited perception of the relationship between foot lesions and increases in the stable fly population. The stable fly may play a role in increasing foot lesions,[4] as severely infested animals usually search for wet or muddy places to minimize the effect of bites. These flies tend to concentrate on the lower parts of the limbs (due to the high number of superficial vessels in these regions) to facilitate hematophagism.[4]

As far as diarrhea is concerned, 12 farmers (24.1%) stated that it occurs during the rainy season, one farmer (1.9%) stated that it occurs in both rainy and the dry seasons, and 39 (74%) farmers did not know or did not answer. Despite the presence of enterobacteria in this fly,[5] the majority of the farmers were not aware of the period of the year in which its frequency is higher.

Regarding tick-borne diseases, seven interviewed (13.5%) stated that the number of cases increases during the rainy season; three (5.8%) stated that the number of cases increases during the dry season, 40 (76.9%) did not know this information, and two (3.8%) did not answer. In this topic, the objective was to ask questions about the frequency of anaplasmosis in the herd, but the majority of the farmers did not know how to differentiate between anaplasmosis and babesiosis. Among those interviewed, the majority stated that the period of higher incidence of anaplasmosis is the rainy season, which coincides with increase in populations of stable fly, a potential vector of anaplasmosis.[6]

Answers related to cattle diseases show a lack awareness about the importance of vectors and the diseases that may affect the studied herd. Manure management practices are a case in point. For manure management in the farm, 39 farmers (75%) of heap manure close to the pen, seven (13.6%) put it into manure tanks, two (3.8%) use it to fertilize forage crops, two use it to fertilize coffee plantations, and two did not answer. It is well known that manure heaps close to cattle pens make it easier for flies to find animals for feeding, since part of their biological cycle occurs on organic matter,[2] but the surveyed farmers showed neither and awareness nor any particular interest in correct manure-management practices.

As far as veterinary assistance is concerned, 35 farmers (67.3%) received visits occasionally, 12 (23.1%) farmers receive permanent assistance, and five (9.6%) do not have any kind of assistance. The farmers reported difficulties in using technologies owing to their low income and to the reduced number of technicians available through governmental services.

In the present study it was noted that there is a the need for better interaction between these technicians and farmers to prevent and control the stable fly, other parasites, and the diseases transmitted by parasites. The epidemiologic studies of the farmers' perceptions about cattle diseases and the related losses in productivity have a fundamental importance for determining the nature of sanitary education provided by the government as well as actions that affect animal health, prophylaxis and disease-control in cattle.[7]

## REFERENCES

1. STEELMAN, C.D. 1976. Effects of external and internal arthropod parasites on domestic livestock production. Ann. Rev. Ent. **21:** 155–178.
2. GUIMARÃES, J.H. 1984. Mosca dos estábulos—Uma importante praga do gado. Agroq. Ciba-Geigy. **23:** 10–14.
3. BITTENCOURT, A.J. & G.E. MOYA-BORJA. 2000. Flutuação sazonal de *Stomoxys calcitrans* em bovinos e eqüinos no município de Espírito Santo do Pinhal, SP/Brasil. Rev. Univ. Rural–Série Ciências da Vida. **22:** 101–106.
4. TODD, D.H. 1964.The biting fly (*S. calcitrans L*) in dairy herds in New Zealand. New Zeal. J. Agric. Res. **7:** 60–79.
5. CASTRO, B.G., S.D. PIRES, B.M. ALMEIDA, *et al.* 2001. Avaliação da capacidade de *Stomoxys calcitrans* (Linnaeus, 1758) em carrear bactérias causadoras de mastite bovina. Anais da XI Jornada de Iniciação científica da UFRRJ—Trabalhos Completos. **11:** 161–164.
6. POTGIETER, F.T., B. SUTHERLAND & H.C. BIGGS. 1981. Attempts to transmit *Anaplasma marginale* with *Hippobosca rufipes* and *Stomoxys calcitrans*. Onderst. J. Vet. Res. **48:** 119–122.
7. ASTUDILLO, V.M. 1979. Encuestas por muestreo para estudios epidemiologicos en poblaciones animales. Serie de manuales didacticos, Organizacion Mundial de la Salud, Centro Panamericano de Fiebre Aftosa, n. 12.

# *Babesia bigemina*

## Sporozoite Isolation from *Boophilus microplus* Nymphs and Initial Immunomolecular Characterization

JUAN MOSQUEDA, JUAN A. RAMOS, ALFONSO FALCON, J. ANTONIO ALVAREZ, VICENTE ARAGON, AND JULIO V. FIGUEROA

*Centro Nacional de Investigacion Disciplinaria en Parasitologia Veterinaria, Instituto Nacional de Investigaciones Forestales, Agrícolas y Pecuarias, Km 11.5 Carretera Federal, Cuernavaca-Cuautla, Colonia Progreso, Jiutepec, Morelos, C.P. 62500, Mexico*

**ABSTRACT: It has been hypothesized that babesial sporozoites express specific antigens that induce protective immunologic responses in cattle. However, they remain uncharacterized, partly for lack of research on the sporozoite stage of *Babesia* spp. This field suffers from complete knowledge of parasite development in the tick salivary gland; limited amounts of sporozoites from ticks, and a lack of protocols for induction and purification of sporozoites. In this work, *Boophilus microplus* larvae infected with *B. bigemina* were fed on susceptible cattle. Nymphs were collected and macerates were separated by a Percoll density gradient. Microscopic analysis of Giemsa-stained smears showed a larger number of sporozoites from nymphs fed for 9 days. Percoll-purified sporozoites were observed in large numbers in groups or individually and free of tick cells. RT-PCR analysis of total RNA extracted from purified sporozoites indicated transcription of the rhoptry associate protein 1 (*rap-1*) genes: *rap-1a*, *rap-b*, *rap-1c*, as well as the heat shock protein 20 (*hsp-20*) gene. Purified sporozoites were cultured *in vitro* analyzed for RAP-1a expression using an immunocytochemistry assay. Erythrocyte-attached sporozoites reacted with a specific RAP-1a monoclonal antibody. This is the first report of *Babesia bigemina* sporozoite antigens. Moreover, purified sporozoites will allow the characterization of stage-specific antigens involved in immunologic protection.**

**KEYWORDS: *Babesia bigemina*; sporozoite; rhoptry-associated protein 1; heat shock protein 20; *Boophilus microplus*; nymph**

## INTRODUCTION

Bovine babesiosis caused by *Babesia bigemina* and *Babesia bovis* is an important disease of tropical and subtropical regions in the world, including the Americas.[1,2] Both species are transmitted transovarially by *Boophilus* ticks, but only tick larvae transmit *B. bovis*, whereas nymphs and adults transmit *B. bigemina*.[3–7] Sporozoite

Address for correspondence: Juan Mosqueda. CENID-PAVET, INIFAP. Apartado Postal 206, CIVAC, Morelos, C.P. 62550. Mexico.
mosqueda.juanjoel@inifap.gob.mx

**Ann. N.Y. Acad. Sci. 1026: 222–231 (2004). © 2004 New York Academy of Sciences.
doi: 10.1196/annals.1307.034**

stages develop in the salivary gland of the feeding tick and, when expelled, immediately attach and invade their target cell—the bovine erythrocyte. Here, they develop into intraerythrocytic, paired merozoites, which escape the cell only to invade and replicate in new erythrocytes causing anemia, hemoglobulinemia, and occasionally hemoglobinuria.[3,4] We have previously hypothesized that a common molecular mechanism of erythrocyte invasion occurs in sporozoites and merozoites of true *Babesia* species, including *B. bovis* and *B. bigemina.*[8] This is based not only on the fact that both sporozoites and merozoites have a common target cell, similar shape, and same apical organelles, but both stages also express molecules with a proposed role in attachment and invasion of erythrocytes. For example, *B. bovis* merozoite surface antigen 1 (MSA-1), merozoite surface antigen 2a, 2b, and 2c (MSA-2a, 2b, 2c), and the rhoptry associated protein 1 (RAP-1), are expressed by sporozoite and merozoites. Attachment and invasion of both stages to erythrocytes can be prevented by specific antibodies.[8–12] However these observations do not rule out the possibility of unique sporozoite antigens, which should be included as targets in a potential vaccine against bovine babesiosis in order to first block the initial sporozoite invasion and then neutralize the subsequent rounds of merozoite multiplication.[13] Identification of unique antigens from sporozoites has been hampered by an incomplete knowledge of the parasite development in the tick salivary gland, limited amounts of sporozoites, and a lack of protocols for induction of sporozoite development and purification. Here, we report a method to obtain pure *B. bigemina* sporozoites from experimentally infected *Boophilus microplus* nymphs using Percoll gradients. Moreover, we used the purified sporozoites to analyze the transcription of the *rap-1a*, *rap-1b*, *rap-1c,* and *hsp-20* genes as well as the expression of the *rap-1a* gene using an immunocytochemistry assay and a specific monoclonal antibody (MAb).

## MATERIALS AND METHODS

### Babesia bigemina *Strain*

A Mexico strain of *B. bigemina* was used as the source of sporozoites. The strain was donated in 1979 by R. Smith, who obtained it from infected *B. microplus* ticks collected from a slaughterhouse in Mexico City in 1975. Since then, it has been kept in liquid nitrogen and passages through ticks and cattle.

### Boophilus microplus *Strain*

A *Babesia*-free colony of *B. microplus* ticks (Media Joya strain) was used. The tick strain was originally collected from cattle from a ranch in Jalisco state, Mexico, in June 2000. Since then it has been maintained under laboratory conditions at the Centro Nacional de Investigación Disciplinaria en Parasitologia Veterinaria (CENID-PAVET), in Morelos state, Mexico.

### *Infection of* B. microplus *Ticks and Induction of Sporozoite Development*

Adult ticks were allowed to feed on an intact steer using skin patches.[14] Adult female ticks start engorgement approximately 21 days after being placed on the bovine.[15] Steers were inoculated with seven vials of a *B. bigemina* stabilate at 14 days post-attachment so that parasitemia, determined by microscopic examination of Giemsa-stained blood smears, was maximal during the final stages of female tick engorgement. Engorged ticks were washed and dried and placed in individual vials during ovoposition.[16] To obtain a high percentage of infected ticks, only those females replete during the period of highest parasitemia were used. Infection of female ticks was determined on day 6 of ovoposition by the hemolymph test.[17] Only eggs from infected females with more than 10 kinetes per hemolymph sample were used. Eggs laid during the first 144 hours post-engorgement were discarded and the rest of the eggs were incubated at 27°C and 92% relative humidity for three weeks. Once the larvae hatched and their cuticles hardened, they were kept at 27°C and 92% relative humidity for an additional 21 days, which enhances infection rates.[18] To stimulate the development of *B. bigemina* sporozoites, infected larvae were fed on an uninfected steer for 8, 9, or 10 days using skin patches. After this period, nymphs were removed and cleaned from hair and skin debris. Some of the nymphs were processed immediately and the rest incubated at 37°C for an additional day. Uninfected nymphs were obtained from the same colony by using the same procedure, except that the adult ticks were fed on an uninfected cow. Temperature and humidity conditions were the same as those used for the infected ticks.

### *Sporozoite Purification by Percoll Gradient*

A 40% iso-osmotic Percoll gradient was prepared using phosphate-buffered saline (PBS): 9 parts of Percoll (Sigma, St. Louis, MO, USA) are mixed with one part of PBS 10×. Four parts of this solution were mixed then with 6 parts of 1× PBS. Approximately 300 infected fed nymphs were ground using a mortar and pestle in 2 mL of VYM medium.[19] In a 10-mL ultracentrifuge tube, 7 mL of 40% iso-osmotic Percoll were poured and centrifuged at 30,000*g* for 30 min at 4°C. Once the gradient was formed, 2 mL of the tick macerate was added carefully on top of the gradient and centrifuged at 30,000*g* for 30 min at 4°C. The superior phase formed was collected carefully and resuspended in PBS to eliminate the remaining Percoll. A final centrifugation step was performed at 2,740*g* for 25 min at 4°C. The supernatant was discarded and the pellet was resuspended in 100 µL PBS. Purified sporozoites were used immediately or stored at –70°C.

### *RNA Extraction and RT-PCR Analysis*

Total RNA was extracted from purified sporozoites using the RNAeasy kit (Qiagen, Valencia, CA, USA). RNA samples were treated with DNase (Qiagen) and RNase inhibitor (Invitrogen, Carlsbad, CA, USA). RNA was reverse-transcribed and processed using a commercial RT-PCR kit (One Step SuperScript RT-PCR, Invitrogen, Carlsbad, CA, USA). For reverse transcription and pre-denaturation, samples were incubated at 50°C for 30 min, followed by 2 min at 94°C. Pairs of primers used for identification of *rap-1a*, *rap-b*, *rap-c*, and *hsp-20* cDNA were designed to amplify fragments of 517, 292, 528, and 530 bp, respectively. The *rap-1a*, *rap-b*, and *rap-c* forward and reverse primers were obtained from previously published sequences[20] with modifications. The forward and reverse primers for *rap-1a* were 5′-GCGCTTCTGGATGCGTTCGAG and 5′-CGTGCTTCATAATCAACTTGGCAGGG, respectively. The forward and reverse primers for *rap-1b* were 5′-

GCAGGAGCGAAATGGAAGCTTGTC and 5′-CGTCCCTTCTTCCACATTTGC-CAAC, respectively. The forward and reverse primers for *rap-1c* were 5′-GCCCCT-GACCACACTGTCGAG and 5′-GGCATCCAGCGACAAGTATGTCTTGTAG, respectively. The forward and reverse primers for *hsp-20* were kindly provided by Junzo Norimine and Wendy Brown (Washington State University, Pullman, WA, USA), have been published elsewhere[21] and were: 5′-ATGTCGTGCATTATGAG-GTGCAA and 5′-TGCCTTGCCGTCGATCTGGA, respectively. For PCR amplification, 30 cycles were performed consisting of denaturation at 95°C for 15 sec, annealing for 30 sec at 59°C (*rap-1a*, *rap-b*, and *rap-c*) or 57°C (*hsp-20*), and extension at 72°C for 45 sec, and then final extension for one cycle of 72°C for 7 min. RNA from *B. bigemina*–infected erythrocytes was used as positive control. To control for DNA contamination, RT-PCR was performed without reverse transcriptase. RNA extracted from uninfected nymphs was used as a negative control.

### *Immunocytochemistry and Analysis of Expression of RAP-1a in Sporozoites*

Purified *B. bigemina* sporozoites were added to uninfected erythrocytes and cultured *in vitro* in M199 complete medium for 5 hours. Smears of the culture were made using Probe-On slides (Fisher, Pittsburgh, PA, USA), were air-dried for 2 h and fixed in methanol for 5 min. Smears were rinsed in 125 mM Tris buffer containing 0.05% Triton X-100. Smears were then blocked with this buffer containing 5% goat serum at 37°C for 10 min. MAb 64/64.5.10 (anti-RAP-1a)[22] was kindly provided by Guy Palmer (Washington State University, Pullman, WA., USA). It was used at a final concentration of 10 μg/mL and was incubated at 37°C for 25 min. Biotinylated goat anti-mouse immunoglobulin (LSAB2 System-HRP, DAKO, Carpentaria, Ca., USA) was incubated for 25 min at 37°C followed by addition of streptavidin-horseradish peroxidase complex and incubation for 20 min at 37°C. Slides were blotted and rinsed 10× between steps. The chromogen AEC (DAKO, Carpinteria, Ca., USA) was added and incubated for 5 min at room temperature to develop the reaction. Smears were then blotted and rinsed 3× in distilled water with a final wash for one minute. Smears were counterstained for 2 min with filtered hematoxylin and rinsed 4× in water. Smears were cover-slipped in aqueous mounting medium. As a positive control, slides of *B. bigemina*–infected erythrocytes were used. Percoll-separated macerates from uninfected *B. microplus* nymphs cultured *in vitro* with erythrocytes were used as a negative control. The MAb CACT138A, a mouse anti-bovine CD4+ IgG1 (Monoclonal Antibody Center, Pullman, WA, USA) was used as negative isotype control.

## RESULTS

### **B. bigemina** *Sporozoites Are Purified from Infected Nymph Macerates Using Percoll Gradients*

To obtain pure sporozoites from infected nymphs without contaminating tick cells, we used a Percoll gradient protocol using 40% iso-osmotic Percoll. After a final centrifugation of the tick macerate in a Percoll gradient, three phases were formed: a thin upper phase, an intermediate phase, and a lower phase. Microscopic

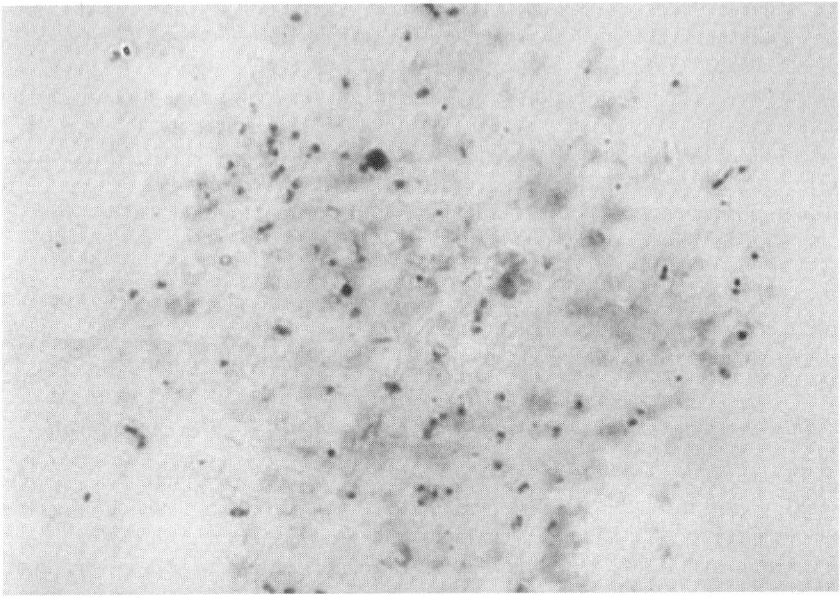

**FIGURE 1.** Light microscopy analysis of *B. bigemina* sporozoites. Black and white image of Percoll-purified sporozoites stained with Giemsa (1,000×). The dark nuclei of the parasites are observed originally stained in purple with pink cytoplasm.

analysis of these phases indicated that the upper phase contained mostly cell membranes and some sporozoites, while the intermediate phase contained most of the sporozoites. Tick cells, midgut contents and cuticle debris were located in the lower phase. Sporozoites were observed in groups or individually (FIG. 1) in macerates of nymphs collected on days 9 and 10, in both immediately processed nymphs or nymphs incubated for an additional day at 37°C. However, fewer numbers of sporozoites were observed in macerates obtained from 10-day-fed nymphs when using the same amount of infected nymphs. Moreover, incubation of 9- or 10-day-fed nymphs for additional 24 h at 37°C decreased the amount of RNA obtained from the samples compared with that of nymphs processed immediately after collection (data not shown). For the rest of the experiments, 9- or 10-day-fed nymphs processed the same day were used.

### Rap-1a, rap-1b, rap-1c, *and* hsp-20 *Are Transcribed in Sporozoites of* B. bigemina

To analyze the transcription of the *rap-1a*, *rap-1b*, *rap-1c*, and *hsp-20* genes, in purified *B. bigemina* sporozoites, an RT-PCR method was performed. Specific primer sets were used to amplify reverse-transcribed cDNA obtained from *B. bigemina*–infected fed nymphs. The *rap-1a* primers were predicted to amplify a fragment from nucleotide 428 to 945; the *rap-1b* primers were designed to amplify a fragment from nucleotide 247 to 539. The *rap-1c* primers amplify a fragment from nucleotide 93 to

621 and the *hsp-20* primers amplify a fragment from nucleotide 1 to 696. The resulting amplicons had the expected size of 517 bp (*rap-1a*), 292 bp (*rap-1b*), 528 bp (*rap-1c*) (FIG. 2), and 530 bp (*hsp-20*) (FIG. 3). Amplicons of similar size to the *rap-1a*, *rap-1b*, *rap-1c*, and *hsp-20* fragments were obtained in cDNA from merozoite samples used as a positive control. No amplification was observed when RNA from uninfected nymphs was used or when reverse transcriptase was omitted, confirming specificity and purity of RNA (FIG. 2, lanes 3, 7, and 11, and FIG. 3, lane 3).

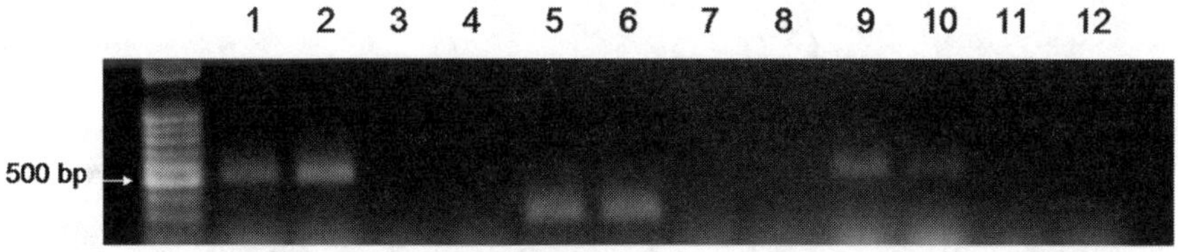


**FIGURE 2.** *rap-1* genes *a*, *b*, and *c* are transcribed in *B. bigemina* sporozoites. Reverse transcriptase PCR (RT-PCR) analysis of total RNA extracted from *B. bigemina*–infected erythrocytes (lanes 1, 5, and 9), *B. bigemina*–purified sporozoites (lanes 2, 6, and 10), *B. bigemina*–purified sporozoites without reverse transcriptase treatment (lanes 3, 7, and 11), and uninfected nymphs (lanes 4, 8, and 12). Amplification used primers specific for *rap-1a* (517 bp, lanes 1–4), *rap-1b* (292 bp, lanes 5–8), and *rap-1c* (528 bp, lanes 9–12). Amplicons were detected by agarose gel electrophoresis and ethidium bromide staining. Molecular size markers are shown on the left.

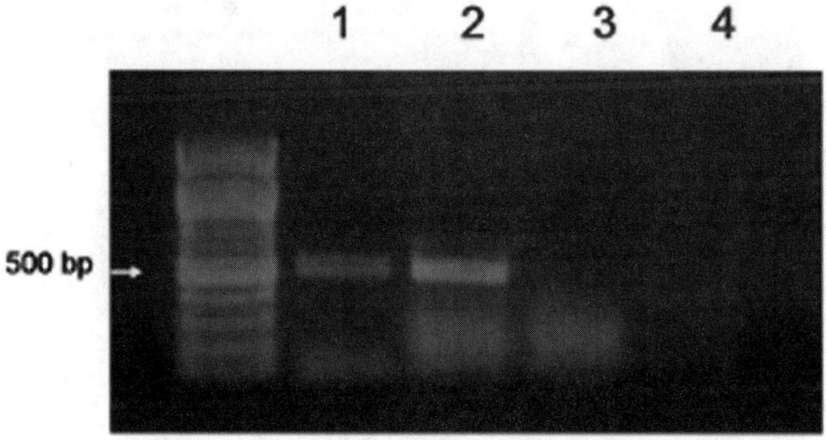


**FIGURE 3.** The *hsp-20* gene is transcribed in *B. bigemina* sporozoites. Reverse transcriptase PCR (RT-PCR) analysis of total RNA extracted from *B. bigemina*–infected erythrocytes (lane 1), *B. bigemina*–purified sporozoites (lane 2), *B. bigemina*–purified sporozoites without reverse transcriptase treatment (lane 3), and uninfected nymphs (lane 4). Amplification used primers specific for *hsp-20* (530 bp). Molecular size markers are shown on the left.

### *RAP-1a Is Expressed in Sporozoites Cultured with Erythrocytes*

To determine if RAP-1a was expressed in infective sporozoite stages at the time of erythrocyte attachment, purified sporozoites obtained from infected nymphs were cultured *in vitro* with erythrocytes for a maximum of 5 h to allow time to bind and infect erythrocytes without developing into merozoites. Smears of these cultures were incubated with a specific antibody anti–RAP-1a. MAb 64/64.5.10 against RAP-1a bound to sporozoites attached to the erythrocyte membrane (FIG. 4A). This MAb reacted also with cultured merozoites used as positive control (FIG 4B), but did not bind erythrocytes cultured alone or with extracts from uninfected larvae (data not shown). Neither sporozoites (FIG. 4C) nor merozoites (FIG. 4D) were bound by negative control MAb CACT138A.

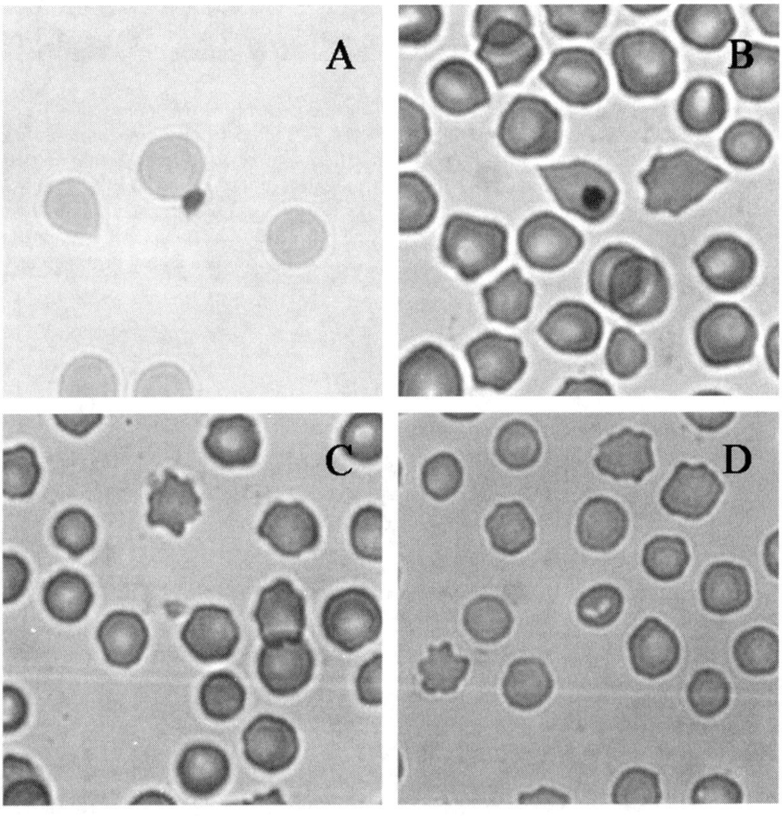


**FIGURE 4.** RAP-1a is expressed in sporozoites attached to erythrocytes. Black and white image of immunocytochemistry of *B. bigemina* cultured with erythrocytes. Smears of erythrocyte cultures initiated with sporozoites (**A** and **C**), and merozoites (**B** and **D**), were incubated with a MAb against RAP-1a (**A** and **B**), or negative control MAb against CACT138A, a mouse anti-bovine CD4+ IgG1 (**C** and **D**). The reaction was visualized with AEC, which results in a dark staining (1,000×).

## DISCUSSION

Merozoites and sporozoites of true *Babesia* species including *B. bigemina*, *B. bovis*, *B divergens*, *B. canis*, *B. caballi*, and *B. ovis,* invade the erythrocyte through an unknown molecular process thought to involve surface coat and apical organelle antigens.[3,4,23,24] Although there is evidence of protective immunologic response to merozoite antigens, there is no information about specific sporozoite antigens and their role in erythrocyte invasion or immune responses in infected cattle.[13,25,26] This is due to several factors, including an incomplete understanding of the parasite life cycle in the vector tick and a lack of protocols to obtain large, pure quantities of sporozoites. Because sporozoites develop in the salivary glands, contamination with tick cells or other *Babesia* stages prevents research on specific transcripts or on stage-specific antigens prior to erythrocyte infection. Here we show that by using 40% iso-osmotic Percoll gradient, purified sporozoites can be obtained from infected nymphs fed for 9 or 10 days. Fewer sporozoites were obtained from nymphs fed for 10 days and none from those fed for 8 days. Eight-day-fed nymphs might not be fully developed, which might prevent sporozoite development. It has been shown before that nymphs could transmit sporozoites as early as day 9 of feeding.[3] Incubating the nymphs for an additional day at 37°C did not increase the amount of sporozoites recovered, probably due to mortality of nymphs during the incubation period. The sporozoites obtained were free of other tick stages of *B. bigemina*, like kinetes or sporonts, as determined by microscopic analysis of Giemsa-stained smears (FIG. 1). These other stages are also present at the time of sporozoite development, but are larger in size; while the kinetes measure $11.0 \times 2.5$ μm and the sporonts reach up to 90 μm in diameter, the sporozoites measure only $2.5 \times 1.2$ μm.[27,28] Although tick cell membranes were observed co-migrating with the sporozoites in both the upper and intermediate phases, these were scarce in the intermediate phase, where most of the sporozoites were observed. Importantly, no tick cells were found in the preparations.

We used the purified sporozoites to analyze transcription of genes previously shown in merozoites. The organization of the *rap-1* locus has been recently reported.[20] It consist of five tandemly arranged copies of the *rap-1a* gene intercalated with five identical copies of the *rap-1b* and a single copy of the *rap-1c* gene at the 3′ end of the locus.[20] Using specific primers for *rap-1a*, *1b*, and *1c*, we detected transcription messages in purified sporozoites as well as in merozoites used as positive control (FIG. 2). This indicates that, similar to *B. bovis*, *rap-1* genes of *B. bigemina* are transcribed in sporozoites and in merozoites.[8] According to this hypothesis, we analyzed the transcription of the *hsp-20* gene, another gene previously shown to be transcribed in merozoites.[21] Transcripts were also present in purified sporozoites as well as in merozoites (FIG. 3). In *B. bovis*, we have shown that *hsp-20* is also transcribed and expressed in sporozoites[29] as well as in merozoites, so the presence of *hsp-20* transcripts in *B. bigemina* merozoites and sporozoites is not surprising. Further determination of HSP-20 expression in sporozoites of *B. bigemina* will confirm the hypothesis that this antigen is also shared by both stages.

Finally, to determine if RAP-1a is expressed by sporozoites, we used a monoclonal antibody specific for this antigen.[22,30] We analyzed the expression of RAP-1a in sporozoites incubated with uninfected erythrocytes. Sporozoites attached to the erythrocyte surface expressed RAP-1a as determined by immunocytochemistry

(FIG. 1A). These results confirm the presence of transcripts in sporozoite RNA and indicate that similar to *B. bovis* RAP-1, *B. bigemina* RAP-1a is transcribed and expressed in sporozoites and merozoites. It was reported that a monoclonal antibody anti–*B. bigemina* RAP-1a inhibited merozoite development *in vitro*.[31]The effect of specific antibodies against RAP-1a in sporozoite initial attachment and invasion remains to be determined, but antibodies have inhibited attachment of sporozoites to erythrocytes and development of merozoites *in vitro* in *B. bovis*.[8,11] Interestingly, in *B. bigemina* erythrocytic stages, *rap-1a, 1b,* and *1c* transcripts are present, but only RAP-1a is expressed. This indicates that expression of *rap-1* genes is regulated at the transcriptional and translational level.[20] Whether *rap-1b* and *rap-1c* are translated in sporozoites of *B. bigemina* remains to be determined and we are currently testing this hypothesis. Together, these results indicate that Percoll-purified sporozoites can be used to analyze transcription and expression of genes present in sporozoites prior to and at the time of erythrocyte invasion; our results from these and previous experiments support the hypothesis that similar surface coat and apical organelle proteins as well as a common mechanism of erythrocyte invasion are conserved in *Babesia* spp. sporozoites and merozoites.

## ACKNOWLEDGMENTS

The MAb 64/64.5.10 was kindly provided by Guy Palmer, and the forward and reverse primers for *hsp-20* were kindly provided by Junzo Norimine and Wendy Brown (Washington State University, Pullman, WA, USA). The technical assistance of Carmen Rojas is greatly appreciated. This work was supported by Consejo Nacional de Ciencia y Tecnologia (Project No. 34473-B).

## REFERENCES

1. MACCOSKER, P.J. 1981. The global importance of babesiosis. *In* Babesiosis. M. Ristic & J.P. Kreier, Eds.: 1–24. Academic Press. New York.
2. RIDDLES, P.M. & I.G. WRIGHT. 1992. Control of intraerythrocytic parasites. *In* Animal parasite control utilizing biotechnology, W.K. Yong, Ed.: 221–240. CRC Press. Boca Raton, FL.
3. HOYTE, H.M.D. 1961. Initial development of infections with *Babesia bigemina*. J. Protozool. **8:** 462–466.
4. HOYTE, M.H.D. 1965. Further observations on the initial development of infections with *Babesia bigemina*. J. Protozool. **12:** 83–85.
5. MAHONEY, D.F. & G.B. MIRRE. 1971. Bovine babesiasis: estimation of infection rates in the tick vector *Boophilus microplus* (Canestrini). Ann. Trop. Med. Parasitol. **65:** 309–317.
6. MAHONEY, D.F. & G.B. MIRRE. 1974. *Babesia argentina*: the infection of splenectomized calves with extracts of larval ticks (*Boophilus microplus*). Res. Vet. Sci. **16:** 112–114.
7. MAHONEY, D.F. & G.B. MIRRE. 1977. The selection of larvae of *Boophilus microplus* infected with *Babesia bovis* (syn B argentina). Res. Vet. Sci. **23:** 126–127.
8. MOSQUEDA, J. *et al.* 2002. *Babesia bovis* merozoite surface antigen 1 and rhoptry-associated protein 1 are expressed in sporozoites, and specific antibodies inhibit sporozoite attachment to erythrocytes. Infect. Immun. **70:** 1599–1603.
9. MOSQUEDA, J., T.F. MCELWAIN & G.H. PALMER. 2002. *Babesia bovis* merozoite surface antigen 2 proteins are expressed on the merozoite and sporozoite surface, and spe-

cific antibodies inhibit attachment and invasion of erythrocytes. Infect. Immun. **70:** 6448–6455.
10. Hines, S.A. *et al.* 1995. Immunization of cattle with recombinant *Babesia bovis* merozoite surface antigen-1. Infect. Immun. **63:** 349–352.
11. Yokoyama, N. *et al.* 2002. Cellular localization of *Babesia bovis* merozoite rhoptry-associated protein 1 and its erythrocyte-binding activity. Infect. Immun. **70:** 5822–5826.
12. Wilkowsky, S.E. *et al.* 2003. *Babesia bovis* merozoite surface protein-2c (MSA-2c) contains highly immunogenic, conserved B-cell epitopes that elicit neutralization-sensitive antibodies in cattle. Mol. Biochem. Parasitol. **127:** 133–141.
13. Palmer, G.H. & T.F. McElwain. 1995. Molecular basis for vaccine development against anaplasmosis and babesiosis. Vet. Parasitol. **57:** 233–253.
14. Hodgson, J.L. 1991. Detection of *Babesia bigemina* infective forms in salivary glands of *Boophilus microplus* using a DNA probe. Department of Veterinary Microbiology and Pathology. Washington State University. Pullman, WA. p. 81.
15. Nuñez, J.L., M.E. Muñoz-Cobeñas & H.L. Moltedo. 1985. *Boophilus microplus*, the common cattle tick. Berlin, Germany: Springer-Verlag.
16. Hodgson, J.L. *et al.* 1992. *Babesia bigemina*: quantitation of infection in nymphal and adult *Boophilus microplus* using a DNA probe. Exp. Parasitol. **74:** 17–26.
17. Riek, R.F. 1964. The life cycle of *Babesia bigemina* (Smith and Kilborne, 1893) in the tick vector *Boophilus microplus* (Canestrini). Austral. J. Agricult. Res. **17:** 247–254.
18. Dalgliesh, R.J. & N.P. Stewart. 1982. Some effects of time, temperature and feeding on infection rates with *Babesia bovis* and *Babesia bigemina* in *Boophilus microplus* larvae. Int. J. Parasitol. **12:** 323–326.
19. Vega, C.A. *et al.* 1985. *In vitro* cultivation of *Babesia bigemina*. Am. J. Vet. Res. **46:** 416–420.
20. Suarez, C.E. *et al.* 2003. Organization, transcription, and expression of rhoptry associated protein genes in the *Babesia bigemina* rap-1 locus. Mol. Biochem. Parasitol. **127:** 101–112.
21. Brown, W.C. *et al.* 2001. A novel 20-kilodalton protein conserved in *Babesia bovis* and *B. bigemina* stimulates memory CD4(+) T lymphocyte responses in *B. bovis*-immune cattle. Mol. Biochem. Parasitol. **118:** 97–109.
22. Vidotto, O. *et al.* 1995. *Babesia bigemina*: identification of B cell epitopes associated with parasitized erythrocytes. Exp. Parasitol. **81:** 491–500.
23. Friedhoff, K.T. 1988. Transmission of *Babesia*. *In* Babesiosis of Domestic Animals and Man. M. Ristic, Ed.: 23–52. CRC Press. Boca Raton, FL
24. Purnell, R.E. 1981. Babesiosis in various hosts. *In* Babesiosis. M. Ristic & J.P. Kreier, Eds.: 25–63. Academic Press. New York.
25. Wright, I.G. *et al.* 1992. The development of a recombinant *Babesia* vaccine. Vet. Parasitol. **44:** 3–13.
26. Brown, W.C. & G.H. Palmer. 1999. Designing blood-stage vaccines against *Babesia bovis* and *B. bigemina*. Parasitol Today **15:** 275–281.
27. Mehlhorn, H. & E. Shein. 1984. The piroplasms: life cycle and sexual stages. Adv. Parasitol. **23:** 37–103.
28. Potgieter, F.T. & H.J. Els. 1976. Light and electron microscopic observations on the development of small merozoites of *Babesia bovis* in *Boophilus microplus* larvae. Onderstepoort J. Vet. Res. **43:** 123–128.
29. Norimine, J. *et al.* 2004. Conservation of *Babesia bovis* small heat shock protein (Hsp20) among strains and definition of T helper cell epitopes recognized by cattle with diverse major histocompatibility class II haplotypes. Infect. Immun. **72:** 1096–1106.
30. Hotzel, I. *et al.* 1996. Dimorphic sequences of rap-1 genes encode B and CD4+ T helper lymphocyte epitopes in the *Babesia bigemina* rhoptry associated protein-1. Mol. Biochem. Parasitol. **81:** 89–99.
31. Figueroa, J.V. & G.M. Buening. 1991. *In vitro* inhibition of multiplication of *Babesia bigemina* by using monoclonal antibodies. J. Clin. Microbiol. **29:** 997–1003.

# Successful Infestation by *Amblyomma pseudoconcolor* and *A. cooperi* (Acari: Ixodidae) on Horses

SAMUEL C. CHACON, JOÃO LUIZ H. FACCINI,
AND VÂNIA R.E.P. BITTENCOURT

*Inst. Veterinária, Depto. Parasitologia Animal, Universidade Federal Rural do Rio de Janeiro, Rio de Janeiro, Brazil*

**ABSTRACT: The host relationships for most species of the genus *Amblyomma* are poorly known in Brazil. The ability of *A. pseudoconcolor* and *A. cooperi* to successfully feed on horses was investigated during ongoing research on the life cycle of these two species, which are primarily associated with wildlife. Results of these experiments suggest that horses are potential hosts for the adult stages of both species.**

**KEYWORDS: *Amblyomma pseudoconcolor*; *A. cooperi*; Ixodidae; life cycle; experimental infestation; horse;**

## INTRODUCTION

Host specificity is one of the paramount topics in the study of the ecology of a given species of parasite. Accordingly, parasites are classified in a gradient from monoxenous (one host species) to polixenous (several host species). As the host range increases, the host – parasite relationship become more complex and, consequently, the control of the involved parasite species.

The host relationships for most species of the tick genus *Amblyomma* Koch are poorly known in Brazil, including the host range. *A. cooperi* Nutall and Warburton, 1908 parasitizes capybaras primarily whereas the host records for *A. pseudoconcolor* Aragão, 1908 include mostly the endangered armadillo species in Northeast Brazil, threatened due to the impact of ecologic changes caused by human activities such hunting. During ongoing research on the life cycle of these ticks under the laboratory conditions at the UFRRJ, the ability of these ticks to successfully feed on domestic animals was investigated.

## MATERIALS AND METHODS

Laboratory colonies of *A. cooperi* and *A. pseudoconcolor* were started with engorged females collected from a naturally infested capybara and armadillo, respec-

Address for correspondence: João Luiz H. Faccini, UFRRJ, Inst. Veterinária, Departmento Parasitologia Animal, Seropédica, RJ, Brazil, 23890-000. Voice/fax: 55-21-2682-1618. faccini@ufrrj.br

**Ann. N.Y. Acad. Sci. 1026: 232–234 (2004). © 2004 New York Academy of Sciences.**
**doi: 10.1196/annals.1307.035**

**TABLE 1. Selected feeding and reproductive parameters of three species of *Amblyomma* Koch experimentally fed on horses**

| Parameters | *A. pseudoconcolor* | *A. cooperi* | *A. cajennense* |
|---|---|---|---|
| Ovipositing females (mg) | 40.5–644 | 91.5–634.2 | 385.3–802.5 |
| | $261.4^{a} \pm 165.3$ | $291.2^{a} \pm 159.8$ | $595.1^{b} \pm 126.1$ |
| | 49 | 22 | 20 |
| Egg mass (mg) | 2.2–423 | 0.3–307 | 100.8–497.4 |
| | $128.4^{a} \pm 118.7$ | $81.4^{a} \pm 72.4$ | $222.9^{b} \pm 88.3$ |
| | 49 | 22 | 20 |
| Reproductive index (REI)[d] | 2.6–66.5 | 0.1– 48.4 | 13.3–62 |
| | $38.9^{a} \pm 19.9$ | $27^{b} \pm 11.4$ | $37.4^{a,b} \pm 10.4$ |
| | 49 | 22 | 20 |
| Female/egg mass weight | r = 0.98 | r = 0.74 | r = 0.57 |
| Eclosion (%) | 40–100 | 2–50 | 3–98 |
| | $83^{a} \pm 13.9$ | $19.7^{b} \pm 14.1$ | $54.6^{c} \pm 29.2$ |
| | 41 | 21 | 36 |

Means followed by different letters in a row are significantly different ( $P < 0.05$).
[d]Reproductive efficiency index (REI). According to G. F. Bennett, 1974. Acarologia **XVI**(1): 52–61.

tively. Thus, each of two horses were infested with 160 adult ticks (80 males and 80 females) either of *A. cooperi* or *A. pseudoconcolor* obtained from larvae and nymphs previously fed on domestic rabbits. The infestation site was delimited by an area of approximately 300 $cm^2$ on the side of the neck[1]. Naturally detached females were weighed and held in an incubator at 27 ± 1°C and 80 ± 10% RH. The ANOVA and the Tukey-Kramer test ( $P < 0.05$) were used to compare means of each parameter.

## RESULTS AND DISCUSSION

For *A. cooperi*, 22/80 ( 27.5%) engorged females oviposited fertile eggs whereas 49/80 (61.3%) of *A. pseudoconcolor* oviposited fertile eggs. Selected feeding and reproductive data of these females and those of *Amblyomma cajennense* are presented as range, mean ± SD and number of observations in TABLE 1.

When these results are compared with those obtained in our laboratory for *A. cajennense* on horses,[2] its primary host in Brazil, they suggest that horses are potential hosts for the adult stage of both species in nature, although statistical differences ( $P < 0.05$ ) were seen (TABLE 1). This situation might have importance in the transmission of a given bioagent such a spotted fever group *Rickettsia*, which probably circulates among *A. cooperi* and capybaras in Brazil.[3]

## REFERENCES

1. SANAVRIA, A. & M.C.A. PRATA. 1996 . Metodologia para colonização de *Amblyomma cajennense* (Fabricius, 1787) (Acari: Ixodidae) em laboratório. Rev. Bras. Parasitol. Vet. **5:** 87–90.
2. CHACON, S.C., P.G. CORREIA, F.S. BARBIERI, *et al.* 2003. Efeito de tres temperaturas constantes sobre a fase não parasitária de *Amblyomma cajennense* ( Fabricius, 1787) (Acari: Ixodidae). Rev. Bras. Parasitol. Vet. **12** (1): 13–20.
3. LEMOS, E.R.D., H.H.B. MELLES, S.C. COLOMBO, *et al.* 1996. Primary isolation of Spotted Fever group Rickettsiae from *Amblyomma cooperi* collected from *Hydrochaeris hydrochaeris* in Brazil. Mem. Inst. O. Cruz **91:** 273–275.

# Microscopic Features of Tick-Bite Lesions in Anteaters and Armadillos

## Emas National Park and the Pantanal Region of Brazil

M.F. LIMA E SILVA,[a] M.P.J. SZABÓ,[a,b] AND G.H. BECHARA[a]

[a]*São Paulo State University-Universidade Estadual Paulista Julio de Mesquita Filho (UNESP), Department of Animal Pathology, Jaboticabal-SP, Brazil*

[b]*University of Franca-Universidade de Franca (UNIFRAN), Franca-SP, Brazil*

**ABSTRACT: The naturally occurring wildlife host associations between ticks and tick-borne pathogens found in the neotropics are poorly described. Understanding tick-bite lesions is important as these are the site of host reaction to and pathogen delivery by ticks. As part of a comprehensive study concerning established and emerging tick-host relationships. the present work describes some aspects of tick-bite lesions in anteaters and armadillos captured at the Emas National Park and the Pantanal region of Brazil. Biopsies were of skin were taken and examine. Tick feeding sites of all animals displayed an eosinophilic homogeneous mass, the cement cone, and, occasionally, a feeding cavity underneath the tick attachment site. At these locations the epidermis was usually thickened due to keratinocyte hyperplasia. The main dermal changes included tissue infiltration with a varying number of inflammatory cells, edema, hemorrhage. and vascular dilatation. Cellular infiltration of the dermis was predominantly composed of mononuclear cells, neutrophils. and eosinophils. Mast cells were also seen in both non-parasitized and parasitized skin but were found in higher numbers at perivascular sites and in parasitized skin. Basophils were not seen at tick attachment sites of anteaters or armadillos.**

**KEYWORDS: anteaters; *Myrmecophaga tridactyla*; armadillo; *Euphractus sexcintus*; ticks; histopathology**

Ticks are a major concern throughout the world for their negative impact on animal production and also on public health. In tropical and subtropical regions of Brazil, but the impact varies according to region. The species of the tick, the tick-borne pathogen, and the host as well as environmental and socioeconomic conditions may generate different problems for each region.[1]

Tick-host relationships are susceptible to evolutionary mechanisms that affect the close association between parasite and host. In these cases, the survival of the parasite depends on host resistance.[2] Once contact between a host and a parasite is established, parasite survival depends on feeding mechanisms. Ticks feeding on hosts

Address for correspondence: M.V. Maria Fernanda de Lima e Silva, Faculdade de Ciências Agrárias e Veterinárias, Universidade Estadual Paulista, 14884-900 Jaboticabal-SP, Brazil. Voice: 55-16-32092662.

mafer.1@zipmail.com.br

**Ann. N.Y. Acad. Sci. 1026: 235–241 (2004). © 2004 New York Academy of Sciences. doi: 10.1196/annals.1307.036**

are also exposed to non-specific host defenses and to acquired resistance mechanisms of the host, an immunological phenomenon.[3] Usually, associations between ticks and laboratory animals are characterized by more prevalent expression of acquired resistance than in the naturally occurring associations between ticks and animals.[4] At the same time, naturally occurring wildlife host associations with ticks and tick-borne pathogens are poorly described within the neotropics. Understanding tick-bite lesions is important as these are the sites of host reaction to and pathogen delivery by ticks. Moreover, wildlife hosts may harbor potential pathogens of domestic animals and humans and, considering tick vectoring capacity, may be involved in the emergence of new diseases.

## MATERIALS AND METHODS

### *Study Site and Animal Capture*

Four anteaters (*Myrmecophaga trydactyla*) and seven armadillos (*Euphractus sexcintus*) were captured at the Alegria Farm (19°08′S, 56°46′W) and some neighboring areas in the Nhecolândia region of the Pantanal, Mato Grosso do Sul State, Brazil and Emas National Park, Goiás State, Brazil, in several scientific expeditions held from September 1996 to September 2000. Animal capture as described was submitted to and authorized by the National Institute of Environment and Natural Resources of Brazil (IBAMA—Instituto Brasileiro do Meio Ambiente e dos Recursos Renováveis).

For capture, animals were located visually and restrained either manually or with the aid of a net. All captured animals were anesthetized with xylazine (Rompum, Bayer do Brasil S.A.) and ketamine hydrochloride (Ketalar, Park Davis do Brasil S.A). When the collection of biological parameters was completed, (sample) animals were inspected, treated for any wounds or lesions, and released. The animals were released only after total recovery from the anesthetic to protect them from predators.

### *Skin Sample Collection and Histological Processing*

Biopsies were taken from anesthetized animals with the aid of a 6-mm diameter punch from tick-feeding sites and from normal skin (control samples). Skin samples were kept for 24 hours in the fixative and then in 70% alcohol. Later, samples were embedded in paraffin and processed according to routine histological techniques. Each biopsy was serially sectioned at a thickness of 6 μm, and stained with hematoxylin-eosin and May Grünwald-Giemsa stains.

### *Section Analysis*

Sections displaying the center of tick-bite lesions were chosen by the presence of the tick's cement cone and were used for histological analyses as described previously by Szabó and Bechara.[4] General features of tick-attachment sites and cellular infiltrate composition were evaluated on, respectively, hematoxylin-eosin and Giemsa-stained sections. General features and infiltrating inflammatory cell numbers were assessed as semiquantitative levels (-, absent; +, discrete; ++, light; +++, moderated; ++++, accentuated).

**TABLE 1. Intensity of histopathological findings in skin from anteater parasitized with *Amblyomma* sp.**

| | Animal | | | |
|---|---|---|---|---|
| Histopathological Findings | 01 | 02 | 03 | 04 |
| Hyperplasia | + | +++ | ++ | ++++ |
| Edema | ++++ | ++++ | +++ | ++ |
| Vasodilatation | +++ | - | - | +++ |
| Hemorrhage | + | ++ | + | +++ |
| Lymphocytes | ++++ | ++++ | ++++ | +++ |
| Plasma cells | +++ | + | + | - |
| Macrophages | ++ | ++ | + | - |
| Neutrophils | + | + | +++ | ++++ |
| Eosinophils | +++ | +++ | ++ | +++ |
| Mast cells | +++ | +++ | ++ | + |
| Basophils | - | - | - | - |

Histopathological findings: -, absent; +, discrete; ++, light; +++, moderate; ++++, accentuated.

## RESULTS

### *Tick Identification*

Identification of tick species infesting anteaters and armadillos has been presented elsewhere.[6,7] Tick attachment sites analyzed in the present work were exclusively lesions induced by adult *Amblyomma* sp. ticks.

### *Histopathology*

Several parasitized skin samples of each host species could not be included in the analysis as the center of lesions was not always found. Results are thus based on sections from four anteaters and two armadillos. Histopathological findings are presented in TABLES 1 and 2.

Normal, uninfested skin of both species, used as control samples, revealed intact epidermis, superficial dermis including blood and lymphatic vessels in moderated quantity, presence of fibroblasts, mast cells, and skin appendices. Some animals presented discrete eosinophilic and mononuclear cell infiltration, possibly related to previous ectoparasitic contact.

The histopathological findings of tick-infested skin samples were similar in both species but with some variation. General features of tick attachment sites included epidermal hyperplasia, discrete inflammatory edema, hemorrhage, and vascular dilation of the dermis. All parasitized animals presented a cement cone inserted under the stratum corneum and extending to granular layer of the epidermis. In some cases a feeding cavity under the tick's attachment site could be seen. These feeding cavities were filled with dead cells, debris, and neutrophils (FIG. 1).

**TABLE 2. Intensity of histopathological findings in skin from armadillo parasitized with *Amblyomma* sp.**

| | Animal | |
|---|---|---|
| Histopathological Findings | 01 | 02 |
| Hyperplasia | ++++ | ++ |
| Edema | - | - |
| Vasodilatation | - | + |
| Hemorrhage | - | - |
| Lymphocytes | +++ | ++++ |
| Plasma cells | - | + |
| Macrophages | ++ | +++ |
| Neutrophils | - | +++ |
| Eosinophils | + | ++ |
| Mast cells | ++ | +++ |
| Basophils | - | - |

Histopathological findings: -, absent; +, discrete; ++, light; +++, moderate; ++++, accentuated.

Inflammatory exudate could be seen surrounding the cement cone and could also be observed at the deep dermis in a perivascular distribution. This exudate was characterized by accentuated influx of lymphocytes, eosinophils and macrophages, and occasional neutrophilic infiltration, but it lacked basophils (FIGS. 2 and 3). Mast cells appeared as perivascular cuffing in moderate numbers, both in normal and parasitized skin. Many mast cells could also be seen close to blood vessels, far from tick attachment sites (FIG. 4).

## DISCUSSION

The present work displays morphological information at tick feeding sites of two wild host species captured in nature. These animals were exposed to many environmental stimuli and among these, repeated tick infestations. It should be stressed that the inflammatory reactions observed in this study may have had an immune response involvement. At the same time, although anteater and armadillo are natural hosts to a variety of species of *Amblyomma*, the significance of the findings described above is unclear, as there is no information available concerning resistance of these hosts to the ticks that caused the lesions. The age of lesions could not be determined since there is no information for the feeding period of ticks from each biopsy. On the other hand the high percentage of lymphocytes and plasma cells indicate an ongoing immune response.

The moderate to accentuated mast cell and eosinophil numbers indicate that these cell types also have a role in the reaction as they do in other parasitic diseases.[8] It is noteworthy, however, that basophils, a cell type associated to resistance to ticks,[5,9] could not be detected in these samples.

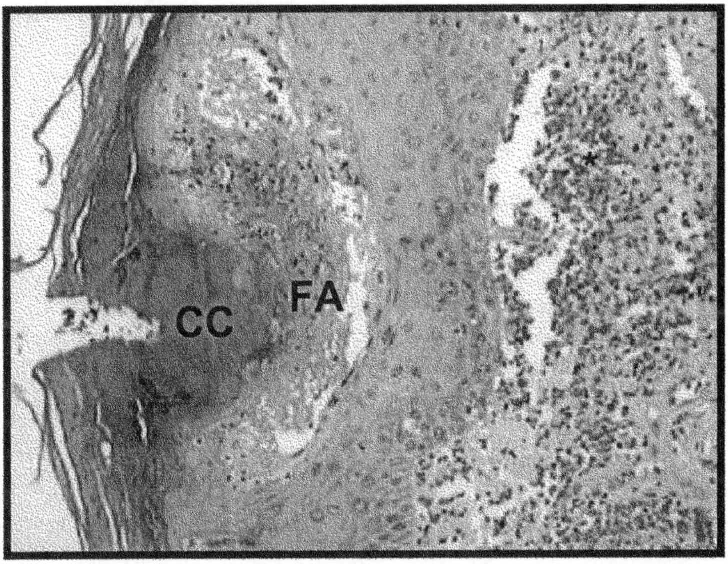


**FIGURE 1.** Anteater skin parasitized by *Amblyomma* sp. Observe cement cone (CC), feeding cavity (FA), and blood in superficial dermis (*). Hematoxylin-eosin stain. (Magnification ×10).

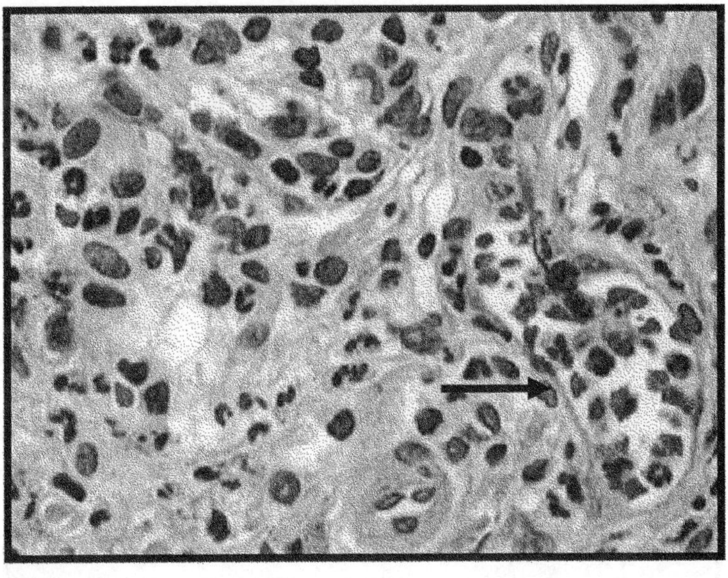

**FIGURE 2.** Anteater skin parasitized by *Amblyomma* sp. Observe inflammatory infiltrate (mononuclear cells and neutrophils) and leukocyte margination in a vessel (*arrow*). Hematoxylin-eosin stain. (Magnification ×40).

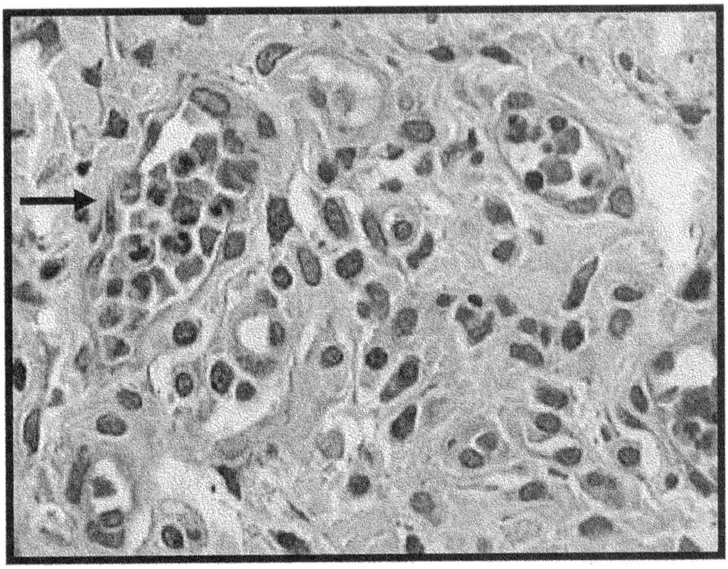

**FIGURE 3.** Armadillo skin parasitized by *Amblyomma* sp. Observe mononuclear cells and leukocyte margination in a vessel (*arrow*). Hematoxylin-eosin stain. (Magnification ×40).

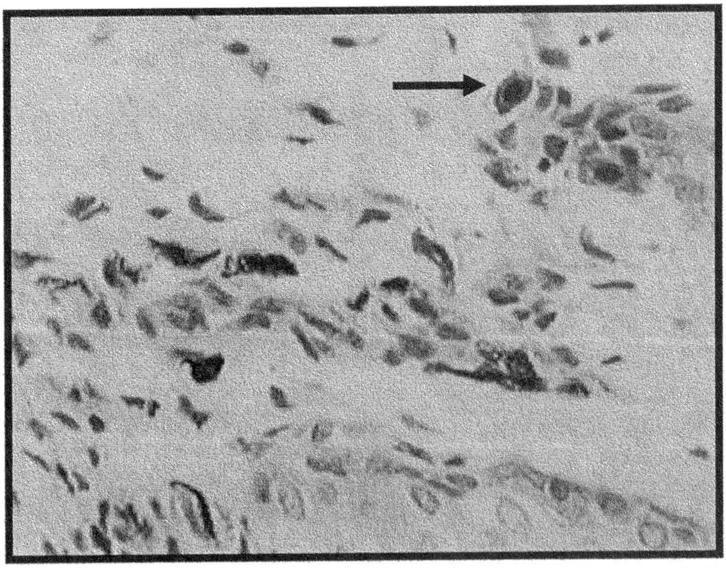

**FIGURE 4.** Armadillo skin parasitized by *Amblyomma* sp. Observe discrete degranulated mast cells (*arrow*). May-Grünwald-Giemsa stain. (Magnification ×40).

On the whole this work provides a glimpse of the morphological features at the tick attachment sites in naturally occurring host-parasite relationships in the wild in these species. Even though it is highly speculative, it may be supposed that the lack of basophils indicates a lack of strong resistance to ticks in these hosts. Under such conditions, pathogen transmission by ticks is enhanced and thus these host-parasite relationships may be associated with pathogen transmission. Whatever the case, more detailed studies would be worthwhile.

## ACKNOWLEDGMENTS

The Instituto Brasileiro do Meio Ambiente e dos Recursos Renováveis (IBAMA) kindly gave permission for the capture of the animals. This research was supported by Fundacao de Amparo a Pesquisa do Estado de Sao Paulo (FAPESP), Conselho Nacional de Desenvolvimento Cientifico e Tecnologico (CNPq), and São Paulo State University.

## REFERENCES

1. VÁSQUEZ, C.C., Z.G. VÁSQUEZ & M.T. QUINTERO. 1995. Avanços en la immunización contra garrapatas del ganado bovino. Vet. Méx. **26:** 251–262.
2. DINEEN, J.K. 1963. Immunological aspects of parasitism. Nature **17:** 268–269.
3. MITCHELL, G.F. 1979. Effector cells molecules and mechanisms in host-protective immunity to parasites. Immunology **38:** 209–223.
4. RIBEIRO, J.M.C. 1989. Role of saliva in tick-host interactions. Exp. Appl. Acarol. **7:** 15–20.
5. SZABÓ, M.P.J. & G.H. BECHARA. 1999. Sequential histopathology at the *Rhipicephalus sanguineus* tick feeding site on dogs and guinea pig. Exp. Appl. Acarol. **23:** 915–928.
6. CAMPOS PEREIRA, M., M.P.J. SZABÓ, G.H. BECHARA, *et al.* 1999. Ticks associated with wild animals in the Pantanal region of Brazil. J. Med. Entomol.**37:** 209–212.
7. BECHARA, GH, M.P.J. SZABÓ, W.V. ALMEIDA FILHO, *et al.* 2002. Ticks associated with armadillo (*Euphractus sexcinctus*) and anteater (*Myrmecophaga trydactyla*) of Emas National Park, State of Goias, Brazil. Ann. NY Acad. Sci. **969:** 290–293.
8. ABBAS, A. K., A.H. LICHTMAN & J.S. POBER. 2000. Cellular and Molecular Immunology, 4th edit. W.B. Saunders Company.
9. BROWN, S.J. 1988. Highlights of contemporary research on host immune responses to ticks. Vet. Parasitol. Amsterdam **28:** 321–334.

# Gene Discovery in *Boophilus microplus*, the Cattle Tick

## The Transcriptomes of Ovaries, Salivary Glands, and Hemocytes

ISABEL K. F. DE MIRANDA SANTOS,[a,b] JESUS G. VALENZUELA,[d] JOSÉ MARCOS C. RIBEIRO,[d] MARILIA DE CASTRO,[a] JULIANA NARDELLI COSTA,[a] ANA MARIA COSTA,[a] EDSON RAMIRO DA SILVA,[a] OLAVO BILAC REGO NETO,[a] CLARISSE ROCHA,[a] SIRLEI DAFFRE,[e] BEATRIZ R. FERREIRA,[b] JOÃO SANTANA DA SILVA,[b] MATIAS PABLO SZABÓ,[c] AND GERVASIO HENRIQUE BECHARA[c]

[a]*Center for Genetic Resources and Biotechnology, Empresa Brasileira de Pesquisa Agropecuária, Brasília, DF, Brazil*

[b]*Ribeirão Preto School of Medicine, Universidade de São Paulo, Ribeirão Preto, SP, Brazil*

[c]*School of Agronomical and Veterinary Sciences, Universidade Estadual Paulista, Jaboticabal, SP, Brazil*

[d]*Section of Vector Biology, National Institute of Allergy and Infectious Diseases, National Institutes of Health, Bethesda, Maryland 20892-6612, USA*

[e]*Institute of Biology, Universidade de São Paulo, São Paulo, SP, Brazil*

**ABSTRACT: The quest for new control strategies for ticks can profit from high throughput genomics. In order to identify genes that are involved in oogenesis and development, in defense, and in hematophagy, the transcriptomes of ovaries, hemocytes, and salivary glands from rapidly ingurgitating females, and of salivary glands from males of *Boophilus microplus* were PCR amplified, and the expressed sequence tags (EST) of random clones were mass sequenced. So far, more than 1,344 EST have been generated for these tissues, with approximately 30% novelty, depending on the the tissue studied. To date approximately 760 nucleotide sequences from *B. microplus* are deposited in the NCBI database. Mass sequencing of partial cDNAs of parasite genes can build up this scant database and rapidly generate a large quantity of useful information about potential targets for immunobiological or chemical control.**

**KEYWORDS: *Boophilus microplus*; transcriptomes; salivary glands; ovary; hemocytes**

Address for correspondence: Isabel K.F. de Miranda Santos, Departamento de Bioquímica e Imunologia, Faculdade de Medicina de Ribeirão Preto, Universidade de São Paulo, Avenida Bandeirantes 3900, 14 049-900 Ribeirão Preto, SP, Brazil. Voice: +55-16-602-3234; fax: +55-16-633-6631.

imsantos@rpm.fmrp.usp.br

**Ann. N.Y. Acad. Sci. 1026: 242–246 (2004). © 2004 New York Academy of Sciences. doi: 10.1196/annals.1307.037**

**TABLE 1. Number of sequences deposited for *Boophilus microplus* in the NCBI database, July 2003**

| Taxonomic Classification | Proteins | Nucleotides (contains EST) |
|---|---|---|
| Arthropoda | 145,159 | 988,798 |
| Insecta | 135,689 | 957,386 |
| Arachnida | 3,527 | 16,624 |
| Ixodida | 683 | 10,044 |
| Boophilus | 108 | 827 |
| *B. microplus* | 98 | 761 |

## INTRODUCTION

Ticks cause serious economic losses to animal production worldwide, in the order of billions of dollars. New alternatives for their control are urgently needed owing to the increasing lack of efficacy and/or of acceptance of the current methods that rely on chemicals.[1] Vaccines are one attractive strategy, but the complexity of the tick/host interface will certainly require the use of multicomponent vaccines. Other technologies may also be available pending new knowledge about the mechanisms of parasitism in ticks and their general biology. New genes and antigens of ticks are, therefore, in great demand. Analysis of gene expression profiles by means of expressed sequence tags (EST) offers insights into the biology of ticks and can pave the way for new control methods for ticks.[2] EST are rapid and informative, and function is inferred by comparative analysis. They can indicate new candidates for immunological, biological, nontoxic chemical, and semiochemical control strategies. TABLE 1 shows the current entries, relative to arthropods, of information available in the database of the National Center for Biotechnology Information of the National Institutes of Health, for genes and proteins of *Boophilus microplus*, the cattle tick. It is clear that development of new technologies depends on far more information than is currently available for this species. In order to identify genes that are involved in oogenesis and development, in defense, in hematophagy, and in parasite escape mechanisms, the present work studied the transcriptomes of ovaries, hemocytes, and salivary glands from rapidly ingurgitating females, and of salivary glands from males of *B. microplus*.

## MATERIALS AND METHODS

### *Construction of cDNA Libraries*

Approximately 50 pairs of salivary glands from males and females, 50 pairs of ovaries, and 500 μL of hemolymph were used to obtain mRNA using the Micro-FastTRack kit (Invitrogen, San Diego, CA). PCR-based libraries were made according to the instructions for the SMART cDNA library construction kit (Clontech, Palo

Alto, CA). Approximately 500 ng of mRNA from each type of tissue were reversed transcribed to cDNA and constitute the libraries, as previously described.[3]

### *Sequencing of the* Boophilus microplus *cDNA Libraries*

Libraries were plated, and resulting plaques were randomly selected and transferred to 96 well plates containing 100 μL of water per well. An aliquot of the phage sample was PCR amplified, and the ESTs of random clones were mass sequenced, as previously described.[3]

### *Sequence Information Cleaning, Searching for Sequence Similarities of the cDNA Sequences and Sequence Clustering*

Bioinformatics treatment of the data was performed with in-house programs written by one of us (J.M.C.R.) in VisualBasic6.0 (Microsoft Corp. Redmond, WA), as detailed previously.[3] The output of the programs are imported into a Microsoft Excel spreadsheet with hyperlinks for each cluster, to the matches in the databases, to FASTA files and CLUSTAL alignments.

## RESULTS AND CONCLUSIONS

A total of 813 EST were generated for ovaries, resulting in 588 unique sequences, with 453 singletons. A BLASTX search of the tags of the nonredundant sequence database gave 292 matches. Of interest are genes of segment polarity and a match with tuftelin, a protein related to biomineralization. EST with similarities to vitellogenins and cathepsins, previously described in *B. microplus*, were also found. For additional EST with similarites of interest, see TABLE 2. Hemocytes have generated 196 EST (157 clusters, with 149 singletons, and 62 with no matches in the database) and contain similarities to a metal-chelating protein, metallothionein, to silk proteins, to Kunitz-type serine protease inhibitors, and to a defensin previously described in hemocytes of *B. microplus*. Female salivary glands have generated 324 EST (188 unique sequences, with 149 singletons). A BLASTX search gave 292 matches. Of interest are clusters similar to metalloproteinases, Kunitz-type protease inhibitors, iron response elements, a major allergen of mites, selenium-binding proteins, and glutathione peroxidases. Interestingly, the transcriptome of male salivary glands has, so far, revealed a different profile of expressed genes related to hematophagy: while it also contains putative metalloproteinases and Kunitz-type protease inhibitors, only male salivary glands have revealed genes with similarities to immunoglobulin- and histamine-binding proteins, and they were abundantly expressed, as determined by the number of clones per cluster (respectively, 13 and 5 clones per cluster). To date, 401 EST have been generated for male salivary glands, of which 89 had no matches to the database. For additional EST with similarities of interest, see TABLE 2.

Approximately 761 nucleotide sequences from *B. microplus* are deposited in the NCBI database, and this number needs to grow in order to provide information that is relevant for tick control strategies. Mass sequencing of partial cDNAs of parasite genes can rapidly generate a large quantity of useful information about potential tar-

**TABLE 2. Putative functions of ESTs of interest from cDNA libraries of *B. microplus*, based on similarities with genes deposited in the NCBI database**

| Ovary | Salivary Gland, Female | Salivary Gland, Male | Hemocytes |
|---|---|---|---|
| Peter pan | Metalloproteinase with thrombospondin motif | Metalloproteinase | Metallthionein |
| Syntaxin | Tropomiosin | Immunoglobulin-binding protein | No–on transient A protein |
| MW domain-binding protein | Selenium-binding protein | Peptidoglycan-recognition protein | Minor ampullate silk protein |
| Catepsin | | Heparanase | Defensin |
| Conglutinin | Dipeptidase | Histamine-binding protein | Kunitz-like protease inhibitor |
| Conglutinin precursor | Actin filament–binding protein | Cement protein from *Rhipicephalus appendiculatus* | Fibroin-like protein |
| Alphacrystallin | Apolipoprotein | Newborn larvae-specific serine protease, trypsin-like | |
| Vitellogenin | Serotonin receptor | Ferritin | |
| Tuftelin | Gluthatione peroxidase | Taste bud-specific protein | |
| Serine protease | Major allergen from mite | Proteoglycan, megakaryocyte stimulatory factor (hempexin domain) | |
| Vitellin envelope protein | Kunitz-type inhibitor of serine protease | Myosin | |
| Enhancer of polycomb | Iron response element | Histamine responsive G protein | |
| | Diphenol oxidase | Silk gland specific protein | |
| Pelo-PI | Selenium-binding protein | Kunitz-like protease inhibitor | |
| Larval glue protein | | Carboxipeptidase | |
| Laminin gamma precursor | | | |
| Yellow protein | | | |
| Leptin receptor | | | |
| Olfactory receptor protein | | | |
| Superoxide desmutase | | | |
| Glutathione peroxidase | | | |

gets for immunobiological or chemical control and can rapidly identify new genes. PCR-based cDNA libraries, that are not normalized *sensu strictu,* can render representative data and, at the same time, furnish information on the relative abundance of expressed genes. The putative functions and the presence of genes in the genome of *B. microplus* must be confirmed, creating a demand for high throughput technologies to evaluate EST in control strategies and other settings.

## ACKNOWLEDGMENT

Work reported here was supported by the Conselho Nacional de Desenvolvimento Cientifico e Tecnologico (CNPg).

## REFERENCES

1. KUNZ, S.E. & D.H. KEMP. 1994. Insecticides and acaricides: resistance and environmental impact. Rev. Sci. Tech. **13:** 1249–1286.
2. VALENZUELA, J.G. 2002. High throughput approaches to study salivary proteins and genes of vectors of disease. Insect Biochem. Mol. Biol. **32:** 1199–1209.
3. VALENZUELA, J.G., V.M. PHAM, M.K. GARFIELD, *et al.* 2002. Toward a description of the sialome of the adult female mosquito *Aedes aegypti*. Insect Biochem. Mol. Biol. **32:** 1101–1122.

# Partial Protection Induced by a BHV-1 Recombinant Vaccine against Challenge with BHV-5

FERNANDO R. SPILKI,[a,b] ALESSANDRA D. SILVA,[a,b] SÍLVIA HÜBNER,[a,b] PAULO A. ESTEVES,[a,b] ANA CLÁUDIA FRANCO,[a,b] DAVID DRIEMEIER,[c] AND PAULO M. ROEHE[a,b]

[a]*Laboratório de Virologia, DM-ICBS/Universidade Federal do Rio Grande do Sul (UFRGS), Porto Alegre, RS, Brazil*

[b]*Equipe de Virologia, Centro de Pesquisas Veterinárias Desidério Finamor, Eldorado do Sul, RS, Brazil*

[c]*Setor de Patologia, Departamento de Patologia e Clínica Veterinária, FAVET, UFRGS, Porto Alegre, RS, Brazil*

ABSTRACT: **Bovine herpesvirus type 5 (BHV-5) is the causative agent of bovine herpetic encephalitis, a major concern for cattle farming in Brazil and Argentina. We recently developed a differential, gE-negative vaccine (265 gE–), based on a Brazilian BHV-1 strain. The present study was carried out to examine whether such a vaccine would confer protection to BHV-5 infections. It was concluded that the recombinant BHV-1 vaccine tested here is not capable of conferring full protection to BHV-5 challenge.**

KEYWORDS: **BHV-1; BHV-5; challenge; protection; vaccine**

## INTRODUCTION

Bovine herpesvirus type 5 (BHV-5), an alphaherpesvirus, is the causative agent of bovine herpetic encephalitis,[1] a major concern for cattle farming in Brazil and Argentina.[2,3] Despite the extensive serological cross-reactions between BHV-5 and infectious bovine rhinotracheitis virus[4] (bovine herpesvirus 1; BHV-1), doubt persists on whether BHV-1 vaccines would be able to protect cattle against BHV-5. Previous experiments using conventional BHV-1 vaccines have shown some degree of protection, but the establishment of disease in nonvaccinated controls was not clear.[5] Under field conditions, in the lack of other control measures, it has been a common practice to vaccinate against BHV-1 when outbreaks of BHV-5 are detected. We have recently developed a differential, gE-negative vaccine (265 gE−), based on a Brazilian BHV-1 strain.[6] The present study was carried out to examine whether such a vaccine would confer protection to BHV-5 infections.

Address for correspondence: Paulo M. Roehe, Laboratório de Virologia, DM-ICBS/Universidade Federal do Rio Grande do Sul (UFRGS), Porto Alegre, RS, Brazil. Voice: +51 481 3711. proehe@ufrgs.br

**Ann. N.Y. Acad. Sci. 1026: 247–250 (2004). © 2004 New York Academy of Sciences. doi: 10.1196/annals.1307.038**

## EXPERIMENTAL DESIGN

Eight calves were vaccinated with the recombinant BHV-1 strain 265 gE– [4 calves intranasally (IN); four calves intramuscularly (IM)]; six calves were kept as controls. Thirty-five days later, both groups were challenged with $10^{8.8}$ 50% tissue culture infectious doses ($TCID_{50}$) of the Brazilian BHV-5 wild-type strain EVI 088/95. The animals were monitored daily until day 21 postchallenge (pc), when 4 animals from each group were culled. Sixty days later, the remaining calves were submit-

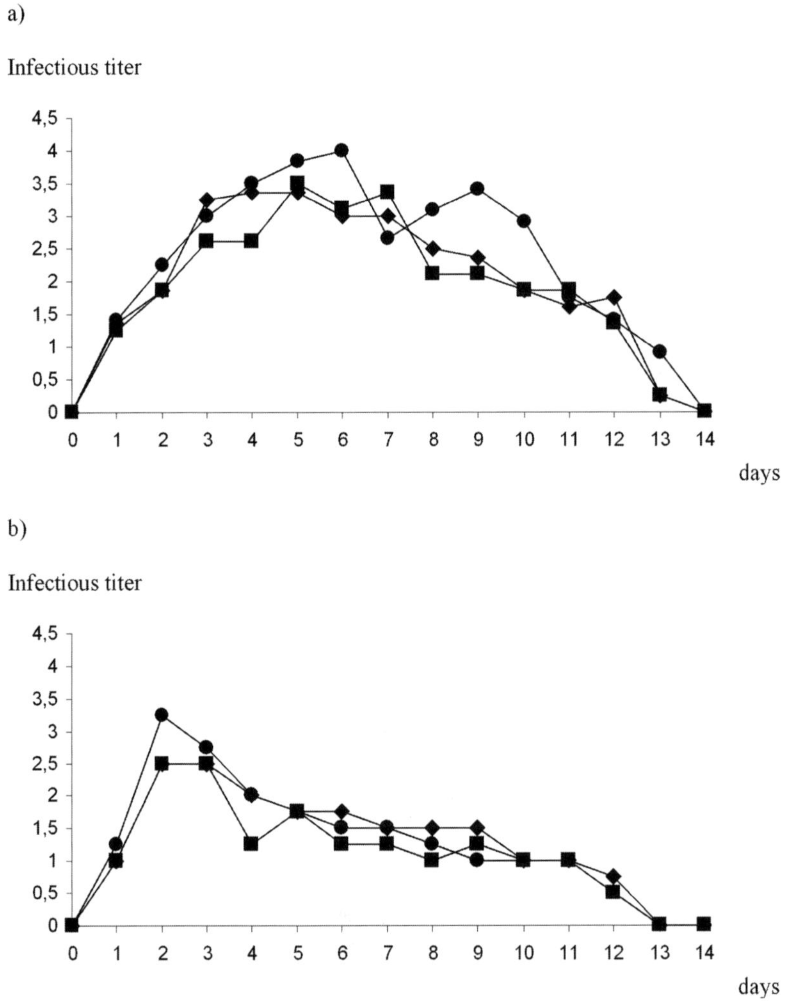


**FIGURE 1.** BHV-5 virus shedding (expressed as log 10): **(a)** after challenge and **(b)** upon corticosteroid-induced reactivation. Note that there are no significant differences between intranasally (*diamonds*) and intramuscularly (*squares*) BHV-1 vaccinated calves and those included on the nonvaccinated group (*circles*). In panel **a**, "days" refers to days postchallenge; in panel **b**, it refers to days after corticosteroid administration.

ted to corticosteroid administration in order to induce virus reactivation. All laboratory and animal care procedures, such as viral isolation and titration, seroneutralization assays, and corticosteroid-induced reactivation, were performed as described previously.[7]

## RESULTS AND DISCUSSION

Calves readily developed antibodies against BHV-1 and cross-reactive neutralizing antibodies against BHV-5; the titers were above 32 at day 35 after vaccination in both IN and IM groups. During acute BHV-5 disease, between days 3 and 11 after infection, mild clinical signs of respiratory disease were evident in both groups of infected calves. In addition, the calves displayed signs of depression of low magnitude. Two calves from the control group were more severely affected and presented pronounced signs of neurological disease (hypersalivation, teeth chewing, recumbency, incoordination, and difficulty in standing) between days 7 and 21 pc, being culled *in extremis*. On day 21 pc, four of the vaccinated calves and two of the remaining calves from the control group were culled. Postmortem examinations revealed that all calves in both groups had typical BHV-5 lesions in the brain, including foci of malacia and sinking areas on the frontal, parietal, and temporal lobes. Mononuclear meningoencephalitis was a consistent microscopical finding. Upon corticosteroid-induced reactivation, no respiratory or nervous signs were noticed, despite evident virus shedding. Atrophic brain lesions were present in three of the vaccinated calves as well as in two calves from the control group. However, it was not possible to determine whether such lesions were consequent to reactivation or primary infection. No significant differences were observed in BHV-5 virus shedding between previously BHV-1 vaccinated and nonvaccinated calves either in the early phase following challenge or during reactivation (FIG. 1).

It is concluded that the recombinant BHV-1 vaccine tested here is not capable of conferring full protection to BHV-5 challenge. Under field conditions, in the lack of other control measures, vaccinations have lead practitioners to recommend BHV-1 vaccination in trying to control BHV-5 outbreaks. Clearly, other BHV-1 vaccines, when applied with this purpose, would have to be equally evaluated; otherwise, such indication may be worthless. Future studies shall be conducted in order to evaluate type-specific vaccines to control BHV-5 infections.

## REFERENCES

1. MURPHY, F.A., E.P. GIBBS, M.C. HORZINEK & M.J. STUDDERT. 1999. Veterinary Virology. Third edition. Academic Press. New York.
2. SALVADOR, S.C., R.A.A. LEMOS, F. RIET-CORREA, *et al.* 1998. Meningoencefalite em bovinos causada por herpesvirus bovino-5 no Mato Grosso do Sul e São Paulo. Pesqui. Vet. Bras. **18:** 76–83.
3. PIDONE, C.L., C.M. GALOSI, M.G. ECHEVERRIA, *et al.* 1999. Restriction endonuclease analysis of BHV-1 and BHV-5 strains isolated in Argentina. Zentralbl. Veterinaermed. B **46**(7)**:** 453–456.
4. BRATANICH, A.C., S.I. SARDI, E.N. SMITSAART & A.A. SCHUDEL. 1991. Comparative studies of BHV-1 variants by *in vivo–in vitro* tests. Zentralbl. Veterinaermed. B **38**(1)**:** 41–48.

5. CASCIO, K.E., E.B. BELKNAP, P.C. SCHULTHEISS, *et al.* 1999. Encephalitis induced by bovine herpesvirus 5 and protection by prior vaccination or infection with bovine herpesvirus 1. J. Vet. Diagn. Invest. **11**(2)**:** 134–139.
6. FRANCO, A.C., F.A.M. RIJSEWIJK, E.F. FLORES, *et al.* 2002. Construction and characterization of a glycoprotein E deletion of bovine herpesvirus type 1.2 strain isolated in Brazil. Braz. J. Microbiol. **33:** 274–278.
7. FRANCO, A.C., F.R. SPILKI, P.A. ESTEVES, *et al.* 2002. A Brazilian glycoprotein E–negative bovine herpesvirus type 1.2a (BHV-1.2a) mutant is attenuated for cattle and induces protection against wild-type virus challenge. Pesqui. Vet. Bras. **22:** 135–140.

# Effect of Various Acupuncture Treatment Protocols upon Sepsis in Wistar Rats

M. V. R. SCOGNAMILLO-SZABÓ,[a] G. H. BECHARA,[a] S. H. FERREIRA,[b] AND F. Q. CUNHA[b]

[a]*Faculty of Agricultural and Veterinary Sciences, São Paulo State University, Jaboticabal–SP, Brazil*

[b]*Faculty of Medicine of Ribeirão Preto, University of São Paulo, Ribeirão Preto–SP, Brazil*

ABSTRACT: **Sepsis is a syndrome characterized by infection and generalized inflammatory response that can lead to organ failure and death. In this study we standardize a model to investigate acupuncture's effects upon sepsis. the objectives were to study the use of acupuncture in the infectious process and to formulate acupuncture's treatment protocol for sepsis. The CLP (cecal ligation and puncture) model in rats was used to induce sepsis through bacterial entrance into the peritoneal cavity. An acupuncture treatment protocol that enhanced survival and reversed the neutrophil impairment migration toward the peritoneal cavity in rats with sepsis was achieved. It seems that acupuncture can be used for the treatment of experimental infectious processes. The effects of acupuncture and related mechanisms are discussed.**

KEYWORDS: **acupuncture; sepsis; rats; neutrophil migration; cecal ligation and puncture**

## INTRODUCTION

Acupuncture is the insertion of needles in cutaneous-specific locations of the body, known as acupoints, for the treatment or prevention of several diseases, including asthma, rhinitis, inflammatory bowel disease, and rheumatoid arthritis.[1–5] Although this technique is increasingly used for the treatment of pain and other conditions, the rational basis underling its use remains unclear. Improved knowledge of the acupuncture therapeutic mechanism is essential to validating it since acupuncture is difficult to test under double-blind and placebo-controlled conditions.[6–11] In fact, few reports concerning the effect of acupuncture on inflammatory/infectious models are available, although clinical trials claim the success of acupuncture in inflammatory disorders.

Address for correspondence: M.V.R. Scognamillo-Szabó, Faculty of Agricultural and Veterinary Sciences, São Paulo State University, Jaboticabal–SP, Brazil. Voice: +55-16-3209-2662; fax: +55-16-3202-4275.
szabo@asbyte.com.br

**Ann. N.Y. Acad. Sci. 1026: 251–256 (2004). © 2004 New York Academy of Sciences.**
**doi: 10.1196/annals.1307.039**

Sepsis and septic shock are intense systemic inflammatory responses with multiple physiological and immunological abnormalities commonly caused by bacterial infection and possibly leading to organ failure and death.[12] Neutrophil migration to the infectious focus is extremely important for the control of bacterial growth and consequently for the prevention of bacterial dissemination. The importance of this phenomenon to the evolution of sepsis has been demonstrated clearly in a cecal ligation and puncture (CLP) model. In lethal CLP, failure of neutrophil migration to the infectious focus is accompanied by an increased number of bacteria in the peritoneal fluid and blood and by a high mortality rate. Conversely, in nonlethal CLP, in which the impairment of neutrophil migration is not observed, the bacterial infection is restricted to the peritoneal cavity, and the animals exhibit increased survival.[13–16]

In this study we standardized a protocol of treatment to investigate the effects of acupuncture on neutrophil migration in sepsis. The cecal ligation and puncture (CLP) model was used in rats to induce the entrance of bacteria into the peritoneal cavity, leading to sepsis.

## MATERIALS AND METHODS

### *Induction of Sepsis*

Male Wistar rats (weight 230–250 g) were used throughout the study. Sepsis was induced by cecal ligation and puncture (CLP), as described elsewhere.[13] Briefly, rats were anesthetized with 2.5% tribromoethanol, and the cecum was exposed and ligated below the ileocecal junction, without causing bowel obstruction, and punctured 4 or 20 times with a 16G gauge needle. Animals given 4 or 20 punctures show 100 or 0% survival, respectively; we called these nonlethal (NL) and lethal (L) CLP, respectively.

### *Experimental Design*

Animals that underwent CLP were submitted to two different protocols of manual acupuncture (test groups). Six hours later, treatment test and control (nontreated) groups were analyzed for neutrophil migration and bacteria count in the peritoneal cavity. Negative control groups consisted of naive animals. All experiments were performed in the morning between 8:00 A.M. and 11:00 A.M. to avoid circadian interference.[17,18]

### *Treatments*

Manual acupuncture was performed immediately after CLP. Detailed descriptions of treatment protocols are shown in TABLES 1 and 2. The acupoints were selected based on traditional treatment of fever-related diseases.[19] For acupuncture sessions, animals were lightly immobilized using hands to minimize stress. The nontreated control group was also lightly immobilized using the same method. Acupuncture points were identified by one of the authors (M.V.R.S.S), an experienced acupuncturist.

**TABLE 1. Acupuncture treatment protocol 1 used in rats with CLP-induced sepsis**

| Time | Acupoint | Acupoint Localization |
|---|---|---|
| Zero | S 36 | Below the cranial crest of the tibia, in the belly of the tibialis cranialis muscle |
| 2 h | GV 01 | Midline between anus and tail |
| 4 h | GV 01 | Midline between anus and tail |

**TABLE 2. Acupuncture treatment protocol 2 used in rats with CLP-induced sepsis**

| Time | Acupoint | Acupoint Localization |
|---|---|---|
| Zero | B 25 | Between the apical ends of the transverse process of L5-L6 vertebrae |
| 1:30 h | GV 01 | Midline between anus and tail |
| | GV 03 | Lumbosacral space |
| 4:30 h | GV 01 | Midline between anus and tail |
| | GV 14 | Space between C7 and T1 vertebrae |
| | Liv 02 | In the distal third of the depression between metatarsal I and II |
| 6 h | Er Jien | On the convex surface of the tip of the ear |
| | GV 14 | Space between C7 and T1 vertebrae |
| | LI 11 | In the depression between the dorso-lateral condyle of humerus and the processes anconaeus |
| 10 h | GV 01 | Lumbosacral space |
| 14 h | GV 01 | Lumbosacral space |

### *Neutrophil Migration and Number of Bacteria in Peritoneal Cavities*

Six hours after CLP the animals were killed in an ether chamber, and the peritoneal cavity was washed by injecting 10 mL of sterile PBS containing 1 mM EDTA. Aliquots of serial log dilutions of the peritoneal lavage fluid were plated on Miller-Hinton agar dishes. Colony-forming units (CFU) were counted after overnight incubation at 37°C, and the results were expressed as the number of CFU per cavity. Total leukocyte counts were made in a cell counter, and differential cell counts were made on cytocentrifuge slides stained by the May-Grünwald-Giemsa method (Rosenfeld). The results are expressed as the number of neutrophils per cavity.

### *Statistical Analyses*

Neutrophil and bacteria counts are reported as mean ± SE. The means between different treatments were compared by analysis of variance. If significance was determined, individual comparisons were subsequently tested with the Bonferroni test for unpaired values. Statistical significance was set at $P < .05$.

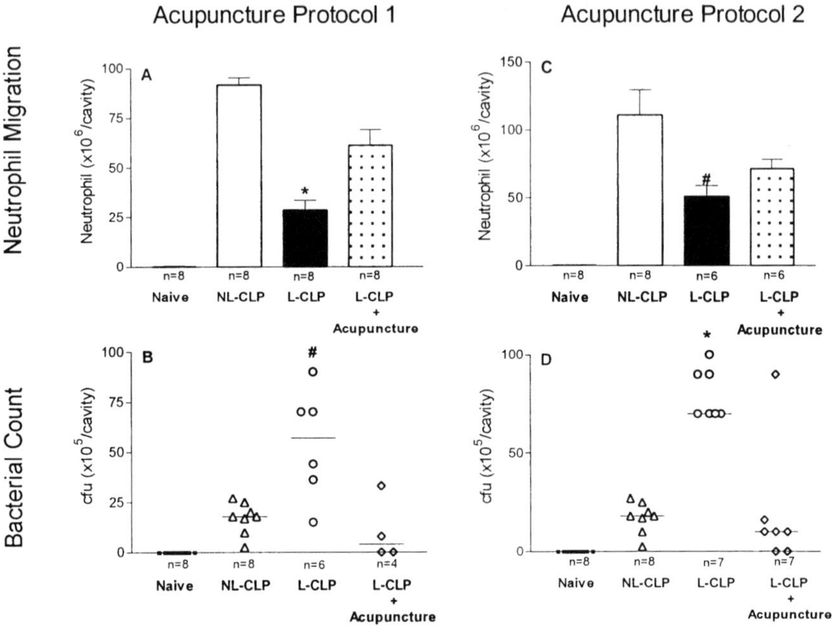


**FIGURE 1.** Effect of acupuncture treatment protocols on neutrophil migration and bacterial counts in the peritoneal cavities of rats 6 h after nonlethal (NL) and lethal (L) cecal ligation and puncture (CLP). Neutrophil migration and quantification of the amount of bacteria were also determined in naive rats. Results are expressed as mean ± SE of neutrophils per cavity and as mean CFU per cavity and are representative of two independent experiments. The number of animals in the different experimental groups are indicated below the bars. #$P$<.05 and *$P$<.001 compared with NL-CLP (analyses of variance, followed by the Bonferroni test).

## RESULTS

The number of bacteria present in the peritoneal cavities of L-CLP rats was 2.5-fold higher than that observed in NL-CLP. Despite this, the L-CLP group presented failure of neutrophil migration into peritoneal cavities, compared with NL-CLP. The impairment of neutrophil migration observed in L-CLP was partially reverted by acupuncture (FIG. 1A and C), and, as a consequence, the number of bacteria present in the peritoneal cavities of acupunctured rats was lower compared with L-CLP non-treated animals (FIG. 1B and D). The reestablishment of neutrophil migration in L-CLP was more evident with protocol 1.

## DISCUSSION

During sepsis several physiopathological events are clearly associated with the presence of the inflammatory mediators in the circulation, which induces accumulation and activation of leukocytes, in important organs, such as the lung. Nevertheless, inhibition of neutrophil migration to infection focus is observed and often associated with uncontrolled bacterial growth and subsequent bacteremia, which may then lead to increased lethality.[20] The relationship found between reduced neutrophil chemotaxis and poor prognosis in human subjects suggests that restoration of the neutrophil chemotatic function could be an appropriate strategy in the septic patient.[14] The high level of cytokines and nitric oxide in the circulation seems to mediate this event. So, pharmacological approaches that reduced the production of systemic cytokines and nitric oxide reestablished the neutrophil migration failure and consequently controlled the infection focus.[14–16] In the present study we observed that acupuncture partially reestablished the neutrophil migration and reduced the number of bacteria in the infectious focus. The effect of acupuncture on the production of systemic cytokines and nitric oxide is under investigation. In this regard, there is evidence in the literature indicating that acupuncture treatment reduces the intensity of the inflammatory process observed in experimental and human inflammatory diseases, including arthritis, epicondylitis, complex regional pain syndrome type 1, vasculitis, and reduces the production of the inflammatory mediators.[3–5,21–27]

## ACKNOWLEDGMENTS

We thank Ana Kátia dos Santos for technical assistance. This work was supported by Fundação de Amparo à Pesquisa do Estado de São Paulo (FAPESP), Brazil.

## REFERENCES

1. SCHOEN, A.M., Ed. 1994. Veterinary Acupuncture: Ancient Art to Modern Medicine. American Veterinary Publications. Goleta, CA.
2. ERNST, E. & A. WHITE. 1999. Acupuncture, a Scientific Appraisal, p. 30–59. Butterworth Heinemann. Oxford.
3. DAVID, J., S. TOWNSEND, R. SATHANATHAN, S. KRISS & C.J. DORE. 1999. The effect of acupuncture on patients with rheumatoid arthritis: a randomized, placebo-controlled cross-over study. Rheumatology **38:** 864–869.
4. TUKMACHI, E. 1999. Acupuncture treatment of osteoarthritis. Acupuncture in Med. **17:** 65–67.
5. ZIJLSTRA, F.J. I. VAN DEN BERG-DE LANGE, F.J.P.M. HUYGEN & J. KLEIN. 2003. Anti-inflammatory actions of acupuncture. Mediators Inflamm. **12:** 59–69.
6. STREITBERGER, K. & J. KLEINHENZ. 1998. Introducing a placebo needle into acupuncture research. Lancet 352: 364–365.
7. TUKMACHI, E. 2000. Acupuncture and rheumatoid arthritis. Rheumatology **39:** 1153–1154.
8. LANGEVIN, H.M., D.L. CHURCHILL, J.R. FOX, *et al.* 2001. Biomechanical response to acupuncture needling in humans. J. Appl. Physiol. **91:** 2471–2478.
9. LANGEVIN, H.M. & J.A. YANDOW. 2002. Relationship of acupuncture points and meridians to connective tissue planes. The Anatomical Record (New Anatomist) **269:** 257–265.
10. CASIMIRO, L., L. BROSSEAU, S. MILNE, *et al.* 2004. Acupuncture and electroacupuncture for the treatment of RA (Cochrane Review). *In* The Cochrane Library. Issue 2, Oxford: Update Software.
11. SHERMAN, K.J., C.J. HOGEBOOM, D.C. CHERKIN & R.A. DEYO. 2002. Description and validation of a nonivasive acupuncture procedure. J. Altern. Complement. Med. **8:** 11–19.

12. SCCM-Society of Critical Care Medicine. [on line] <http://www.sccm.org/estatistics>, February 28, 2003.
13. Wichtermann, K.A., A.E. Baue & I.H. Chaundry. 1980. Sepsis and septic shock: a review of laboratory models and a proposal. J. Surgery Res. **29:** 189-201.
14. Tavares-Murta, B.M., M. Zaparoli, R.B. Ferreira, *et al.* 2002. Failure of neutrophil chemotatic function in septic patients. Crit. Care Med. **30:** 1056-1061.
15. Benjamim, C.F., J.S. Silva, Z.B. Fortes, *et al.* 2002. Inhibition of leukocyte rolling by nitric oxide during sepsis leads to reduced migration of active microbicidal neutrophils. Infect. Immun. **70:** 3602-3610.
16. Crossara-Alberto, D.P. A.L.C. Darini, R.Y. Inoue, *et al.* 2002. Involvement of NO in the failure of neutrophil migration in sepsis induced by *Staphylococcus aureus*. Br. J. Pharmacol. **136:** 645-658.
17. Chen, X. & Z. He. 1989. A quantitative study of circadian variations in mast cell number in different regions of the mouse. Acta Anatomica 136: 222-225.
18. Storch, K.F. 2002. Extensive and divergent circadian gene expression in liver and heart. Nature **417:** 78-83.
19. Maciocia, G. 1999. [Fundamentals of Chinese Medicine]. Editora Roca. Rio de Janeiro.
20. Teixeira, M.M., F.Q. Cunha & S.H. Ferreira. 2001. Editorial: Response to an infectious insult (e.g., bacterial or fungal infection). Braz. J. Med. Biol. Res. **34**(5): preceding 555.
21. Sin, Y.M., M.S. Gwee & M.S. Loh. 1984. Electric acupuncture on carrageenan-pleurisy: comparative study using various body regions for stimulation. Am. J. Acupunct. **12:** 355-358.
22. Ceccherelli, F., G. Gagliardi, R. Visentin, *et al.* 1999. The effects of parachlorophenylalanine and naloxone on acupuncture and electroacupuncture modulation of capsaicin-induced neurogenic edema in the rat hind paw. A controlled blind study. Clin. Exp. Rheumatol. **17:** 655-662.
23. Zhang, Y.Q., G.C. Ji, G.C. Wu & Z.Q. Zhao. 2002. Excitatory amino acid receptor antagonists and electroacupuncture synergetically inhibit carrageenin-induced behavioral hyperalgesia and spinal fos expression in rats. Pain **99:** 525-535.
24. Kwon, Y.B., H.J. Lee, H.J. Han, *et al.* 2002. The water-soluble fraction of bee venom produces antinociceptive and anti-inflammatory effects on rheumatoid arthritis in rats. Life Sci. **31:** 191-204.
25. Kwon, Y.B., J.D. Lee, H.J. Lee, *et al.* 2001. Bee venom injection into an acupuncture point reduces arthritis associated edema and nociceptive responses. Pain **90:** 271-280.
26. Song, X., Z. Tang, Z. Hou & S. Zhu. 1993. An experimental study on acupuncture anti-hemorrhagic shock. J. Tradit. Chin. Med. **13:** 207-210.
27. Niimi, H., S. Yamaguchi, Q-H. Hu & F-Y. Zhuang. 2001. Microvascular vasodilatatory responses to electric acupuncture in rat brain under acute hemorrhagic hypotension. Clin. Hemorheol. Microcirc. **23:** 191-195.

# Immunization of Bovines Using a DNA Vaccine (pcDNA3.1/MSP1b) Prepared from the Jaboticabal Strain of *Anaplasma marginale*

G. M. DE ANDRADE,[a] R. Z. MACHADO,[a] M. C. VIDOTTO,[b] AND O. VIDOTTO[b]

[a]*Universidade Estadual Paulista–UNESP, Jaboticabal, SP, Brazil*

[b]*Universidade Estadual de Londrina, UEL, Londrina, PR, Brazil*

**ABSTRACT: Anaplasma is a tick-borne ehrlichial pathogen of cattle that causes the disease, anaplasmosis. In the present study, a total of 11 *Anaplasma marginale* seronegative calves were assigned into two groups: one immunized (G1, *n* = 6) and one nonimmunized-control (G2, *n* = 5). Six calves were immunized by using a DNA vaccine containing the gene of a major surface protein, MSP1b, encoded by the plasmid identified as pcDNA3.1/MSP1b. Calves received three intramuscular inoculations of 100 μg of pcDNA3.1/MSP1b at a 20-day interval. The control group received buffer phosphate at the same schedule as the experimental group. The immune response elicited by immunization with pcDNA3.1/MSP1b was evaluated in mice and calves. Twenty days following initial immunization, specific serum antibody from four BALB/c mice bound MSP1b in immunoblots. Sixty days after the last immunization, all calves were challenged with cryopreserved *A. marginale* at a dose of $10^4$ parasites/mL/animal by intravenous injection. Results of packed cell volume (PCV) and detection of infected erythrocytes in all experimental groups revealed that the decrease of PCV and detection of infected erythrocytes occurred at 28 to 42 days after challenge. Mean temperature values did not increase over 39.85°C. Antibodies developed by immunized bovines from G2 were detected 14 days after challenge. MSP1b was characterized during the immunization period and MSP2 was the most predominant polypeptide at the challenge period. DNA of *A. marginale* was detected in all groups just after challenge by nested PCR assay. It can be concluded that all immunized bovines were partially protected against homologous challenge.**

**KEYWORDS: *Anaplasma marginale*; MSPs; DNA; bovines; immunization**

## INTRODUCTION

Anaplasmosis is a tick-transmitted hemoparasite disease caused by the ehrlichial pathogen, *Anaplasma marginale*.[1] The disease causes significant morbidity and mortality in tropical and subtropical regions worldwide. Clinical disease is characterized by severe anemia, weight loss, fever, abortion, decreased milk production,

Address for correspondence: R. Z. Machado, Universidade Estadual Paulista–UNESP, Jaboticabal, SP, Brazil.
zacarias@fcav.unesp.br

**Ann. N.Y. Acad. Sci. 1026: 257–266 (2004). © 2004 New York Academy of Sciences.**
**doi: 10.1196/annals.1307.040**

and death.[2] Cattle that recover from acute disease become persistently infected and are lifelong carriers, serving as a reservoir for the transmission of *A. marginale*, either biologically, by ixodid ticks, or mechanically, by blood, contaminated needles, or biting flies.[3] Infected cattle are protected from challenge infection with homologous strains and are partially protected from challenge with heterologous strains.[3] In Brazil, bovine anaplasmosis and babesiosis are important tick-borne diseases of cattle, and the principal vector of both diseases is the tick, *Boophilus microplus*, causing substantial livestock losses estimated as over 2 billion dollars per year.[4] Losses due to mortality and morbidity, limited upgrading of local herds by restricted importation of genetically superior cattle, and the loss of exportation and other sale markets are significant.[5] Control methods for anaplasmosis include premunition by infection of cattle with infected blood from *A. marginale* carrier cattle, and infection with attenuated live or less pathogenic *A. centrale* live vaccines.[6] Immunization of cattle using killed whole organisms or purified outer membranes has been shown to induce partial protection against high-level rickettsemia and severe disease.[7] The use of inactivated *A. marginale* immunization to protect cattle has been reported and attempts to make improved vaccine have been made by purifying and using specific surface membrane proteins.[3] Six major surface proteins (MSPs) from *A. marginale* have been well characterized, which include MSP1a, MSP1b, MSP2, MSP3, MSP4, and MSP5.[8] MSP1a is encoded by a single-copy gene containing a neutralization sensitive epitope,[9] whereas MSP1b is encoded by a polymorphic gene family among isolates.[10] MSP1a was shown to be an adhesin for bovine erythrocytes and tick cells, while the MSP1b was shown to be an adhesin for bovine erythrocytes.[11] Immunization of cattle with affinity-purified native MSP1 complex (a heterodimer containing MSP1a and a 100-kDa protein designated MSP1b) induced protective immunity against challenge with homologous and heterologous strains of *A. marginale.*[3] Previous work demonstrated that bovines immunized with a DNA-mediated vaccine using the plasmid pVCL/MSP1a, which encoded the complete *msp*1a gene *A. marginale*, had developed antibody titers.[12] Since relatively few studies using DNA vaccines have been performed with large animals, the present study aimed to immunize *A. marginale* sera-negative calves using a DNA vaccine (pcDNA3.1/MSP1b) followed by challenge with a virulent *A. marginale* strain.

## MATERIALS AND METHODS

### *Construction and Purification of pcDNA3.1/MSP1b for Immunization*

Genomic DNA was extracted from stabilates of the Jaboticabal strain of *A. marginale* using a Puregene (Gentra) DNA extraction kit according to manufacturer's recommendation. The *msp*1b gene was amplified by PCR using primers 5′-CACCATGACAGAAGACGACAA-3′ and 5′-CCTAGACCAACCAGAAGACTGC-3′ (MSP1b: GeneBank accession number M59845) for insertion into the DNA expression vector plasmid pcDNA3.1 (Invitrogen) and transformed into *Escherichia coli* One Shot Top 10 (Invitrogen) according to the manufacturer's instructions. Restriction enzyme digestion with BamH I and Xho I identified a construct containing the correct orientation, and this plasmid was designated pcDNA3.1/MSP1b. Purification of pcDNA3.1/MSP1b for vaccination was carried out by alkali lysis methods using Qiagen

Plasmid Midi and Maxi kit according to instructions in the manual. DNA was ethanol-precipitated, resuspended in sterile PBS, and maintained at –20°C. The purity of DNA was confirmed by measuring the optical density at 260 nm, followed by 1% agarose gel electrophoresis and ethidium bromide staining.

### *Immunization of Mice with pcDNA3.1/MSP1b*

Six 6-week-old male BALB/c mice were injected intramuscularly with 10 μg of the pcDNA3.1/MSP1b vaccine ($n = 4$) or phosphate buffer saline (PBS) as controls ($n = 2$) by using an insulin syringe and a 27-gauge needle (Becton Dickinson). Mice were immunized at days 0 and 21 with 10 μg of pcDNA3.1/MSP1b. Pre- and post-immune serum samples were collected from the tail vein and stored at –20°C until they were tested for MSP1b-specific antibody by Western blotting.[13]

### *Immunization of Calves with pcDNA3.1/MSP1b and Parasite Challenge*

Eleven Holstein calves were used in this experiment. The calves were shown to be negative for antibody to *A. marginale* by ELISA[14] and by cELISA (VMRD, Inc.) prior to and during all immunization periods. A group of six calves (G1), 4–6 months old, were immunized intramuscularly with 100 μg/1.6 mL of pcDNA3.1/MSP1b using a 22-gauge needle. Calves were injected at days 0, 20, and 40. Control group calves (G2, $n = 5$) were injected with an equivalent amount of PBS. All three immunizations were given in the neck muscle. Calves were challenged 60 days after the first immunization with cryopreserved *A. marginale* at a dose of $10^4$ parasites/mL/animal. Calves were bled during vaccine application days and at 0, 14, 21, 28, 35, 42, and 47 days after challenge. During the entire experimental period, calves were evaluated by rectal temperature, packed cell volume (PCV), and rickettsemia. Serum samples were tested for the presence of antibodies by ELISA[14] and cELISA (VMRD, Inc.) assays. Characterization of proteins was by Western blotting.[13] Blood samples were used for detection of *A. marginale* by nested PCR.[15]

### *Antibody Analysis by ELISA and cELISA*

ELISA was carried out as described.[14] Immunolon 2 microtiter plates (Dynatech Lab, Inc.) were coated with 100 μL per well of a 5 μg/mL solution of MSP antigens and incubated overnight at 4°C. Plates were blocked with 5% skim milk for 1 h at 37°C with carbonate/sodium bicarbonate buffer (pH 9.6). Plates were washed five times with PBST (0.5% Tween 20 in PBS). Sera were diluted 1:400 with PBST-added 5% rabbit normal sera. The plates were incubated with the sera for 90 min at 37°C, washed five times with PBST, and then incubated with rabbit antibovine IgG-alkaline phosphate conjugate (Sigma) diluted 1:10,000 in PBST-added 5% skim milk for 90 min at 37°C. Plates were washed again, developed with pNPP (Sigma) for 50 min, and finally stopped with 25 μL of 3.2 M NaOH, and read on an ELISA reader (Dynex) at 405 nm. Antibody titers were determined as the maximum dilution of the serum that yielded an OD value that was at least twice and a half as high as the control negative serum.

A competitive enzyme-linked immunosorbent assay (cELISA) kit based on antibody binding to recombinant *A. marginale* MSP5 was performed as previously described following manufacturer's recommendations (VMRD, Inc.). Aliquots of sera were added at 70 μL per well, and plates (plate B) were incubated for 30 min at RT. Fifty μL of each serum was then transferred to plate A and incubated for 1 h at RT. Wells were then washed twice with 200 μL of wash buffer per well. Fifty μL of Mab ANAF16C1 conjugated to horseradish peroxidase was added to each and incubated for 20 min at RT. After wells had been washed, 50 μL of substrate solution was added to each well and then incubated for 20 min at RT. Reaction was finally stopped with 50 μL of stop solution. Plates were read on an ELISA reader (Dynex) at 620 nm. To interpret the results, the % inhibition was calculated as follows:

$$100 - [(\text{sample OD} \times 100) \div (\text{mean negative control OD})] = \%\ \text{inhibition}.$$

Therefore, samples having <30% inhibition are negative and samples having ≥30% inhibition are positive.

### *SDS-PAGE and Western Blotting*

SDS-PAGE and Western blotting were carried out essentially as described elsewhere.[13] All sera were used at 1:50 dilution. Rainbow marker (Amersham), high-molecular-weight range, was used on each immunoblot.

### *Nested PCR for Detection of* A. marginale

Nested PCR was performed as described[15] to confirm the presence of *A. marginale*. The primers, obtained from the DNA sequence of the *msp*5 gene of Florida *A. marginale* (MSP5: GeneBank M3392), were external forward 5′-CATAGCCTCCCCCTCTTTC-3′, external reverse 5′-TCCTCGCCTTGCCCCTCAGA-3′, and internal forward 5′-TACACGTGCCCTACCGACTTA-3′. PCR was performed in a final volume of 25 μL with a Genius thermocycler (Techne) for 5 min at 95°C, followed by 35 cycles at 95°C for 1 min, 65°C for 2 min, and 72°C for 1 min, with a final extension at 72°C for 10 min followed by cooling at 4°C. A 100-bp ladder (Invitrogen) was coelectrophoresed to serve as a molecular size standard. A DNA band of 345-bp size was visualized with ultraviolet light and photographed with Eagle Eye Jr equipment (Stratagene).

## RESULTS

### *Antibody Responses in BALB/c Mice Immunized with pcDNA3.1/MSP1b*

Twenty days following initial immunization, *A. marginale*–specific antibodies in sera collected at day 21 from four BALB/c mice bound MSP1b in immunoblotting (FIG. 1). Control mice serum samples did not shown any reactivity to MSP1b at day 21 (FIG. 1).

### *Clinical Parameters at Challenge Exposure*

Four out of six calves (G1) presented clinical signs of anaplasmosis that included pale mucus, mild depression, and low food intake. Results of rickettsemia, temperature, and PCV are presented (TABLE 1). Initial rickettsemia was observed at 28 days after challenge (DAC) ranging from 0.8% to 1.34%. Mean temperature ranged from

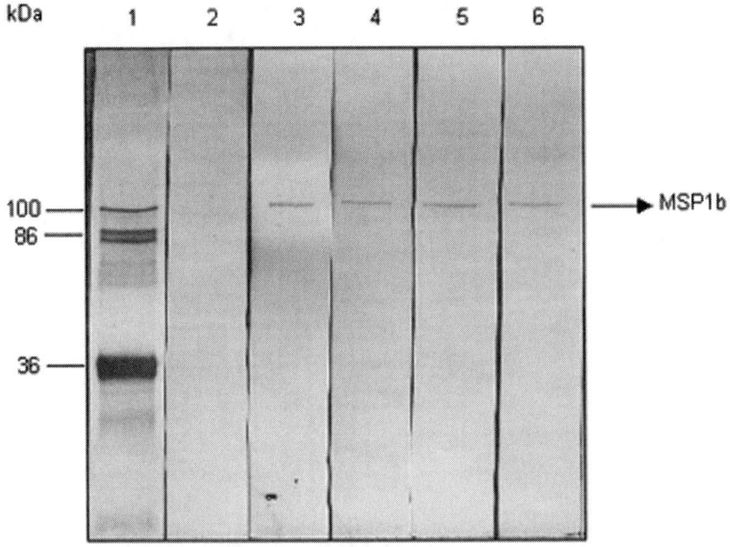


**FIGURE 1.** Recognition of MSP1b by antibodies in sera from mice ($n = 4$) at 21 days after the first immunization with pcDNA3.1/MSP1b. Each lane contains 5 μg of *A. marginale* homogenate: molecular mass marker (*lane 1*); nonimmunized mice (*lane* 2); mice immune sera (*lanes 3–6*).

38.8°C to 39.85°C at challenge exposure. There was a maximal PCV loss of 36.06% at 35 DAC. Four calves were treated against anaplasmosis at 35 DAC. Likewise, at 35 DAC, calves from G2 ($n = 4$) presented clinical signs of anaplasmosis similar to the immunized group. Rickettsemia observed from 35 to 47 DAC ranged from 0.07% to 1.53% (TABLE 1). Mean temperature ranged from 38.2°C to 38.7°C at challenge exposure. There was a maximal PCV loss of 18.31% at 35 and 42 DAC.

### *Antibody Responses in Calves Immunized with pcDNA3.1/MSP1b during the Experimental Period*

Antibody levels against *A. marginale* were detected 20 days after the last immunization in calves from G1, and an ELISA mean optical density (OD) value for serum samples of 0.45 was obtained. Furthermore, at challenge exposure, mean ODs ranged from 0.46 to 0.71 as detected by ELISA, and values ranging from 58.75% to 86.25% inhibition were obtained by cELISA. Additionally, antibodies developed by nonimmunized calves were detected 14 DAC. Mean ODs from serum ranged from 0.32 to 0.50 detected by ELISA, and values ranging from 61.25% to 86.25% inhibition were detected by cELISA at challenge exposure (data not shown).

Twenty days following the third immunization and 42 DAC, serum antibody from calves (G1) bound MSP1b in immunoblots (FIG. 2). Forty-two DAC, serum antibodies from calves bound MSP1, MSP2, and MSP3, whereas the protein of 36 kDa (MSP2) was a dominant band as it reacted strongly with the sera of all the animals (FIG. 2).

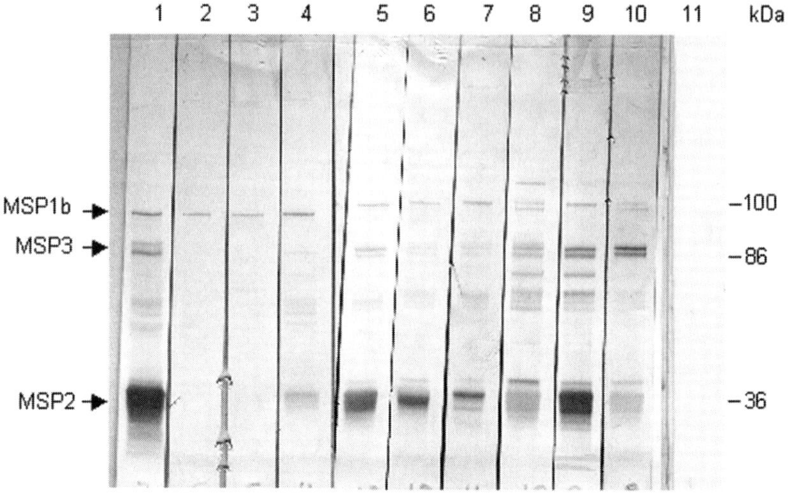


**FIGURE 2.** Initial body polypeptides of the Jaboticabal strain of *A. marginale* recognized by sera from calves immunized with pcDNA3.1/MSP1b (G1). Sera were collected from calves at 20 days after the third dose and 42 days after challenge diluted 1:50, and reacted with *A. marginale* antigen (40 µg per lane) in immunoblots: positive control (*lane 1*); negative control (*lane 11*); sera from calves after the third vaccine (*lanes 2–4*); sera from immunized calves and at 42 days after challenge (*lanes 5–10*).

**TABLE 1. Parameters of protection in cattle vaccinated with pcDNA3.1/MSP1b from *Anaplasma marginale* and challenged with homologous strain**

| | Rickettsemia (%) | | Mean temperature (°C) | | PCV reduction (%) | |
|---|---|---|---|---|---|---|
| Period | V | C | V | C | V | C |
| Before | 0 | 0 | 38.6 | 38.6 | — | — |
| After challenge | | | | | | |
| 14 DAC | 0 | 0 | 39.1 | 38.2 | 11.47 | 12.67 |
| 21 DAC | 0 | 0 | 38.6 | 38.2 | 14.76 | 15.49 |
| 28 DAC | 0.8–1.34 ($n$ = 4) | 0 | 39.8 | 38.3 | 26.13 | 11.26 |
| 35 DAC | 0.9 ($n$ = 2)* | 1.23–1.53 ($n$ = 2)* | 38.5 | 39.4 | 36.06 | 18.31 |
| 42 DAC | 1.12 ($n$ = 1) | 0.1 ($n$ = 1) | 39.3 | 38.2 | 26.22 | 18.32 |
| 47 DAC | 0 | 0.07 ($n$ = 1) | 38.8 | 38.7 | 18.03 | 9.85 |

TERMS: V, vaccinated; C, control; *, day of treatment; *n*, number of calves; —, no reduction.

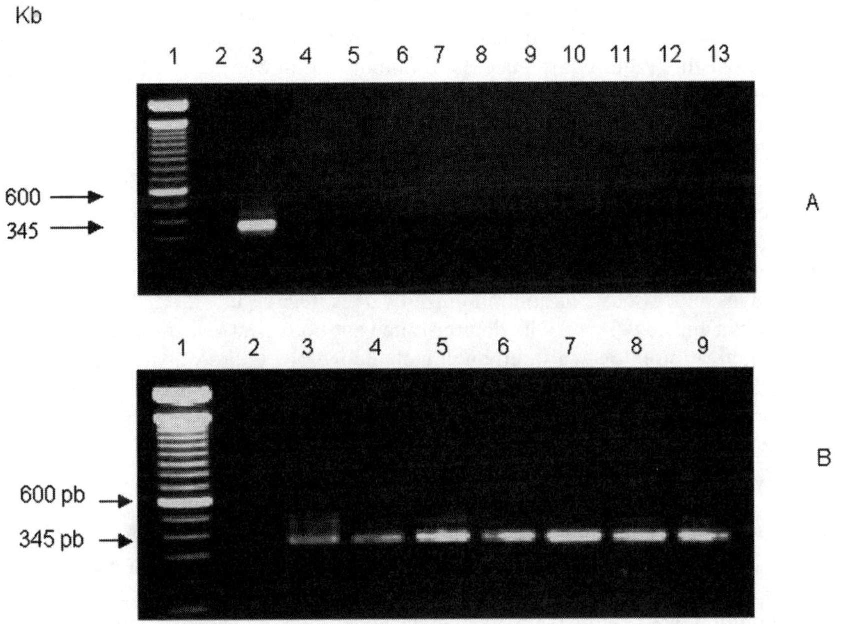


**FIGURE 3.** Detection of *A. marginale* DNA in blood samples obtained from calves immunized with pcDNA3.1/MSP1b (G1) before immunization (*lanes 4–13*) (**A**) and 35 days after challenge (*lanes 4–9*) (**B**). For panels **A** and **B**: molecular size markers, 100 bp (*lane 1*); negative control (*lane 2*); positive control, Jaboticabal strain (*lane 3*). The size of the nested PCR products (345 bp) is indicated in the left margin.

### *PCR Assay*

Direct detection of *A. marginale* by nested PCR in blood of calves before immunization, during the process of immunization, and after challenge showed that DNA of *A. marginale* was detected in both groups just after challenge (FIG. 3).

## DISCUSSION

This study demonstrates, for the first time, the induction in cattle of antibody immune response to a rickettsial gene delivered as DNA vaccine in postimmunization and postchallenge periods. Both BALB/c mice and calves (G1 and G2) developed immune responses induced by the pcDNA3.1/MSP1b, which expressed MSP1b. Mice immunized sera demonstrated specific reactivity to only MSP1b, as demonstrated in Western blotting. This result shows that muscle cells release antigen locally, which is then processed by antigen-presenting cells (APCs). Plasmid DNA may be taken up and protein expressed directly by APCs, which are attracted to the immunization site by inflammatory mediators.[16] Furthermore, immunization of plasmid DNA within transfected dendritic cells enhanced the immune response to plasmid-encoded proteins.[17] In addition, the number of inoculations of a DNA-encoding plasmid required for successful immunization was apparently variable. In our experiments, four mice received two inoculations of pcDNA3.1/MSP1b, whereas all mice developed antibody against MSP1b. In contrast, mice immunized four times with plasmid encoding MSP1b-F2 did not develop detectable antibody titers, despite two mice immunized with plasmid encoding MSP1b-F3 VR4 who developed detectable antibodies.[18] Arulkanthan and coworkers[12] reported that six mice inoculated with plasmid encoding MSP1a developed antibody against the immunogen. In our experiments, antibody was first detected 20 days after the third dose in immunized calves and 14 DAC in nonimmunized calves. Our results differ from those of Arulkanthan and coworkers, who demonstrated antibody first detection after the second immunization in one calf, although a challenge procedure was not carried out. The MSP1 complex of *A. marginale* has been shown to play an important role in the infection of host cells.[19] Both MSP1a and MSP1b were adhesins for bovine erythrocytes, suggesting that these sites be targeted in the development of subunit vaccines against anaplasmosis.[20] Although MSP1b was demonstrated not to be an adhesin for tick cells, the binding of MSP1b antibody appears to effectively block the adhesion of *A. marginale* to tick cells and may be due to the close association of the two surface proteins.[21] When recombinant MSP1a and MSP1b are presented separately to the bovine immune system rather than as a complex, it appears to allow for recognition of all epitopes involved in the generation of the neutralizing antibody response.[22] In our experiment, immunization of calves (G1) induced partial protective response against a virulent homologous challenge. *A. marginale*–specific antibodies in the sera from calves immunized with pcDNA3.1/MSP1b were detected by ELISA at 20 days after the last immunization, and serum antibodies from calves immunized recognized the MSP1b by Western blotting. After challenge, serum antibodies from all calves recognized MSP1, MSP2, and MSP3. Different studies carried out in Brazil, approaching antigenic characterization of geographically distinct Brazilian *A. marginale* isolates, have revealed common and different peptides found among the isolates. Kano and coworkers,[23] comparing six Brazilian isolates from Mato Grosso do Sul (MS), Minas Gerais (MG), São Paulo (SP), Paraná (PR-Li and PR-HV), and Rio Grande do Sul (RS) states, showed that MSP1a, MSP4, and MSP5 were conserved in all isolates of *A. marginale*. The MSP2 was identified in isolates from MG, PR-Li, PR-HV, and RS; furthermore, MSP3 was identified in isolates from MG and PR-Li. This antigenic diversity among Brazilian isolates should be considered for future studies involving the development of an ideal vaccine against anaplasmosis. The DNA vaccine seemed to prevent high rickettsemia levels, although clinical signs of anaplasmosis were observed in both groups at day 35 after challenge, which resulted in treatment of four calves in G1 and four calves in G2 against anaplasmosis. Furthermore, serum antibody titers in calves detected by serological assays were not enough to achieve a total protection against anaplasmosis, although high titers of antibodies were reported as responsible for protection level.[7] In our experiment, it was not significant at first sight since antibodies developed by immunized calves were directed toward MSP1b and not to other MSPs from *A. marginale*. Moreover, the lack of protection observed in immunized calves in comparison with nonvaccinated calves could be explained by the polymorphism encoded in the variant MSP1b proteins since just one protein was cloned. Last, DNA vaccines against anaplasmosis need to be further studied considering some limitations, cited in the literature, such as (a) presentation of the antigen to the bovine immune system to achieve optimal protective responses and (b) possible dependence of the immunogenicity of MSPs on the conformation of protein.[24]

## ACKNOWLEDGMENTS

We thank the Fundação de Amparo a Pesquisa do Estado de São Paulo–FAPESP, process numbers 99/04495-5 and 99/10488-1, for financial support.

## REFERENCES

1. DUMLER, J.S., A.F. BARBET, C.P.J. BEKKER *et al.* 2001. Reorganization of genera in the families Rickettsiaceae and Anaplasmataceae in the order *Rickettsiales*; unification of some species of *Ehrlichia* with *Anaplasma*, *Cowdria* with *Ehrlichia*, and *Ehrlichia* with *Neorickettsia*, descriptions of six new species combinations and designation of *Ehrlichia equi* and "HGE agent" as subjective synonyms of *Ehrlichia phagocytophila*. Int. J. Syst. Evol. Microbiol. **51:** 2145–2165.
2. RICHEY, E.J. 1981. Bovine anaplasmosis. *In* Current Veterinary Therapy Food Animal Practice, pp. 767–772. Saunders. Philadelphia.
3. PALMER, G.H. 1989. *Anaplasma* vaccines. *In* Veterinary Protozoan and Hemoparasites Vaccines, pp. 1–29. CRC Press. Boca Raton, FL.
4. GRISI, L., C.L. MASSARD, G.E. MOYA-BORJA & J.B. PEREIRA. 2002. Impacto econômico das principais ectoparasitoses em bovinos no Brasil. Hora Vet. **125:** 8–10.
5. BRAZ JUNIOR, C.J., L.M.F. PASSOS & D.J. LIMA. 1995. Utilização do teste ELISA para detecção de anticorpos anti *Anaplasma marginale*. *In* Seminário Brasileiro de Parasitologia Veterinária, Campo Grande. Anais, p. 173.
6. RISTIC, M. & C.A. CARSON. 1977. Methods of immunoprophylaxis against bovine anaplasmosis with emphasis on use of the attenuated *Anaplasma marginale* vaccine. *In* Immunity to Blood Parasites of Animals and Man: Advances in Experimental Medicine and Biology, pp. 151–188. Plenum. New York.
7. TEBELE, N., T.C. MCGUIRE & G.H. PALMER. 1991. Induction of protective immunity by using *Anaplasma marginale* initial body membranes. Infect. Immun. **59:** 3199–3204.
8. OBERLE, S.M. & A.F. BARBET. 1993. Derivation of the complete msp4 gene sequence of *Anaplasma marginale* without cloning. Gene **136:** 291–294.
9. ALLRED, D.R., T.C. MCGUIRE, G.H. PALMER *et al.* 1990. Molecular basis for surface antigen size polymorphisms and conservation of a neutralization-sensitive epitope in *Anaplasma marginale*. Proc. Natl. Acad. Sci. USA **87:** 3220–3224.
10. BARBET, A.F. & D.R. ALLRED. 1991. The msp1b multigene family of *Anaplasma marginale*: nucleotide analysis of an expressed copy. Infect. Immun. **59:** 971–976.
11. MCGUIRE, T.C., W.C. DAVIS, A.L. BRASSFIELD *et al.* 1991. Identification of *Anaplasma marginale* long-term carrier cattle by detection of serum antibody to isolate MSP-3. J. Clin. Microbiol. **29:** 788–793.
12. ARULKANTHAN, A., W.C. BROWN, T.C. MCGUIRE & D.P. KNOWLES. 1999. Biased immunoglobulin G1 isotype responses induced in cattle with DNA expressing msp1a of *Anaplasma marginale*. Infect. Immun. **67:** 3481–3487.
13. LAEMMLI, U.K. 1979. Cleavage of structural proteins during assembly of the head of bacteriophage, $T_4$. Nature **227:** 680–685.
14. MACHADO, R.Z. 1995. Emprego do ensaio imunoenzimático indireto (ELISA teste) no estudo da resposta imune humoral de bovinos importados e premunidos contra a tristeza parasitária. Rev. Bras. Parasitol. **4**(2)**:** 217.
15. TORIONI DE ECHAIDE, S., D.P. KNOWLES, T.C. MCGUIRE *et al.* 1998. Detection of cattle naturally infected with *Anaplasma marginale* in a region of endemicity by nested

PCR and a competitive enzyme-linked immunosorbent assay using recombinant major surface protein 5. J. Clin. Microbiol. **36:** 777–782.
16. DONELLY, J.J., J.F. ULMER, J.W. SHIVER & M.A. LUI. 1997. DNA vaccines. Annu. Rev. Immunol. **15:** 617–648.
17. MANICKAN, E., S. KANANGAT, R.J. ROUSE *et al.* 1997. Enhancement of immune response to naked DNA vaccine by immunization with transfected dendritic cells. J. Leukocyte Biol. **61:** 125–132.
18. CAMACHO-NUEZ, M.L.M., C.E. SUAREZ & T.C. MCGUIRE. 2000. Expression of polymorphic msp 1b genes during acute *Anaplasma marginale* rickettsemia. Infect. Immunol. **68:** 1946–1952.
19. BOWIE, M.V., J. DE LA FUENTE, K.M. KOCAN *et al.* 2002. Conservation of major surface protein 1 genes of the ehrlichial pathogen *Anaplasma marginale* during cyclic transmission between ticks and cattle. Gene **282:** 95–102.
20. MCGAREY, D.J. & D.R. ALLRED. 1994. Characterization of hemagglutinating components on the *Anaplasma marginale* initial body surface and identification of possible adhesions. Infect. Immun. **62:** 4587–4593.
21. DE LA FUENTE, J., K.M. KOCAN, J.C. GARCIA-GARCIA *et al.* 2002. Vaccination of cattle with *Anaplasma marginale* derived from tick cell culture and bovine erythrocytes followed by challenge-exposure with infected ticks. Vet. Microbiol. **89:** 239–251.
22. BLOUIN, E.D., J.T. SALIKI, J. DE LA FUENTE *et al.* 2003. Antibodies to *Anaplasma marginale* major surface proteins 1a and 1b inhibit infectivity for cultured tick cells. Vet. Parasitol. **11:** 247–260.
23. KANO, F.S., O. VIDOTTO, R.C. PACHECO & M.C. VIDOTTO. 2002. Antigenic characterization of *Anaplasma marginale* isolates from different regions of Brasil. Vet. Microbiol. **87:** 131–138.
24. KOCAN, K.M., E.F. BLOUIN & A.F. BARBET. 2000. Anaplasmosis control: past, present, and future. Ann. N.Y. Acad. Sci. **916:** 501–509.

# The Immunology of *Leishmania* Infection and the Implications for Vaccine Development

YANNICK VANLOUBBEECK AND DOUGLAS E. JONES

*Department of Veterinary Pathology, College of Veterinary Medicine, Iowa State University, Ames, Iowa 50011, USA*

**ABSTRACT: *Leishmania* parasites are vector-borne protozoal pathogens found in tropical and subtropical regions of both the Old and New World. These parasites can cause visceral or cutaneous disease, and the pathology of the infection is determined by both host immune factors and species/strain differences of the parasite. Dogs are an important reservoir for maintaining the population of *Leishmania* parasites that can lead to visceral leishmaniasis in humans, and a vaccination approach may be an effective method for reducing the numbers of infected dogs. Resistance to leishmaniasis has been consistently associated with a T helper 1 immune response, characterized by the production of IFN-gamma by the antigen-specific lymphocyte population. The development of this Th1 response has been shown to be dependent upon both cytokines and dendritic cells during T cell activation. However, the development of a *Leishmania* vaccine effective in preventing these chronic diseases has proven to be a challenge. Vaccine trials have focused on whole-killed or subunit vaccines with adjuvants. Newer experimental strategies involve the attenuation of the *Leishmania* parasite via gene deletion technologies or the expression of specific *Leishmania* peptides within attenuated organisms, such as Bacillus Calmette Guérin. DNA vaccines and dendritic cell potentiators, such as CpG oligodeoxynucleotides and Flt-3 ligand, are also in the early stage of development. In addition, as part of blocking the transmission cycle of leishmaniasis, several laboratories are also exploring the possibility of immunomodulating the host toward the bite of the sand fly.**

**KEYWORDS: *Leishmania*; T cell; vaccine; immune**

*Leishmania* parasites are vector-borne protozoal pathogens found in tropical and subtropical regions of both the Old and New World. These parasites can cause visceral or cutaneous leishmaniasis, which are estimated to be responsible for 500,000 and 1,500,000 new human infections annually, respectively. Untreated visceral leishmaniasis is often fatal, and there is a significant amount of morbidity associated with the cutaneous disease.[1] Both the response of the host and the species/strain of the infectious parasite determine the clinical outcome that is associated with the infection. For example, *Leishmania donovani* and *L. infantum* are usually associated

Address for correspondence: Douglas E. Jones, Department of Veterinary Pathology, College of Veterinary Medicine, Iowa State University, Ames, Iowa 50011. Voice: 515-294-3282; fax: 515-294-5423.

jonesdou@iastate.edu

**Ann. N.Y. Acad. Sci. 1026: 267–272 (2004).** **© 2004 New York Academy of Sciences.**
**doi: 10.1196/annals.1307.041**

with visceral leishmaniasis, whereas *L. major*, *L. amazonensis*, and *L. tropica* are usually associated with the cutaneous forms of the disease. However, *L. tropica* and *L. amazonensis* have also been reported to be associated with visceral disease.[2,3] Similarly, dogs and humans infected with *L. infantum* and *L. donovani* may demonstrate both visceral and cutaneous symptoms during disease progression.

Dogs have been implicated as an important reservoir for maintaining the population of *Leishmania* parasites that can lead to visceral leishmaniasis in humans, and they can suffer significant morbidity and mortality in association with these diseases. Control of canine leishmaniasis is proposed to be central for limiting the endemic transmission cycle of this zoonotic disease. As culling of infected dogs has had mixed results in influencing transmission rates and is often unpalatable to dog owners, a vaccination approach would appear to be an effective method in reducing the numbers of infected dogs (reviewed in refs. 4 and 5).

Vaccination strategies are based on our current understanding of the characteristics of an effective anti-*Leishmania* immune response, as they have been determined from human and murine studies (reviewed in refs. 6 and 7). Briefly, resistance to leishmaniasis has been associated with a predominant IFN-gamma production from the antigen-specific $CD4^+$ T lymphocyte population—termed a T helper 1 (Th1) immune response. These cells are then effective in promoting macrophage activation at the site of the lesion, and the intracellular *Leishmania* are killed in a nitric oxide–dependent manner. In addition, activation of the $CD8^+$ T cell population has been shown to play a protective role after *L. major* infection after both low-dose primary infection and during a memory immune response.[8,9] $CD8^+$ T cells have also been shown to be important for effective vaccination in experimental murine leishmaniasis.[10,11] By contrast, T helper 2 immune responses, which are characterized by IL-4 production from the antigen-specific $CD4^+$ T cell population, are associated with susceptibility and exacerbation of the disease. It should be emphasized that the immune response is a complex reaction to infection, and both Th1 and Th2 phenotypic cells can almost always be found during the immune response. The biological phenotype of the immune response is, therefore, determined in a large part by the predominance of one cell type over the other, not simply the presence or absence of Th1- or Th2-type immune cells.

Over the past several years it has become recognized that dendritic cells (DCs), such as the Langerhans cells in the skin, are uniquely suited to promote the activation of naive $CD4^+$ T cells toward one of the two polarized phenotypic responses (Th1 vs. Th2).[12] Dendritic cells express a large repertoire of pattern-recognition receptors (PRR) on their cell surface. These receptors include the toll-like receptors (TLR) that recognize specific, conserved, pathogen-related structures (reviewed in ref. 13). As part of the body's immunosurveillance system, immature dendritic cells typically encounter foreign antigen in the context of pathogen structures that bind one or more of these PRR. Activation of the dendritic cell through the PRR sets in motion the migration of the DCs to the nearest lymph node and the DC maturation into an effective antigen-presenting cell capable of activating naive T cells (reviewed in ref. 14). In addition to their ability to present foreign antigen to $CD4^+$ T cells in the context of MHC class II, DCs have a unique ability to present such antigens in the context of MHC class I and to stimulate antigen-specific $CD8^+$ T cell populations (reviewed in ref. 15). This ability is termed cross-priming and may play a significant role in directing an effective immune response to intracellular pathogens, such as

*Leishmania*. How DCs are able to influence T helper cell polarization is not fully understood. While IL-12 production by DCs induces Th1 responses, the DC factors responsible for Th2 development are not known and may involve production of IL-10.[12] Therefore, the phenotype of the T helper response is determined by a myriad of factors, including host genetics (the propensity of the host DCs and T cells to promote a Th1 response) and parasite factors (e.g., the ability of the parasite to influence DC functions).

During clinical leishmaniasis in both humans and dogs, there is often no detectable Th1 cell–mediated immune response, or the immune response seems to be of a mixed phenotype (reviewed in refs. 4 and 16). An apparent inability to produce an effective IFN-gamma response is also observed in mouse models of visceral leishmaniasis and cutaneous leishmaniasis caused by *Leishmania amazonensis*.[17,18] In the latter model, infected mice develop chronic cutaneous lesions, and the immunoreactive T cells are not skewed toward a Th1 or Th2 response (ref. 19 and our unpublished observations). Interestingly, we have found that attempts at promoting a Th1 polarization of the $CD4^+$ T cells, using the Th1-inducing cytokine IL-12 as an adjuvant, were ineffective in promoting resistance to *L. amazonensis* in this experimental system.[20] In addition, vaccination using *Leishmania* antigen-pulsed DCs and IL-12 was only transiently protective. However, using coinfection experiments in mice, we have found that the immune response generated after resolution of *L. major* infection is capable of promoting a healing response toward *L. amazonensis* (our unpublished observations). Altogether, these results highlight the complexity of this biological system, suggesting that a healing cross-protective response can be developed, but that administration of Th1-promoting factors alone may not be sufficient at generating a long-lasting effective immune response toward all *Leishmania* parasites.

In summary, the current available data suggest that protection to *Leishmania* infection requires the effective activation of several cell populations, including dendritic cells, antigen-specific $CD4^+$ and $CD8^+$ T cells, and macrophages. Given the central role of DCs in shaping the phenotype of a pathogen-specific immune response, vaccine candidates should be aimed at efficiently stimulating DCs, which, in turn, should activate the different arms of the cellular immune response in order to provide broad, cross-reactive, and long-lasting protection.

Several vaccination strategies for both cutaneous and visceral leishmaniasis are being developed. Most of the vaccines target the host DCs with adjuvants, such as Bacillus Calmette Guérin (BCG), *C. parvum,* or, more recently, with nonmethylated CpG oligodeoxynucleotides, which mimic nucleotide sequences common to bacterial DNA (reviewed in refs. 4 and 21). These adjuvants stimulate DCs through various PRR, in particular the TLR, and provide pathogen signals recognized by the DC population to promote maturation and a Th1 bias of the responding antigen-specific T cells. Other adjuvant strategies being developed include the administration of Th1 promoting cytokines, such as IL-12, within the vaccine.

In clinical trials of both humans and dogs, whole-killed vaccines with BCG as an adjuvant promote predominantly a cell mediated–type immune response. However, the efficacy of these vaccines is not clear (reviewed in refs. 4 and 21). Presumably they stimulate a predominant $CD4^+$ T cell response through endocytosis and antigen presentation on MHC class II molecules. It is possible that their limited efficacy may result from inadequate cross-presentation and little $CD8^+$ T cell activation. In addition, as different individual microbial antigens can induce DCs to promote either Th1 or Th2 responses,[22] it is probable that the immune response to a relatively complex antigenic mixture would have some aspects of both a Th1 and a Th2 response. Therefore a whole-killed vaccine may not be able to target the appropriate population of T cells to consistently generate an effective Th1 response.

A refinement of the whole-killed vaccine strategy is to identify pathogen peptides or other components of the pathogen that promote a Th1 response. For example, the parasite antigen LeIF (*Leishmania* elongation initiation factor) induces IL-12 production from the antigen-presenting cells and promotes a skewing of the antigen-specific recall response toward a Th1 response.[23] Therefore, certain peptide-based vaccines may be better at exclusively expanding a population of cells that are highly skewed toward a Th1 response.[24,25] These vaccine candidates are used in combination with a wide array of adjuvants (reviewed in refs. 4 and 21).

Other vaccination strategies exploit the specific interaction of infectious pathogens with DCs to target a *Leishmania* antigen to the DCs compartment. These vaccines involve the expression of specific *Leishmania* peptides as part of other attenuated organisms, such as salmonella or BCG (reviewed in refs. 4 and 21). In this system, the DC response is determined by the DC interaction and maturation pathway evoked by the "carrier" organism, and the *Leishmania*-specific T cell response is determined by the parasite-specific protein or peptide that has been placed into the attenuated carrier organism. Other vaccines using *Leishmania* parasites attenuated through genetic deletions may effectively mimic a natural infection, which is known to promote an adequate memory immune response in the majority of people infected with *L. major* (reviewed in refs. 4 and 21). However, *Leishmania* parasites can influence DC function, and these vaccines may not necessarily be able to consistently skew the immune phenotype.

In addition, expansion of the host DC population is also achievable. DCs themselves can be grown *in vitro* from individual animals, pulsed with pathogen-specific antigen/peptides and reinjected into the individual to promote an adequate immune response.[24] Also, Flt-3 ligand (a compound promoting the expansion of mature DCs *in vivo*) has been shown to provide partial protection to *L. major* and may enhance overall DC functions.[26,27]

Of particular interest for effective vaccination strategies is the potential of DNA vaccines. These consist of bacterial plasmid DNA that encodes the peptide or protein of interest. Such plasmids have intrinsic adjuvant properties as they contain non-methylated CpG motifs. In addition, gene sequences of cytokines or any other adjuvants that promote a Th1 response can be included within the plasmid DNA. DNA vaccination results in cellular uptake of the DNA and transport to the nucleus. The cells that take up the DNA and express the proteins encoded in the vaccine can be varied and include muscle cells, skin cells, and DCs themselves. Therefore, transfected DCs can present DNA-encoded antigen directly through the class I pathway, enhancing $CD8^+$ T cell responses, or the DCs can endocytose dead transfected cells and express the encoded antigen through class I (via cross-presentation) and/or class II pathways.[28] One of the *Leishmania* DNA vaccines described in the literature promoted a $CD8^+$ T cell response as well as a $CD4^+$ T cell response, suggesting that the immune response to DNA vaccination has a broader repertoire. In addition, the vaccine was shown to promote immunity for longer periods of time.[10]

As part of blocking the transmission cycle of leishmaniasis, the possibility of immunomodulating the host toward the bite of the sand fly is also being explored. Several potent immunomodulating substances have been identified in sand fly saliva (reviewed in ref. 29). Recently it was shown that prior exposure of mice to bites of uninfected sand flies conferred protection against subsequent infection via sand fly feeding.[29] In addition, in murine studies, a DNA vaccine encoding a single salivary protein from *Phlebotomus papatasi* was able to afford significant resistance to a low-dose intradermal challenge with *L. major* and salivary gland homogenate.[30] Again, the promotion of the adequate and consistent immune response to these salivary proteins will be enhanced if they are targeted to maximize DC–T cell interaction and the generation of an effective immune response that inhibits the establishment of the parasite in the host.

In conclusion, vaccination strategies that will be long lasting and effective against multiple species of *Leishmania* parasites will target broad aspects of the immune response. Maximizing this response will use our growing knowledge of the importance of the DC population in instructing and/or selecting the appropriate $CD4^+$ and $CD8^+$ T cell phenotype. The ability to target DC in many different mammalian species will promote the effective vaccination of canine populations that are susceptible to infection from *Leishmania* and implicated as reservoir hosts for this protozoal parasite.

## REFERENCES


1. Center for Disease Control and Prevention, N.C.F.I.D., Division of Parasitic Diseases, *Leishmania* Infection; Fact Sheet. 2000. www.CDC.gov.
2. Magill, A.J., M. Grogl, R.A. Gasser Jr., *et al.* 1993. Visceral infection caused by *Leishmania tropica* in veterans of Operation Desert Storm. N. Engl. J. Med. **328:** 1383–1387.
3. Barral, A., D. Pedral-Sampaio, G. Grimaldi Junior, *et al.* 1991. Leishmaniasis in Bahia, Brazil: evidence that *Leishmania amazonensis* produces a wide spectrum of clinical disease. Am. J. Trop. Med. Hyg. **44:** 536–46.
4. Gradoni, L. 2001. An update on antileishmanial vaccine candidates and prospects for a canine *Leishmania* vaccine. Vet. Parasitol. **100:** 87–103.
5. Davies, C.R., P. Kaye, S.L. Croft & S. Sundar. 2003. Leishmaniasis: new approaches to disease control. Br. Med. J. **326:** 377–382.
6. Sacks, D. & N. Noben-Trauth. 2002. The immunology of susceptibility and resistance to *Leishmania major* in mice. Nat. Rev. Immunol. **2:** 845–858.
7. Melby, P.C. 2002. Recent developments in leishmaniasis. Curr. Opin. Infect. Dis. **15:** 485–490.
8. Belkaid, Y., E. Von Stebut, S. Mendez, *et al.* 2002. $CD8^+$ T cells are required for primary immunity in C57BL/6 mice following low-dose, intradermal challenge with *Leishmania major*. J. Immunol. **168:** 3992–4000.
9. Muller, I., P. Kropf, J.A. Louis & G. Milon. 1994. Expansion of gamma interferon-producing $CD8^+$ T cells following secondary infection of mice immune to *Leishmania major.* Infect. Immun. **62:** 2575–2581.
10. Mendez, S., S. Gurunathan, S. Kamhawi, *et al.* 2001. The potency and durability of DNA- and protein-based vaccines against *Leishmania major* evaluated using low-dose, intradermal challenge. J. Immunol. **166:** 5122–5128.
11. Colmenares, M., P.E. Kima, E. Samoff, *et al.* 2003. Perforin and gamma interferon are critical CD8(+) T-cell-mediated responses in vaccine-induced immunity against *Leishmania amazonensis* infection. Infect. Immun. **71:** 3172–3182.
12. Moser, M. & K.M. Murphy. 2000. Dendritic cell regulation of TH1-TH2 development. Nat. Immunol. **1:** 199–205.

13. MEDZHITOV, R. 2001. Toll-like receptors and innate immunity. Nat. Rev. Immunol. **1:** 135–145.
14. BANCHEREAU, J., F. BRIERE, C. CAUX, *et al.* 2000. Immunobiology of dendritic cells. Annu. Rev. Immunol. **18:** 767–811.
15. GUERMONPREZ, P., J. VALLADEAU, L. ZITVOGEL, *et al.* 2002. Antigen presentation and T cell stimulation by dendritic cells. Annu. Rev. Immunol. **20:** 621–667.
16. RUSSO, D.M., M. BARRAL-NETTO, A. BARRAL & S.G. REED. 1993. Human T-cell responses in Leishmania infections. Prog. Clin. Parasitol. **3:** 119–144.
17. KAYE, P.M., A.J. CURRY & J.M. BLACKWELL. 1991. Differential production of Th1- and Th2-derived cytokines does not determine the genetically controlled or vaccine-induced rate of cure in murine visceral leishmaniasis. J. Immunol. **146:** 2763–2770.
18. AFONSO, L.C. & P. SCOTT. 1993. Immune responses associated with susceptibility of C57BL/10 mice to *Leishmania amazonensis.* Infect. Immun. **61:** 2952–2959.
19. JI, J., J. SUN, H. QI & L. SOONG. 2002. Analysis of T helper cell responses during infection with *Leishmania amazonensis.* Am. J. Trop. Med. Hyg. **66:** 338–345.
20. JONES, D.E., L.U. BUXBAUM & P. SCOTT. 2000. IL-4-independent inhibition of IL-12 responsiveness during *Leishmania amazonensis* infection. J. Immunol. **165:** 364–372.
21. HANDMAN, E. 2001. Leishmaniasis: current status of vaccine development. Clin. Microbiol. Rev. **14:** 229–243.
22. MANICKASINGHAM, S.P., A.D. EDWARDS, O. SCHULZ & C. REIS E SOUSA. 2003. The ability of murine dendritic cell subsets to direct T helper cell differentiation is dependent on microbial signals. Eur. J. Immunol. **33:** 101–107.
23. SKEIKY, Y.A., M. KENNEDY, D. KAUFMAN, *et al.* 1998. LeIF: a recombinant *Leishmania* protein that induces an IL-12-mediated Th1 cytokine profile. J. Immunol. **161:** 6171–6179.
24. BERBERICH, C., J.R. RAMIREZ-PINEDA, C. HAMBRECHT, *et al.* 2003. Dendritic cell (DC)-based protection against an intracellular pathogen is dependent upon DC-derived IL-12 and can be induced by molecularly defined antigens. J. Immunol. **170:** 3171–3179.
25. COLER, R.N., Y.A. SKEIKY, K. BERNARDS, *et al.* 2002. Immunization with a polyprotein vaccine consisting of the T-cell antigens thiol-specific antioxidant, *Leishmania major* stress-inducible protein 1, and *Leishmania* elongation initiation factor protects against leishmaniasis. Infect. Immun. **70:** 4215–4225.
26. MARASKOVSKY, E., K. BRASEL, M. TEEPE, *et al.* 1996. Dramatic increase in the numbers of functionally mature dendritic cells in Flt3 ligand-treated mice: multiple dendritic cell subpopulations identified. J. Exp. Med. **184:** 1953–1962.
27. KREMER, I.B., M.P. GOULD, K.D. COOPER & F.P. HEINZEL. 2001. Pretreatment with recombinant Flt3 ligand partially protects against progressive cutaneous leishmaniasis in susceptible BALB/c mice. Infect. Immun. **69:** 673–680.
28. STEINMAN, R.M. & M. POPE. 2002. Exploiting dendritic cells to improve vaccine efficacy. J. Clin. Invest. **109:** 1519–1526.
29. KAMHAWI, S. 2000. The biological and immunomodulatory properties of sand fly saliva and its role in the establishment of *Leishmania* infections. Microbes Infect. **2:** 1765–1773.
30. VALENZUELA, J.G., Y. BELKAID, M.K. GARFIELD, *et al.* 2001. Toward a defined anti-Leishmania vaccine targeting vector antigens: characterization of a protective salivary protein. J. Exp. Med. **194:** 331–342.

# Local Utilization of Metacresolsulfonic Acid Combined with Streptomycin in the Treatment of Actinomycosis

L. A. F. SILVA, M. C. S. FIORAVANTI, K. S. OLIVEIRA, I. B. ATAYDE, M. A. ANDRADE, V. S. JAYME, R. E. RABELO, A. F. ROMANI, AND E. G. ARAÚJO

*Departament of Veterinary Medicine, School of Veterinary Medicine, Federal University of Goias, P. O. Box 131, 74001-970, Goiânia, Goiás, Brazil*

**ABSTRACT: The effectiveness of combining metacresolsufonic acid with streptomycin in the treatment of actinomycosis, diagnosed either clinically or in the laboratory, was evaluated in 12 bovines and 2 equines. Eighty-seven percent of treated animals were considered clinically cured and did not show any signs of relapse after a six-month follow-up period. Therapeutic diagnosis by clinical observation was the procedure of choice when it was not possible to obtain laboratory diagnosis.**

**KEYWORDS: actinomycosis; bovine; diagnosis; metacresolsufonic acid**

## INTRODUCTION

Actinomycosis is a chronic infectious disease, granulomatous in character, caused by *Actinomyces bovis. A. bovis* is the primary agent, but other bacteria also take part in this process.[1] *Actinomyces bovis* was first observed by Bollinger in 1876, described and named by Harz in 1877, and isolated and cultivated by Wolf and Israel in 1891.[2] The pathology is cosmopolitan and most frequently affects cattle, but it can be observed in other ruminants, as well as in horses, pigs, dogs, and cats; cases of human infection are rare.[3–5]

The process usually begins at the molar teeth and is triggered by an injury caused by foreign bodies. Maxillary soft tissues are rarely involved.[1,2]

Traumas and low immunological resistance can also predispose animals to actinomycosis.[6] Chewing and swallowing food becomes difficult because of the involvement of the jaw, maxilla, and teeth,[1,7] a condition that can lead to weight loss[4] and, in more severe cases, death.[2]

The course of treatment recommended for actinomycosis[1] includes the use of parenteral administration of streptomycin combined with local applications of isoniazide, as well as surgical removal of the involved teeth. It is also recommended to use parenteral administration of penicillin, streptomycin, tetracycline, cephalosporin

Address for correspondence: L.A.F. Silva, Rua 18-A, 591 apt. 502 Ed. Acauã. Setor Aeroporto. 74 070-060 Goiânia-GO, Brazil. Voice: +55-62-521-1572; fax: +55-62-521-1566. lafranco@vet.ufg.br

**Ann. N.Y. Acad. Sci. 1026: 273–276 (2004). © 2004 New York Academy of Sciences. doi: 10.1196/annals.1307.042**

or lincomycin, combined with local or parenteral application of potassium iodine, as well as surgery.[2] Further recommendation says that lesions should be subjected to curettage and washed with organic iodine.[4] Radiotherapy, which can temporally reduce the lesions, is also prescribed.[7] A cure rate of 100% has been reported after two to three months of treatment with parenteral administration of benzatine-penicillin, combined with oral isoniazide.[6]

The aim of this paper is to describe the use of metacresolsulfonic acid combined with streptomycin sulfate in the treatment of actinomycosis in bovines and equines with positive clinical or laboratory diagnosis.

## MATERIALS AND METHODS

Twelve bovines and two equines from different breeds ranging from two to six years old and showing enlarged mandibular swelling with a fistula were treated.

Material for laboratory tests was harvested from five animals, one equine and four bovines. The cattle were tranquilized with a solution of xylazin hydrochloride at 2% (Rompun, Bayer, Brazil), at the proportion of 0.1 to 0.2 mg/kg, followed by local anesthesia with lidocaine at 2% (Xilocaína, Lepetit, Brazil). For the equines, a combination of chlorpromazine (Amplictil, Rhodia, Brazil) and prometazine (Fenergan, Rhodia, Brazil) at the proportion of 25 mg/100 kg was used as a preanesthesia. Full anesthesia was induced by 50 g of guaiacol glyceryl ether (Éter gliceril Guaiacólico, Rhodia, Brazil) administered with 2 g of sodium thiopental (Thionembutal, Abbot, Brazil) diluted in a 1.000 mL 5% glucose solution (Solução Glicosada, Equipex, Brazil).[8] After anesthesia, drilling was performed on all animals, and one had the first molar extracted. Material was harvested under aseptic conditions, deep in the cavity, with a swab that was then taken to the bacteriology laboratory.[9] Once in the laboratory, the identification process began by releasing the yellow granules present in the exudates, also known as "sulfur granules," by washing them twice in 0.85% saline. The granules were then placed on a glass slide, pressed by a cover slip and examined on a light microscope. Part of the remaining sample was Gram-stained, and the remaining part was bacteriogically processed, according to previously established methods.[2,5]

Animals with clinical suspicion and/or laboratory diagnosis were treated with 10% metacresolsulfonic acid (albocresil Solução, BYK, Brazil) diluted in 0.9% saline[10] instilled in the lesion through a cannula attached to the skin, and intramuscular streptomycin (Estreptomicina, Fort Dodge Saúde Animal, Brazil) at aproportion of 10 mg/kg of body weight[11] for 20 consecutive days. The cannula was removed at the end of treatment.

## RESULTS

Clava-like structures suggesting "druses" could be seen through direct observation on a light microscope. Filamentous gram-positive rods with morphologic and staining features similar to actinomyces, as described in the literature,[12] were identified. Small, clear, and bright colonies, isolated in agar-blood were subjected to the following biochemical tests: catalase, urea, glucose, lactose, nitrate–nitrite, indole, methyl red, manitol, and motility. All animals tested yielded positive indications for *Actinomyces bovis*.

All 14 animals clinically suspected of actinomycosis showed, through clinical examination, unilaterally enlarged swelling in the medial portion of the mandible, with a fistula in the abscess and intermittent suppuration either toward the inside or the outside of the mouth. These signs are similar to those reported by others.[1,2,4,6,7] Analyzed material collected from four bovines and one equine was positive for actinomycosis. At a later stage, the remaining animals with a potential clinical diagnosis were also confirmed by therapeutic diagnosis. However, two cases did not respond to this treatment.

After the cannula was removed, the skin lesion healed. This was considered a sign of recovery, although no regression of the enlarged bone volume was observed. The average time between the beginning of treatment and recovery of the animals was 35 days.

## DISCUSSION

Clinical cure was seen in ten bovines and two equines, and follow-up lasted approximately six months; no relapses were reported. From the two cases that did not respond to treatment, one animal was discarded and the other died.

The cases that did not respond to therapy were probably due to the fact that either the animals had not been really contaminated with actinomycosis or that the treatment failed. On the other hand, the death of one of the animals at an advanced stage of the disease could have been caused by starvation due to difficulties in chewing and swallowing.

Teething in younger animals and change of dentition in elders represent a predisposing factor for infection by *Actinomyces bovis*. Thus, it is necessary to adopt a preventive approach in order to avoid feeding coarse food to those animals at such crucial times. It also important to look out for the presence of foreign materials in their environment. Similar information has already been reported.[1,2]

The substitution of the usual therapy for metacresolsulfonic acid is justified and recommended[13] because of its cytostatic, antiseptic, healing, and selective actions, which help to remove necrotic or pathologically affected tissue, without compromising healthy tissue.[14]

The fact that nine animals were considered infected at clinical examination, and that in seven cases therapeutic diagnosis was confirmed, demonstrates the importance of strict clinical screening. The ability to recognize the characteristic signs of this disease is important in regions where the possibility of sending suspected material for laboratory tests is not a practical option.

The 85.71% cure rate in the animals who underwent treatment, although smaller than the 100% reported in literature,[6] reinforces the effectiveness of this therapeutic approach. It is worth noticing that the success previously reported[6] can be related to previous laboratory confirmation, while in this study, treatment was not restricted to animals with a positive laboratory diagnosis.

For most of the animals, body weight score was below ideal, probably due to difficulties in getting, chewing, and swallowing food, as noted by other researchers.[2] After treatment, the animals recovered from this physical deficiency, especially in the case of the bovine that was submitted to tooth extraction, an approach that is also recommended.[4]

After this study had been completed, it was possible to conclude that (1) the use of metacresolsulfonic acid in combination with streptomycin was effective in treating actinomycosis, (2) careful clinical practice is very important to diagnosing this disease, (3) therapeutic diagnosis associated with clinical observation is recommended when laboratory diagnosis is not possible, and (4) tooth extraction from animals whose alveoli are affected contributes to recovery.

## REFERENCES


1. RADOSTITS, O.M., D.C. BLOOD & C.C. GAY. 1994. Veterinary Medicine. A Textbook of the Diseases of Cattle, Sheep, Pigs, Goats and Horses, 8th edit., pp. 851–852. Baillière Tindall. London.
2. GUERREIRO, M.G., S.J., OLIVEIRA, D. SARAIVA, *et al.* 1984. Bacteriologia Especial: com interesse em saúde animal e saúde pública, pp. 448–454. Ed. Sulina. Porto Alegre.
3. JUBB, K.V.F, P.C. KENNEDY & N. PALMER. 1985. Pathology of Domestic Animals. V. I, 3rd edit., p. 67. Academic Press. San Diego.
4. SMITH, B.P. 1993. Tratado de medicina interna de grandes animais. V. I.: 724–726. Manole. São Paulo.
5. CARTER, G.R., M.M. CHENGAPPA & A.W. ROBERTS. 1995. Essentials of Veterinary Microbiology, 5th edit., pp. 214–216. Willians & Wilkins. Média.
6. CORRÊA, W.M. & C.N.M. CORRÊA. 1992. Emfermidades infecciosas dos mamíferos domésticos. 2nd edit., pp. 351–354. Médica e Científica. Rio de Janeiro.
7. FRASER, C.M. 1988. El Manual Merk de veterinária, 3rd edit., pp. 413–415. Merck & Co. Madrid.
8. SILVA, L.A.F., M.I. CARNEIRO, M.C.S. FIORAVANTI, *et al.* 1995. Técnica de circuncisão com encurtamento do pênis para obtenção de rufiões equinos. Arquivo Brasileiro de Medicina Veterinária e Zootecnia. **47:** 789–798.
9. TURNER, A.S. & G.W. MCILWRAITH. 1985. Técnicas cirúrgicas em animais de grande porte, pp. 211–215. Ed. Roca, São Paulo.
10. FIORAVANTI, M.C.S., L.A.F. SILVA, D. EURIDES, *et al.* 1996. Ácido metacresolsulfônico associado a nitrofurazona e enrofloxacina no tratamento de sinusite em bovinos. Veterinária Notícias Uberlândia **2:** 31–35.
11. ANDREI, E. 2002. Compêndio veterinário, 32nd edit., pp. 262. Andrei. São Paulo.
12. NICOLET, J. 1984. Compendio de bacteriologia médica veterinária, pp. 200–202. Ed. Acribia. Zaragoza.
13. SILVA, L.A.F., R.J. DEL CARLO, G.H. TONIOLLO, *et al.* 1984. Àcido metacresolsulfonico associado à extirpação cirúrgica do tumor venéreo canino. Revista Brasileira de Reprodução Animal **8:** 63–68.
14. LIMA, D.R. 1993. Manual de farmacologia clínica, terapêutica e toxicologia, pp. 611. Ed. Guanabara Koogan. Rio de Janeiro.

# Field Challenge of Cattle Vaccinated with a Combined *Babesia bovis* and *Babesia bigemina* Frozen Immunogen

J. ANTONIO ALVAREZ,[a] JUAN A. RAMOS,[a] EDMUNDO E. ROJAS,[a] JUAN J. MOSQUEDA,[a] CARLOS A. VEGA,[a] ANDREA M. OLVERA,[b] JULIO V. FIGUEROA,[a] AND GERMINAL J. CANTÓ[a,b]

[a]*CENID-Parasitología Veterinaria, INIFAP, Morelos, C.P. 62500, Mexico*

[b]*Escuela de Medicina Veterinaria, Universidad Autónoma de Querétaro*

**ABSTRACT: To determine the optimal dose of a combined, frozen immunogen containing *in vitro* culture-derived strains of *Babesia bovis* and *Babesia bigemina*, twenty-four 14-month-old *Bos taurus* steers from a *Boophilus microplus*–free area in Northern Mexico were used in this experiment. Cattle were randomly allocated into six groups with four animals each, and were intramuscularly inoculated as follows: group 1 (control animals) were administered with normal bovine erythrocytes; group 2 received $1 \times 10^7$ *B. bovis*– and *B. bigemina*–infected erythrocytes as a combined fresh immunogen. Groups 3–6 were inoculated with a combined frozen immunogen containing (previous to cryopreservation at −196°C) $1 \times 10^7$, $5 \times 10^7$, $1 \times 10^8$, and $5 \times 10^8$ infected erythrocytes of each parasite species, respectively. Four months after immunization, principal and control animals were translocated to a bovine babesiosis endemic zone for field challenge. This was carried out by introducing the experimental cattle to tick-infested pastures for 30 days without ixodicide treatment. Cattle were monitored from day 8 postintroduction to the field (PIF) by recording the manifestation of clinical disease, rectal temperature values (RT), packed cell volume index (PCV), and percent of parasitized erythrocytes (PPE). At challenge, all experimental cattle became infected with both *Babesia bovis* and *B. bigemina*. However, except for two animals from group 6, none of the vaccinated animals showed signs of acute clinical babesiosis; therefore, no treatment was instituted. Out of six animals showing acute clinical babesiosis (four group 1 controls and two group 6 vaccinates), two animals (one from each group) died, despite babesiacide treatment, as they manifested classical cerebral babesiosis caused by *B. bovis*. Regardless of the dose or type of immunogen used (combined fresh or frozen), 90% of vaccinated cattle were determined to be protected against the virulent *Babesia* sp. field isolates. Nevertheless, by evaluating clinical parameters, such as average of maximum drop in PVC index (28.5%), average duration of parasitemia (3 days for *B. bovis*; 8.5 days for *B. bigemina*), and average duration of RT values ≥ 39.5 °C (2 days), animals receiving $1 \times 10^8$ infected erythrocytes, as combined frozen immunogen, were more efficaciously protected against challenge with virulent *B. bovis* and *B. bigemina* field isolates.**

Address for correspondence: J. Antonio Alvarez, CENID-PAVET, INIFAP. Apartado Postal 206, CIVAC, Morelos, C.P. 62500. Mexico. Voice: +52-777-3192850; fax: +52-777-3192850 ext. 129.

alvarez.jesus@inifap.gob.mx

**Ann. N.Y. Acad. Sci. 1026: 277–283 (2004).** **© 2004 New York Academy of Sciences.**
**doi: 10.1196/annals.1307.043**

KEYWORDS: *Babesia bovis*; *Babesia bigemina*; babesiosis vaccine

## INTRODUCTION

Bovine babesiosis is one of the diseases that causes major economic hardship to cattle raisers in the tropical regions of México. This disease is as prevalent in the Pacific region as in the Gulf of Mexico region.[1] A variety of activities have been considered in the effort to control babesiosis, such as vector control, cattle transport control, chemoprophylaxis, cattle resistant to tick infestation, and immunization.[2] Of the diverse types of vaccines studied up to now, only live attenuated vaccines that are available commercially in such countries as Australia, Colombia, and Israel have shown acceptable effectiveness. Mexico has two attenuated strains available, one of *Babesia bovis* and the other of *B. bigemina*; the BOR strain of *B. bovis* was cloned, irradiated, and maintained *in vitro,*[3] and has been used in susceptible animals where it has shown immunoprotector potential.[4,5] With reference to *B. bigemina*, a strain named BIS is available,[6] which has been maintained *in vitro* for an indefinite number of steps and has presented biological characteristics of reduced virulence to the inoculation of susceptible animals. In addition, the BIS sample has induced protection to heterologous challenge with infected ticks,[7] and with infected blood.[8] Up the present, satisfactory results have been obtained with both species of *Babesia,* applying them as a mixed inoculum.[9,10] Nevertheless, infected red blood cells have been manipulated as fresh inoculum produced *in vitro*. In both cases, excellent protective ability has been induced to the presentation of confrontations with heterologous samples in controlled and field conditions.

With regard to the way in which live vaccines are handled and their stability is debated,[11] this study considers the necessity for evaluating the use of these attenuated samples as frozen material, which could mean optimization of production, maintenance, and distribution of the immunogen.

## MATERIALS AND METHODS

### *Clinical Tracking of the Animals*

During the immunization period as well as during the challenge, the animals were clinically followed with a daily individual register of rectal temperature (RT), packed cell volume index (PCV), and percentage of parasitized erythrocytes (PPE). The clinical tracking continued through 21 days postimmunization (phase I) and 30 days postintroduction to an infected pasture (PIF) (phase II).

### *Phase I Bovine Immunization*

#### *Experimental Animals*

Twenty-four young bull heifers from a *Boophilus microplus*–free area were used; the animals were housed at a high plateau, in a tick vector–free ranch in Querétaro, Mexico. Six groups of four animals each were formed at random. Groups III, IV, V, and VI received the combined frozen immunogen of *B. bovis* and *B. bigemina* containing $1\times10^7$ $5\times10^7$, $1\times10^8$, and $5\times10^8$ infected erythrocytes (IE) of each parasite species, respectively. Group II received a fresh inoculum mixture with doses of $1\times10^7$ IE of each *Babesia* species. Group I served as control and received uninfected red blood cells (NRBC).

### *Phase II Field Challenge of Vaccinated Animals*

Four months after immunization, the experimental animals were transported to an endemic area in Paso del Toro, Veracruz, Mexico. They were immediately placed in *Boophilus microplus*–infested pastures. The field challenge was done in a natural way through the exposure of the cattle to the tick vector for a minimum of 30 days before the application of an ixodicide product.

## RESULTS

During phase I (immunization) no significant changes were observed in the evaluated parameters, although there was a decrease in PCV values with respect to basal value in one of the groups. In this way, the safety of frozen doses and the fresh inoculum was confirmed. With respect to phase II (field challenge with field strains), evaluated parameter data are presented in TABLE 1. It was observed that all animals showed parasitemia for several consecutive days (11–17 days, postchallenge), although it was never greater than 0.7%. All animals were infected with both species of *Babesia*, except for two animals vaccinated with the frozen mixture where no

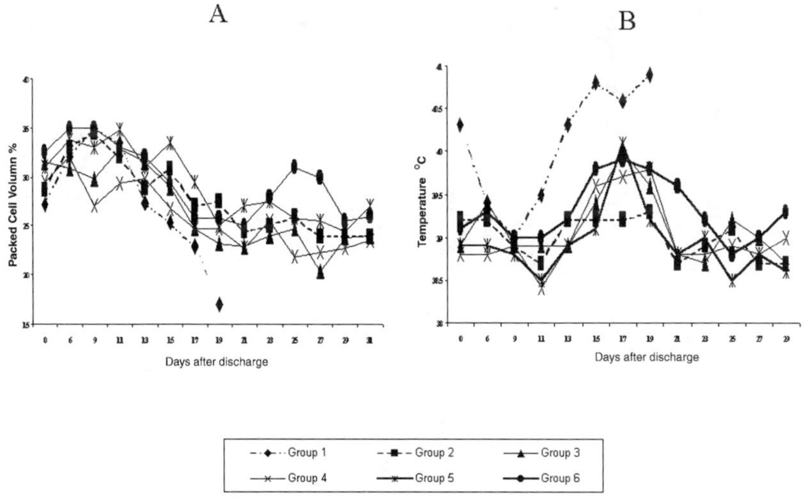


**FIGURE 1.** Mean packed cell volume (**A**) and rectal temperature (**B**) values in groups of animals immunized with the combined *in vitro* culture-derived fresh and frozen *Babesia* vaccine and challenged with tick-transmitted field isolates.

**TABLE 1. Monitoring of animals immunized with a frozen immunogenic mixture of *Babesia bovis* and *Babesia bigemina* during field challenge**

| | | | | | | | | Decrement PCV | | Decrement PCV >50% | | | Percent parasitized erythrocytes | | | Rectal temperature >39.5°C |
|---|---|---|---|---|---|---|---|---|---|---|---|---|---|---|---|---|
| *N*=4 | Tx | RIP | Hb | *B. bovis* infection | *B. bigemina* infection | NS | Pbc | Percent | Days postchallenge | Animals per group | Minimum PCV% | Group animal | *B. bovis* | *B. bigemina* | Total | Days |
| NRBC | 4 | 1 | 2 | 4 | 4 | 1 | 4 | 57.1 | 18 | 4 | 14.7 | 12 | 6.2 | 4.0 | 11 | 7.5 |
| 1 | 0 | 0 | 0 | 4 | 4 | 0 | 0 | 34.3 | 27 | 0 | 22.5 | 18 | 3.5 | 7.5 | 17 | 3.0 |
| 2 | 0 | 0 | 0 | 4 | 4 | 0 | 0 | 33.7 | 27 | 1 | 22.7 | 16 | 4.0 | 7.5 | 16 | 3.7 |
| 3 | 0 | 0 | 0 | 4 | 4 | 0 | 0 | 35.2 | 27 | 1 | 22.2 | 16 | 5.2 | 6.2 | 14 | 4.2 |
| 4 | 0 | 0 | 0 | 3 | 4 | 0 | 0 | 28.5 | 30 | 0 | 24.5 | 21 | 3.0 | 8.5 | 16 | 3.5 |
| 5 | 2 | 1 | 1 | 4 | 4 | 1 | 2 | 35.8 | 26 | 2 | 22.0 | 10 | 3.0 | 5.5 | 14 | 4.0 |

Tx = treatment; RIP = dead animals; Hb = animals with hemoglobinury; NS = animals with nervous symptoms; Pbc = animals with poor body condition; PCV = packed cell volume.

*B. bovis* was found. On average, for each experimental group, PCV values were very similar in all the vaccinated groups (22–24%), being greater than the control group (14.7%), with a maximum PCV decrease of 28–35%, compared to 57.1% observed in the control group on day 18 postchallenge (TABLE 1, FIG. 1A). Statistical differences were observed between the control and all the vaccinated groups ($P < .05$); however, no differences were found among the vaccinated groups ($P > .05$). Most of the animals presented fever, temperature above 39.5°C (FIG. 1B), for a minimum of three days, on average, and statistical differences were found between the control and immunized groups ($P < .05$). The four animals of the control group and two animals from group 6 required specific treatment. None of the animals from groups II, III, IV, or V and two animals from group VI were sick enough with acute clinical babesiosis to need any specific treatment. Of the six animals seriously sick with acute babesiosis (four controls; two from group VI), two animals (one from each group) died, in spite of having received the specific treatment, thus showing classic nervous signs from *B. bovis*. We determined that vaccinated animals received 90% protection, independent of doses or type of immunogen used.

After comparing the five vaccinated groups, it was observed that the animals vaccinated with the frozen immunogen at a dose of $1 \times 10^8$ had the lower decrement of PCV. Animals from this group as well those from the freshly vaccinated group were the only groups in which none of the animals had a PCV >50%, and also where the two groups in which the lower number of days with a temperature >39.5°C were observed (TABLE 1).

## DISCUSSION

The study corroborated, initially, the innocuity and protective capacity of the *B. bovis*– and *B. bigemina*–attenuated immunogen against field challenge; secondly it defined the dose of the combined frozen immunogen ($1 \times 10^8$) that confers similar or better protection to that conferred by the fresh inoculum at a dose of $1 \times 10^7$ IE. Thus, the frozen vaccine at a dose of $1 \times 10^8$ probably is good enough, considering that some of the immunogenic parasites would die during freezing and thawing. In Brazil, using a frozen live vaccine attenuated by passages in calves at a dose of $1 \times 10^7$ showed similar results in terms of clinical findings, and solid protection was demonstrated in both artificial and natural challenging.[14] An important advantage for the frozen vaccines has been reported, as that allows us to verify quality control before releasing.[15]

*B. bovis* was observed in blood stained smears of the control group animals 11 days after the animals were placed in the tick-infested pastures, which indicated that the challenge occurred the first week after arrival. Previous studies in the same pastures, in which the combined fresh immunogen was used under a field challenge, had also shown the high-tick infestation in the endemic area, thus a high probability risk of *Babesia* spp. infection.[12,13] The same studies showed a 100% and 70% protection of the freshly combined immunogen after challenge, depending on whether the animals were vaccinated in a free or endemic area, which is similar to the 87.5% protection observed in the present study with the frozen immunogen. In relation to the PCV decrease after challenge, the same authors found a 32% and 40% decrease in the free- and endemic area–immunized animals, respectively, which is also similar to the 33% PCV decrement observed in the frozen immunogen–vaccinated animals and the 34% decrease found in the fresh immunogen–vaccinated group. However, the control groups in the first two experiments had a 40% and 46% PCV decrement against the 57% observed in the present study. These results could indicate that the animals in the latter study were exposed to a higher tick infection rate. Using live vaccines derived from multiple passage in splenectomized calves caused some reactions in older animals and minimum risk for young animals.[14] All immunized animals, with the exception of two from group IV, were able to resist the field challenge, which was so severe that it produced evident signs of babesiosis, such as recumbency, pallor, marked weight loss, and nervous signs in the control animals. If the physical condition of the animals, which was always evaluated by the same veterinarian, could have been numerically measured and these values statistically analyzed, high statistical differences could have been shown. Significant difference was observed for RT when the control group was compared to the vaccinated animals ($P < .05$) from 14–19 days postchallenge.

In this study we demonstrated that the use of live frozen vaccines induces a strong protective level against a heterologous strain after field challenge. However, additional studies should be conducted to evaluate the duration of immunity, as well as to study the role for T cell subpopulations, macrophages, and molecules involved in the protective immune response.

## REFERENCES


1. Álvarez, J.A. & G.J. Cantó. 1985. Epidemiología de la babesiosis. *In* Parasitología. Vol. Conmemorativo, pp. 55–72. Sociedad Mexicana de Parasitología. S.C. México, D.F.
2. FAO. 1989. Review strategies for the control of ticks and tick-borne diseases and their vectors. FAO expert consultation, pp. 1–15. Rome, Italy.
3. Rodríguez, S.D., G.M. Buening, T.J. Green & C.A. Carson. 1983. Cloning of *Babesia bovis* by *in vitro* cultivation. Infect. Immun. **42:** 15–19.
4. Salas, T.E., J. Garcia, J.A. Ramos, *et al.* 1988. Patogenia de una clona irradiada de *Babesia bovis* obtenida de cultivo *in vitro*. Tec. Pecu. Mex. **26:** 36–42.
5. Cantó, G.J., J.V. Figueroa, J.A. Alvarez, *et al.* 1996. Capacidad inmunoprotectora de una clona irradiada de *Babesia bovis* derivada de cultivo *in vitro*. Tec. Pecu. Mex. **34:** 127–134.
6. Vega, C.A., G.M. Buening, T.J. Green & C.A. Carson. 1985. *In vitro* cultivation of *Babesia bigemina*. Am. J. Vet. Res. **46:** 416–420.
7. Hernández, R., J.A. Álvarez, G.M. Buening, *et al.* 1990. Diferencias en la virulencia y en la inducción de protección de aislamientos de *Babesia bigemina* derivados de cultivo *in vitro*. Tec. Pecu. Mex. **28:** 51–56.
8. Figueroa, J.V., G.J. Cantó, J.A. Alvarez, *et al.* 1998. Capacidad protectora de una cepa de *Babesia bigemina* derivada de cultivo *in vitro*. Tec. Pecu. Mex. **36:** 95–101.
9. Cantó, G.J., J.V. Figueroa, J.A. Ramos, *et al.* 1999. Evaluación de la patogenicidad y capacidad protectora de un inmunógeno fresco combinado de *Babesia bigemina* y *Babesia bovis*. Vet. Mex. **30:** 215–220.
10. Cantó, G.J., J.A. Ramos, E.E. Rojas, *et al.* 2002. Evaluación de la inocuidad y protección de un inmunógeno derivado de cultivo *in vitro* de *Babesia bovis* y *Babesia bigemina* multiplicado en bovinos. Tec. Pecu. Mex. **40:** 127–138.
11. Gil, L.A., B. Higuera & J. Orrego. 1986. Vacuna experimental contra *Babesia bovis* y *B. bigemina*, producida con atenuación de los parásitos por irradiación. Rev. Med. Zoot. **39:** 49–57.

12. CANTÓ, G.J., J.A. ALVAREZ, E.E. ROJAS, *et al.* Protección contra babesiosis bovina con una vacuna mixta de *Babesia bovis* y *Babesia bigemina* derivada de cultivo *in vitro* bajo una confrontación de campo. I. Inmunización en un área libre de la enfermedad. Vet. Mex. In press.
13. CANTÓ, G.J., E.E. ROJAS, J.A. ALVAREZ, *et al.* Protección contra babesiosis bovina con una vacunamixta de *Babesia bovis* y *Babesia bigemina* derivada de cultivo *in vitro* bajo una confrontación de campo. II. Inmunización en un área endémica. In press.
14. DE VOS, A.J. & R.E. BOCK. 2000. Vaccination against bovine babesiosis. Ann. N.Y. Acad. Sci. **967:** 540–545.
15. VIDOTTO, O., C.S. BARBOSA, G.M. ANDRADE, *et al.* 1998. Evaluation of a frozen trivalent attenuated vaccine against babesiosis and anaplasmosis in Brazil. Ann. N.Y. Acad. Sci. **849:** 420–423.

# Immunization of Bovines with Concealed Antigens from *Haematobia irritans*

CARLOS R. BAUTISTA,[a] ISABEL GILES,[b] NATIVIDAD MONTENEGRO,[b] AND JULIO V. FIGUEROA[a]

[a]*CENID-PAVET, INIFAP, Jiutepec Estado de Morelos, Mexico*

[b]*CNSCSA-SAGARPA, Jiutepec, Estado de Morelos, Mexico*

**ABSTRACT: To evaluate an immunization procedure using antigens from *Haematobia irritans* intestine (AgHiI), four bovines (group I) were inoculated with AgHiI mixed with Freund's incomplete adjuvant containing *Lactobacillus casei*, three bovines (group II) received AgHiI, and three bovines (group III) received saline solution. At day 35, blood was collected from each animal to feed H. irritans flies. There was no difference in the fly mortality observed in the three groups. The percentage of reduction of eggs oviposited by each female in 8 days (%RE), as compared with group III, was 29.45 for group I and 11.02 for group II. Antibody levels (AbL) to AgHiI were higher in group I than in groups II and III. A high correlation between %RE and AbL was observed.**

**KEYWORDS: *Haematobia irritans*; concealed antigens; immunization; *Lactobacillus casei***

## INTRODUCTION

The horn fly, *Haematobia irritans*, is an haematophagous parasitic fly that causes severe economic losses in cattle throughout the American continent,[1,2] producing an estimated annual loss of $1 billion dollars in North America.[3] Control is carried out mainly by the use of insecticides, but the indiscriminate use of these has induced the development of insecticide resistance in horn fly populations.[4–6] For this reason new control alternatives, such as vaccination and new adjuvants, are needed. In this respect, the use of concealed antigens from the gut of parasitic arthropods to immunize domesticated animals is one promising control measure;[7–10] however, there is little information regarding the control of *H. irritans* by this approach[11] and on the use of the lactic acid bacteria, *Lactobacillus casei*, as adjuvant.[12,13]

In this context, the aim of the present study was to evaluate the effect of an immunization procedure in bovines with a crude antigen from *Haematobia irritans* intestine, mixed with Freund's incomplete adjuvant, containing viable *Lactobacillus casei,* on the survival and oviposition of *H. irritans*.

Address for correspondence: Carlos R. Bautista, CENID-PAVET, INIFAP, Jiutepec Estado de Morelos, Mexico. Voice: +52-777-319-2850; fax: +52-777-320-5544.
bautista.carlos@inifap.gob.mx

**Ann. N.Y. Acad. Sci. 1026: 284–288 (2004).** **© 2004 New York Academy of Sciences.**
**doi: 10.1196/annals.1307.044**

## MATERIALS AND METHODS

### *Animals*

Ten 12-month-old female bovines free of tuberculosis, brucellosis, babesiosis, and anaplasmosis were allocated at random in three groups, in pens with a concrete floor: four in group I, three in group II, and three in group III. Animals were fed hay and commercial food, and had drinking water *at libitum*.

### *Antigen from* Haematobia irritans *Intestine (AgHiI)*

Intestines were dissected from recently blood-fed *H. irritans* flies; then they were washed several times with cold phosphate-buffered saline solution (PBS) to eliminate red blood cells. Afterwards the intestines were homogenized with a glass homogenizer (Ten Broeck) containing cold PBS and a protease inhibitor cocktail (Complete, Mini, Boehringer Mannheim, Germany). This solution was gently stirred using a magnetic stirrer overnight in the cold, and then it was centrifuged at 3000 × *g* at 4°C for 30 minutes. The supernatant (AgHiI) was collected, dispensed into vials, and frozen at –70°C. The protein content of the AgHiI was determined by the Lowry method.[14]

### Lactobacillus casei

In this study the strain ATCC7469 of *L. casei*, grown in MRS media, was used (Sigma, St Louis MO, USA).[15] The adjuvant used for immunization (FIA+Lc) consisted of 1 mL of Freund's incomplete adjuvant (Sigma, St. Louis MO, USA) and $1.8 \times 10^9$ *Lactobacillus casei* cfu, on the basis of previous studies.[15]

### Haematobia irritans

The *H. irritans* flies used to establish the colony at Jiutepec, Morelos, Mexico were kindly donated by the Knipling-Bushland U.S. Livestock Insects Research Laboratory, USDA-ARS, Kerrville, Texas, USA. This *H. irritans* colony has been maintained since 1995 following the same rearing techniques.[16,17]

### *Experimental Design*

Each animal in group I was inoculated intramuscularly (im) with 1 mg of antigen from *H. irritans* (AgHiI) mixed with 1 mL of FIA+Lc. Each bovine in group II received im 1 mg of AgHiI, while each animal in group III received 1 mL of PBS. The days of treatment for the three groups were 0, 7, 14, 21, and 28. On these same days blood samples were collected to obtain sera to determinate antibodies to intestines of *H. irritans* by an indirect ELISA[18] and anti-bovine IgG-peroxidase conjugate (Sigma, St Louis MO, USA); readings were carried out in an ELISA reader (Multiskan Plus, Labsystems, Finland) at an optical density of 492 nm; a serum was considered positive when it showed an O.D. higher than 1.0. On day 35, blood was collected from each animal, defibrinated with glass beads in a sterile container, and treated with kanamycin sulfate (250 mg/mL) and mycostatin (250 units/mL), prior to refrigeration at 4°C. Groups of 100 *H. irritans*, maintained in cages (one cage per bovine), were fed twice a day (8:00 h and 16:00 h) using one obstetrical sanitary napkin soaked in 100 mL of prepared blood per cage. On the basis of the fact that more than 90% of the total number of eggs was laid by flies in eight days, the fly mortality and the number of eggs oviposited per female was assessed in this period. The correlation between anti-AgHiI antibody levels obtained at day 35 and the total percentage of reduction of eggs oviposited per female in eight days was calculated.[19]

## RESULTS

The mean antibody level (AbL) to *H. irritans* intestine was higher in bovines from group I (vaccinated with AgHiI+FIA+Lc) as compared with those antibody levels observed in groups II and III. At day 35 the mean AbLs were 1.850, 0.841, and 0.812 for groups I, II, and III, respectively (FIG. 1). There was no difference in the mortality rate observed in the three groups (data not shown). On the basis of actual egg numbers to estimate $P$ values, the overall percentage of reduction of eggs oviposited per female in eight days (% RE), as compared with the control group, was 29.45 for group I ($P < .01$) and 11.02 for group II (not significant) (FIG. 2). The % RE in group I as compared with group II was 20.54 ($P < .01$). A high correlation between % RE and AbL was observed ($r = .9377$).

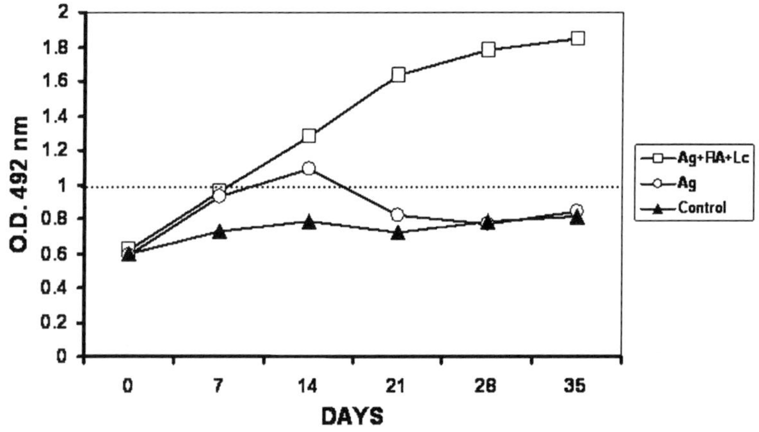


**FIGURE 1.** Average of IgG antibody levels to *Haematobia irritans* intestine as determined by ELISA in sera from immunized bovines. □: antigen from *Haematobia irritans* intestine (AgHiI) + Freund's incomplete adjuvant + *Lactobacillus casei*; ○: AgHiI; ▲: control. Each point represents the average of three to four bovines.

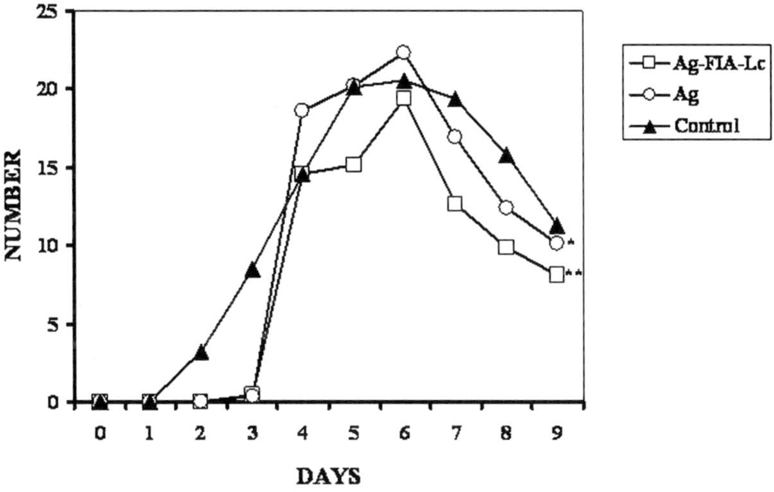


**FIGURE 2.** Daily average of eggs oviposited per *Haematobia irritans* female in cages containing flies (1 cage/1 bovine) fed with blood collected from immunized bovines. □: antigen from *Haematobia irritans* intestine (AgHiI) + Freund's incomplete adjuvant + *Lactobacillus casei*; ○: AgHiI; ▲: control. Each point represents the average of eggs oviposited per female fly from three to four cages. *11.02% total reduction with respect to the control. **29.45% total reduction with respect to the control.

## DISCUSSION

Results suggest that the immunization procedure used induced the buildup of antibodies to the intestines of *H. irritans* in the bovines, which affected the oviposition of the flies but not their survival. In this context, in another study there were no detrimental effects to any life cycle stages of adult horn flies fed with sera from immunized bovines with peritrophins from *H. irritans.*[11] The contrast between our results and those of the later study might be due to the fact that in our study we used a crude antigen from the intestine that probably contained more relevant epitopes to induce protective responses than the antigen used in the later study.

On the basis of the results obtained under the particular conditions of the study, we concluded that the antigen from the *H. irritans* intestine, mixed with FIA containing *L. casei* and inoculated in bovines, induced specific antibodies to this antigen, and when these antibodies were ingested by the adult flies they affected their oviposition but not their survival. However, more studies are needed to demonstrate and isolate the protective components in the antigen used, to corroborate the adjuvant effect of *L. casei*, and to understand the mechanism by which the antibodies induced affect the physiology of the horn fly.

## AKNOWLEDGMENTS

This study was carried out with funds from CONACYT, contract number 31524-B.

## REFERENCES

1. MENDES, J. & A.X. LINHARES. 2002. Cattle dung breeding Diptera in pastures in southeastern Brazil: diversity, abundance and seasonallity. Mem. Inst. Oswaldo Cruz **97:** 37–41.
2. GUGLIELMONE, A.A., E. GIMENO, *et al.* 1999. Skin lesions and cattle hide damage from *Haematobia irritans* infestations. Med. Vet. Entomol. **13:** 324–329.
3. CUPP, E.W., M.S. CUPP, *et al.* 1998. Blood-feeding strategy of *Haematobia irritans* (Diptera: Muscidae). J. Med. Entomol. **35:** 591–595.
4. KUNZ, S.E., M.O. ESTRADA & H.F. SANCHEZ. 1995. Status of *Haematobia irritans* (Diptera: Muscidae) insecticide resistance in northeastern Mexico. J. Med. Entomol. **32:** 726–729.
5. SHEPPARD, D.C. & P.R. TORRES. 1998. Onset of resistance to fenvalerate, a pyrethroid insecticide in Argentine horn flies (Diptera: Muscidae). J. Med. Entomol. **35:** 175–176.
6. GUGLIELMONE, A.A., M.E. CASTELLI, *et al.* 2001. Toxicity of cypermethrin and diazinon to *Haematobia irritans* (Diptera: Muscidae) in its American southern range. Vet. Parasitol. **101:** 67–73.
7. SUKARSIH, S. PARTOUTOMO, *et al.* 2000. Vaccination against the Old World screwworm fly (*Chrysomya bezziana*). Parasite Immunol. **22:** 545–552.
8. WILLADSEN, P. 1990. Perspectives for subunit vaccines for the control of ticks. Parassitologia **32:** 195–200.
9. WILLADSEN, P., P. BIRD, *et al.* 1995. Commercialisation of a recombinant vaccine against *Boophilus microplus*. Parasitology **110** (Suppl)**:** S43–50.
10. HEATH, A.W., A. ARFSTEN, *et al.* 1994. Vaccination against the cat flea *Ctenocephalides felis felis.* Parasite Immunol. **16:** 187–191.
11. WIJFFELS, G., S. HUGHES, *et al.* 1999. Peritrophins of adult dipteran ectoparasites and their evaluation as vaccine antigens. Int. J. Parasitol. **29:** 1363–1377.
12. DE WAARD, R., J. GARSSEN, *et al.* 2001. Enhanced antigen-specific delayed-type hypersensitivity and immunoglobulin G2b responses after oral administration of viable *Lactobacillus casei* YIT9029 in Wistar and Brown Norway rats. Clin. Diagn. Lab Immunol. **8:** 762–767.
13. PLANT, L.J. & P.L. CONWAY. 2002. Adjuvant properties and colonization potential of adhering and non-adhering *Lactobacillus* spp. following oral administration to mice. FEMS Immunol. Med. Microbiol. **34:** 105–111.
14. LOWRY, O.H., N.J. ROSEBROUGH, A.L. FARR & R.L. RANDALL. 1951. Protein measurement with the Folin-phenol reagent. J. Biol. Chem. **193:** 265–268.
15. BAUTISTA-GARFIAS, C.R., O. IXTA-RODRIGUEZ, *et al.* 2001. Effect of viable or dead *Lactobacillus casei* organisms administered orally to mice on resistance against *Trichinella spiralis* infection. Parasite **8:** S226–S228.
16. SCHMIDT, C.D., R.L. HARRIS & R.A. HOFFMAN. 1967. Mass rearing of the horn fly, *Haematobia irritans* (Diptera: Muscidae), in the laboratory. Ann. Entomol. Soc. Am. **60:** 508–510.
17. SCHMIDT, C.D., R.L. HARRIS & R.A. HOFFMAN. 1968. New techniques for rearing horn flies at Kerrville, 1967. Ann. Entomol. Soc. Am. **61:** 1045–1046.
18. DUMENIGO, B.E., A.M. ESPINO, *et al.* 2000. Kinetics of antibody-based antigen detection in serum and faeces of sheep experimentally infected with *Fasciola hepatica.* Vet. Parasitol. **89:** 153–161.
19. SWINSCOW, T.D.V. *et al.* 1978. Statistics at Square One. British Medical Association. London.

# Protection of Dairy Cows Immunized with Tick Tissues against Natural *Boophilus microplus* Infestations in Thailand

SATHAPORN JITTAPALAPONG,[a] WEERAPHOL JANSAWAN,[a]
ASWIN GINGKAEW,[b] OMAR O. BARRIGA,[c] AND ROGER W. STICH[d]

[a]*Department of Parasitology, Faculty of Veterinary Medicine, Kasetsart University, Bangkok 10903, Thailand*

[b]*Rajamongala Institute of Technology, Chanthaburi Campus, Chanthaburi, Thailand*

[c]*Instituto de Ciencias Biomedicas, Facultad de Medicina, Universidad de Chile, Santiago, Chile*

[d]*Department of Veterinary Preventive Medicine, College of Veterinary Medicine, Ohio State University, Columbus, Ohio 43210, USA*

**ABSTRACT: *Boophilus microplus* has a major impact on cattle production, and an antitick vaccine would be a valuable tool for control of this important ectoparasite in Thailand. Previous work has shown that immunization of hosts with different tick tissues has different implications regarding tick feeding and fecundity under experimental conditions. The purpose of this study was to assess the effects of immunization of dairy cattle with *B. microplus* salivary gland or midgut extracts on natural infestations by this tick species. The different antigen extracts (1 mg total protein) or equivalent amounts of adjuvant alone were injected intradermally every two weeks for a total of three times before allowing cattle to graze in a tick-contaminated pasture. Animals were checked daily, and engorged female ticks collected, counted, weighed, and maintained in tick incubators to observe tick performance parameters, including engorged weight, egg mass weight, nonviable eggs, mortality, oviposition period, egg incubation period, and $F_1$ larval weight. After six months, each group was reimmunized with the same antigen and/or adjuvant, and ticks were again collected and evaluated. Immunization of cattle with salivary gland preparations resulted in reductions in mean tick counts and in engorged female weights. Immunization with midgut antigens reduced tick oviposition and reduced egg mass weights. In addition, more ticks recovered from midgut-immunized cows produce nonviable eggs. This investigation indicates that a vaccine based on these antigen preparations could induce a lasting, protective immune response against *B. microplus* that would be expected to provide a safe nontoxic means of tick control.**

**KEYWORDS: *Boophilus microplus*; cows; eggs; immunization; ticks**

Address for correspondence: Sathaporn Jittapalapong, Department of Parasitology, Faculty of Veterinary Medicine, Kasetsart University, Bangkok 10903, Thailand.
fvetspj@ku.ac.th

**Ann. N.Y. Acad. Sci. 1026: 289–297 (2004). © 2004 New York Academy of Sciences.
doi: 10.1196/annals.1307.045**

## INTRODUCTION

The tropical cattle tick, *Boophilus microplus*, is an important ectoparasite of cattle in tropical and subtropical countries. These ticks cause economic losses due to direct effects on the preferred hosts and by the pathogens they transmit. Therefore, tick control is a continuing global priority.[1–3] Traditional control methods include the use of chemicals, with partially successful results, but acaricides are expensive and have adverse effects such as a high incidence of resistance among tick populations, as well as harmful effects on vertebrate hosts, human beings, and the environment.[4–6] Alternative measures include biological control methods such as pasture spelling[7] and artificial selection for tick-resistant cattle.[8,9] Such steps can reduce tick burdens, but enhancement of host resistance through immunization would constitute a major advance in control.[10–13] Several approaches have been used to actively immunize cattle against ticks,[14–22] and it has been reported that protective host immune responses can damage female ticks, reducing tick populations on immunized cattle by up to 70%.

Vaccines against *B. microplus*, containing the recombinant Bm86 antigen, were registered in Australia (TickGARD, Hoechst Animal Health, Australia) and Cuba (Gavac, Heber Biotech S.A., Havana, Cuba) in the early 1990s. The major effect of these vaccines is a successive reduction in tick numbers due to reduction of adult female tick fertility.[23] Considering the complexity of the tick-host-environment relationship, these antitick vaccines seem to work.

*B. microplus* is a one-host tick, almost monospecific for cattle, that is distributed throughout Thailand, except for small areas around forest and mountain regions.[24–26] Resistance to this tick is rare among European breeds of cattle, even after exposure to the tick for months or even years.[27] Crossbred Holstein-Friesian dairy cows are abundant in Thailand, which is a concern owing to the high vulnerability of these breeds to ticks and tick-borne diseases compared to the native breed.[9] There are substantial losses in milk production associated with tick infestation of dairy cows in Thailand. For example, it has been estimated that each engorging adult female tick is responsible for the loss of 8.9 mL of milk and 1 g of live weight gain.[28]

An antitick vaccine is now being considered as another method of cattle tick control in Thailand. The purpose of this research was to investigate the effects of immunizing dairy cows, with extracts of midgut and salivary glands from adult female *B. microplus*, against tick infestations under field conditions. The efficacy of the immunization was evaluated by analyzing tick feeding and fecundity parameters of these treatments compared to the control groups.

## MATERIALS AND METHODS

### *Antigen Preparation*

Adult female *B. microplus* that had fed for 3–7 days were collected from laboratory cattle used to rear and maintain a clean *B. microplus* colony. Ticks were placed into 0.15 M phosphate buffered saline (PBS), pH 7.4, and opened along their dorsal surface. Salivary glands (SG) were removed, dissected free of other tissues, placed into PBS at 4°C, and then suspended in 1% SDS and 5% 2-mercaptoethanol prior to incubation in a water bath at 56°C overnight and boiling for 5 min. The solution was cooled to room temperature, transferred into a 12,000–14,000 molecular weight cut-off dialysis tube (Spectra/Por7, Denver, CO), immersed in 1 L of PBS, and left at 4°C on a magnetic stirrer overnight (PBS was changed every 4–6 h). This mixture (0.5 mL, 2 mg of protein/mL) was filtered and mixed with 0.5 mL of complete or incomplete Freund's adjuvant H37Ra (Difco Laboratories, Detroit, MI) immediately prior to immunization of the hosts.

Tick midgut (MG) samples were also removed at 4°C in PBS. These organs were disrupted for 30 s in PBS at 4°C with a tissue homogenizer (KIKA T8.01, KIKA LABORTECHNIK, Germany) followed by sonication for 15 s (Branson Sonifier 450, USA), set between 30% and 50% duty cycle for 50–100 output power, for a total of 10 times. The homogenates were dialyzed in PBS at 4°C overnight and centrifuged at 16,000 × *g* for 30 min at 4°C. MG preparations were sterilized with a 0.45-μm filter (Millipore, Bedford, MA), and 0.5 mL (2 mg of protein/mL) mixed thoroughly by sonication with 0.5 mL of complete or incomplete Freund's adjuvant prior to immunization. The protein concentration of the extract was estimated by the method of Bradford.

### *Animals*

A total of sixteen 6- to 13-year-old lactating dairy cows, with some previous exposure to *B. microplus*, were used in this study. Before the experiments were started, undetectable antibody titers to tick antigens were confirmed by ELISA and Western blots with tick salivary gland antigens.[29,30] Each animal was randomly allocated to immunized and control groups of four cows each. Vaccinated animals were immunized using extracts derived from adult female *B. microplus* ticks. The first group was immunized with salivary gland extract (SG), the second group with midgut (MG), and the third with adjuvant (ADJ) only, and the last group was injected with PBS only as a control (CTR). Animals were housed in pens and were fed with hay, concentrates, and water *ad libitum* throughout the experimental period until the conclusion of the immunization period, after which they were turned out to the pasture to acquire natural tick infestation.

### *Immunization*

The protocol involved three intradermal immunizations with 1 mg of the extracted protein antigen plus Freund's complete or incomplete adjuvant. The second and third immunizations were given 2 weeks after the first and second ones, respectively. The adjuvant and the control group respectively received adjuvant and PBS only. After the last immunization, animals were allowed to graze in the tick-contaminated pasture. After 6 months, each group was again subjected to a single immunization with the same antigen and adjuvant as described above before being released again into the contaminated pasture and monitored for ticks. This pasture has been observed for tick burden during grazing period for 1 year before the experiment started. Thirty to 100 engorged females were found on cattle within a month, depending upon location and season.

### *Tick Parameters*

Animals were checked daily, and recovered ticks were counted, weighed, and isolated in tick chambers. Tick parameters were measured as described.[31] Engorged ticks were collected twice a day during milking at 6:00 A.M. and 4:00 P.M. Ticks were maintained in tick incubators to observe tick performance parameters such as engorged weight, egg mass weight, nonviable eggs, mortality, oviposition period, egg incubation period, and larval weight. After 14–20 days, when oviposition was completed, individual egg masses were placed in test tubes with a gauze top to monitor hatching.

### *Data Analysis*

The effect of treatment (SG, MG, and ADJ versus CTR) on measurements of tick fecundity and tick feeding parameters was determined by analysis of variance (ANOVA). A multivariate analysis technique was used to descriptively show the pattern of the covariation of the various biological parameters among the animals so that the different variables could be reduced to a smaller set of independent components. This pattern was then compared between the groups of immunization. ANOVA was used to determine significant differences between immunized group means for each of the variables. Pairwise comparisons of least-squares (LS) means were accomplished with the Tukey-Kramer method. All analyses were carried out using the SAS program.

## RESULTS

### *Engorgement*

Effects of vaccination on tick engorgement are summarized in TABLE 1. Among SG-immunized animals, the number and weight of engorged female ticks were significantly reduced compared to the controls ($P < 0.01$). The mean engorged weight of ticks collected from cattle in the MG group was also reduced ($P < 0.05$). The total tick counts for SG, MG, ADJ, and CTR groups were 613, 1032, 1331, and 1445, respectively, during the experimental period.

### *Oviposition*

Ticks that engorged on immunized cows produced eggs whose viability was significantly lower than that of eggs laid by ticks fed on CTR cows (TABLE 2). While immunization with either SG or MG significantly reduced the number of females that oviposited, immunization with SG had the most pronounced reduction. The mean egg mass weights for the MG and CTR groups were also statistically different ($P < 0.01$). Although immunization with either SG or MG significantly reduced egg mass weight, MG had the most pronounced effect. No statistical differences were found between ADJ and CTR groups for this parameter.

**TABLE 1. Performance of engorged female *B. microplus* recovered from immunized and control cows**

| Treatment | Number collected (LS means ± SE) | % Reduction in tick number | Engorged female weight (mg) (LS means ± SE) | % Reduction in tick weight |
|---|---|---|---|---|
| *Initial immunizations* | | | | |
| SG | $328.56 \pm 12.31^a$ | 52.50 | $126.4 \pm 7.33^a$ | 49.38 |
| MG | $505.41 \pm 14.64^b$ | 26.93 | $171.5 \pm 7.49^b$ | 31.32 |
| ADJ | $634.14 \pm 18.57^c$ | 8.32 | $238.8 \pm 9.58^c$ | 4.37 |
| CTR | $691.72 \pm 15.29^c$ | 0 | $249.7 \pm 10.67^c$ | 0 |
| *Booster immunizations* | | | | |
| SG | $285.31 \pm 10.66^a$ | 62.14 | $112.79 \pm 6.34^a$ | 56.98 |
| MG | $526.98 \pm 11.04^b$ | 30.06 | $181.45 \pm 5.63^b$ | 30.79 |
| ADJ | $697.11 \pm 15.67^c$ | 7.49 | $247.81 \pm 7.81^c$ | 5.47 |
| CTR | $753.52 \pm 13.94^c$ | 0 | $262.16 \pm 11.15^c$ | 0 |

$^{a,b,c}$For each column, LS mean values with different superscripts differ ($P < 0.05$).

**TABLE 2. Oviposition by *B. microplus* female ticks collected from immunized and control cows**

| Treatment | Number of females that oviposited | % Oviposition | Egg mass weight (mg) | % Reduction in egg mass weight |
|---|---|---|---|---|
| *Initial immunizations* | | | | |
| SG | $292.91 \pm 8.73^a$ | 89.15 | $55.96 \pm 1.17^b$ | 32.19 |
| MG | $300.27 \pm 9.82^b$ | 59.41 | $26.25 \pm 1.14^a$ | 68.19 |
| ADJ | $599.44 \pm 15.84^c$ | 94.53 | $80.22 \pm 2.37^c$ | 2.79 |
| CTR | $658.53 \pm 15.37^c$ | 95.20 | $82.52 \pm 2.41^c$ | 0 |
| *Booster immunizations* | | | | |
| SG | $264.82 \pm 9.67^a$ | 92.82 | $54.15 \pm 1.06^b$ | 36.91 |
| MG | $289.57 \pm 9.82^b$ | 54.95 | $25.91 \pm 1.74^a$ | 69.81 |
| ADJ | $662.38 \pm 13.30^c$ | 95.02 | $81.04 \pm 2.93^c$ | 5.58 |
| CTR | $721.72 \pm 14.09^c$ | 95.78 | $85.83 \pm 3.25^c$ | 0 |

$^{a,b,c}$For each column, LS mean values with different superscripts differ ($P < 0.05$).

### *Offspring*

The percent nonviable eggs was significantly higher ($P < 0.05$) in ticks recovered from cows in the MG group than those treated with SG, ADJ, or CTR (TABLE 3). Larval weights from the MG group were less than that of CTR, but no statistically significant differences were found among these groups (TABLE 3). Oviposition and egg incubation periods were not affected (TABLE 4).

**TABLE 3. Egg viability of *B. microplus* from immunized and control cows**

| Treatment | Nonviable egg masses[a] | % Nonviable egg masses | Larval weight (μg)[a] | % Reduction in larval weight |
|---|---|---|---|---|
| *Initial immunizations* | | | | |
| SG | 10 ± 2.27 | 3.41 | 25.12 ± 4.68 | 13.59 |
| MG | 37 ± 2.64 | 12.32 | 20.01 ± 5.07 | 31.17 |
| ADJ | 24 ± 2.71 | 4.00 | 27.55 ± 4.86 | 5.23 |
| CTR | 17 ± 1.98 | 2.58 | 29.07 ± 4.37 | 0 |
| *Booster immunizations* | | | | |
| SG | 12 ± 1.83 | 4.53 | 27.34 ± 4.23 | 10.68 |
| MG | 41 ± 2.56 | 14.15 | 19.27 ± 5.83 | 37.05 |
| ADJ | 34 ± 3.31 | 5.13 | 28.39 ± 4.19 | 7.25 |
| CTR | 24 ± 1.94 | 3.33 | 30.61 ± 4.94 | 0 |

[a]Reported as LS mean values ± SE.

**TABLE 4. Oviposition and egg incubation periods of *B. microplus* from immunized and control cows**

| Treatment | Oviposition period (days)[a] | % Increase in oviposition period | Egg incubation period (days)[a] | % Increase in egg incubation period |
|---|---|---|---|---|
| *Initial immunizations* | | | | |
| SG | 18.3 ± 4.85 | 41.86 | 19.1 ± 4.99 | 18.63 |
| MG | 16.6 ± 4.07 | 28.68 | 22.4 ± 5.84 | 39.13 |
| ADJ | 13.7 ± 4.37 | 6.20 | 17.3 ± 5.64 | 7.45 |
| CTR | 12.9 ± 3.69 | 0 | 16.1 ± 4.75 | 0 |
| *Booster immunizations* | | | | |
| SG | 18.1 ± 4.94 | 33.09 | 18.8 ± 5.18 | 11.90 |
| MG | 17.4 ± 4.34 | 27.94 | 22.1 ± 5.47 | 31.55 |
| ADJ | 13.8 ± 3.44 | 1.47 | 17.4 ± 4.56 | 3.57 |
| CTR | 13.6 ± 3.73 | 0 | 16.8 ± 4.82 | 0 |

[a]Reported as LS mean values ± SE.

## DISCUSSION

The results reported in this paper confirm that immunization against *B. microplus* is possible under experimental conditions. The degree of immunity produced was relatively high compared to previous works[14,16] and effective against natural challenge infestations for more than 6 months. Immunization of dairy cows with SG reduced the engorgement number of female ticks by 52%, which corroborates another report that showed a reduction in tick numbers by 73% with a crude extract.[16] Similar results have been demonstrated on cattle vaccinated with crude extracts of adult *B. microplus*, which reduced tick populations on vaccinated cattle by 70% compared to controls.[14] Others have reported immunization with tick antigens to reduce the number of engorged female ticks by 75–81%,[23] 33%,[32] 56%,[33] and 35%.[34] Reduced engorged weights of ticks infesting SG- and MG-vaccinated cows also confirmed the results of other groups.[14,18,23,32,34,35]

Not only did immunization of cows with adult *B. microplus* tissues result in reduced numbers of ticks found on these animals, but it also significantly reduced the ticks' reproductive capacity as demonstrated by the reduction of the average number of eggs laid (egg mass weight) and their lower hatchability (nonviable eggs). Therefore, immunization of dairy cows with MG reduced the reproductive efficiency of cattle ticks.

The reduction in both feeding and fecundity performances after challenge tick infestation in this investigation indicated that a trend of protective immunity was developing in cows immunized with tick tissues. Development of resistance due to MG immunization appeared to be more pronounced than immunization with SG, but this is not surprising because the efficacy of novel antigens from MG has long been documented.[14,18–21] These effects demonstrated that ingestion of blood from vaccinated cattle led to the destruction of the tick midgut digestive cells.[36] In our study, ticks collected from vaccinated cattle were lighter in weight than those collected from controls. This finding suggested damage to the tick midgut as a result of immune reactions developed by the host, as supported by previous investigations.[14,18,20]

Interference with the transmission of tick-borne pathogens is another important consideration in vaccination with tick tissues because pathogen transmission might be prevented through interference with salivary gland function. Further work to define the mechanisms of vertebrate immune resistance to tick infestation and to identify SG and MG antigens associated with reduced tick performance is warranted.

## ACKNOWLEDGMENTS

We thank the Kasetsart University Research Development Institute (KURDI-KIP.90.42) for financial support of this research. We also acknowledge the Faculty of Veterinary Medicine, Kasetsart University, and the Rajamongala Institute of Technology, Chanthaburi Campus, whose faculty and students provided the animal facilities and assistance.

## REFERENCES

1. Snelson, J.T. 1975. Animal ectoparasites and disease vectors causing major reductions in world food supplies. FAO Plant Protect. Bull. **13:** 103–114.
2. Steelman, C.D. 1976. Effects of external and internal arthropod parasites on domestic livestock production. Annu. Rev. Entomol. **21:** 155–178.
3. McCosker, P.J. 1979. Global aspects of the management and control of ticks of veterinary importance. Recent Adv. Acarol. **2:** 45–53.
4. Wikel, S.K. 1988. Immunological control of hematophagous arthropod vectors: utilization of novel antigens. Vet. Parasitol. **29:** 235–264.

5. WILLADSEN, P. & D.H. KEMP. 1988. Vaccination with "concealed" antigens for tick control. Parasitol. Today **4:** 196–198.
6. NOLAN, J., J.T. WILSON, P.E. GREEN & P.E. BIRD. 1989. Synthetic pyrethroid resistance in field samples of the cattle tick (*Boophilus microplus*). Aust. Vet. J. **66:** 179–182.
7. SUTHERST, R.W., G.A. NORTON, N.D. BARLOW *et al.* 1979. An analysis of management strategies for cattle tick (*Boophilus microplus*) control in Australia. J. Appl. Ecol. **16:** 359–382.
8. WHARTON, R.H., K.B.W. UTECH & H.G. TURNER. 1970. Resistance to the cattle tick, *Boophilus microplus*, in a herd of Australian Illawarra Shorthorn cattle: its assessment and heritability. Aust. J. Agric. Res. **21:** 163–181.
9. WHARTON, R.H. & K.R. NORRIS. 1980. Control of parasitic arthropods. Vet. Parasitol. **6:** 135–164.
10. RIEK, R.F. 1962. Studies on the reactions of animal infestation with ticks. VI. Resistance of cattle to infestation with the tick *Boophilus microplus*. Aust. J. Agric. Res. **13:** 532–550.
11. ROBERTS, J.A. 1968. Acquisition by the host of resistance to the cattle tick, *Boophilus microplus* (Canestrini). J. Parasitol. **54:** 657–662.
12. BARRIGA, O.O., S.S. DA SILVA & J.S.C. AZEVEDO. 1993. Inhibition and recovery of tick functions in cattle repeatedly infested with *Boophilus microplus*. J. Parasitol. **79:** 710–715.
13. BARRIGA, O.O., S.S. DA SILVA & J.S.C. AZEVEDO. 1995. Relationships and influences between *Boophilus microplus* characteristics in tick-naïve or repeatedly infested cattle. Vet. Parasitol. **52:** 225–238.
14. JOHNSTON, L.A.Y., D.H. KEMP & R.D. PEARSON. 1986. Immunization of cattle against *Boophilus microplus* using extracts derived from adult female ticks: effects of induced immunity on tick populations. Int. J. Parasitol. **16:** 27–34.
15. PANDA, D.N., M.Z. ANSARI & B.N. SAHAI. 1993. Immunization of cattle using *Boophilus microplus* adult larval extracts: feeding, survival, and reproductive behavior of ticks on immunized cattle. Indian J. Anim. Sci. **63:** 123–127.
16. KHALAF-ALLAH, S.S. & L. EL-AKABAWY. 1996. Immunization of cattle against *Boophilus annulatus* ticks using adult female tick antigen. Dtsch. Tierärzt. Wschr. **103:** 219–221.
17. JOHNSTON, T.H. & H.J. BRANCROFT. 1918. A tick-resistant condition in cattle. Proc. R. Soc. Queensl. **30:** 219–317.
18. KEMP, D.H., R.I.S. AGBEDE, L.A.Y. JOHNSTON & J.M. GOUGH. 1986. Immunization of cattle against *Boophilus microplus* using extracts derived from adult female ticks: feeding and survival of the parasite on vaccinated cattle. Int. J. Parasitol. **16:** 115–120.
19. OPDEBEECK, J.P., J.Y.M. WONG, L.A. JACKSON & C. DOBSON. 1988. Vaccines to protect Hereford cattle against the cattle tick, *Boophilus microplus*. Immunology **63:** 363–367.
20. WILLADSEN, P., R.V. MCKENNA & G.A. RIDING. 1988. Isolation from the cattle tick, *Boophilus microplus*, of antigenic material capable of eliciting a protective immunological response in the bovine host. Int. J. Parasitol. **18:** 183–189.
21. RAND, K.N., T. MOORE, A. SRISKANTHA *et al.* 1989. Cloning and expression of a protective antigen from the cattle tick *Boophilus microplus*. Proc. Natl. Acad. Sci. USA **86:** 9657–9661.
22. WONG, J.Y.M., J.H. DUFTY & J.P. OPDEBEECK. 1990. The expression of bovine lymphocyte antigen and response of Hereford cattle to vaccination against *Boophilus microplus*. Int. J. Parasitol. **20:** 677–679.
23. RODRIGUEZ, M., M.L. PENICHET, A.E. MOURIS *et al.* 1995. Control of *Boophilus microplus* populations in grazing cattle vaccinated with a recombinant Bm 86 antigen preparation. Vet. Parasitol. **57:** 339–349.
24. TANSKUL, P., H.E. STARK & I. INKAM. 1993. A checklist of ticks of Thailand (Acari: Metastigmata: Ixodoidea). J. Med. Entomol. **20:** 330–341.
25. SARATAPHAN, N., D. TUNTASUVAN, S. BOONCHIT & Y. ITO. 1998. Survey on ticks (Acari: Ixodidae) of cattle and buffalo in Thailand. J. Thai Vet. Med. Assoc. **49:** 47–56.
26. PATTANATANANG, K., N. PINYOPANUWAT, B. NIMSUPHAN *et al.* 2001. Rearing and maintaining colony of cattle ticks (*Boophilus microplus*) on the experimental animals and

in the laboratory in Thailand. Proceedings of the 39th Kasetsart University Annual Conference (Feb. 5–7, 2001).
27. UTECH, K.B.W., G.W. SEIFERT & R.H. WHARTON. 1978. Breeding Australian Illawarra Shorthorn cattle for resistance to the *Boophilus microplus*. I. Factors affecting resistance. Aust. J. Agric. Res. **29:** 411–422.
28. JONSSON, N.N. & A.L. MATSCHOSS. 1998. Attitudes and practices of Queensland dairy farmers to the control of the cattle tick, *Boophilus microplus*. Aust. Vet. J. **76:** 746–751.
29. JITTAPALAPONG, S., R.W. STICH, J.C. GORDON *et al.* 2000. Humoral immune response of dogs immunized with salivary gland, midgut, or repeated infestations with *Rhipicephalus sanguineus*. Ann. N.Y. Acad. Sci. **916:** 283–288.
30. JITTAPALAPONG, S. 1999. Immune resistance to *Rhipicephalus sanguineus* in dogs. Ph.D. dissertation. Ohio State University, Columbus, OH.
31. JITTAPALAPONG, S., R.W. STICH, J.C. GORDON *et al.* 2000. Performance of *Rhipicephalus sanguineus* (Acari: Ixodidae) fed on dogs exposed to multiple infestations or immunization with tick salivary gland or midgut tissues. J. Med. Entomol. **37:** 601–611.
32. DE ROSE, R., R.V. MCKENNA, G. COBON *et al.* 1999. Bm 86 antigen induces a protective immune response against *Boophilus microplus* following DNA and protein vaccination in sheep. Vet. Immunol. Immunopathol. **71:** 151–160.
33. JONSSON, N.N., A.L. MATSCHOSS, P. PEPPEER *et al.* 2000. Evaluation of TickGARD$^{PLUS}$, a novel vaccine against *Boophilus microplus*, in lactating Holstein-Friesian cows. Vet. Parasitol. **88:** 275–285.
34. RODRIGUEZ, M., R. RUBIERA, M. PENICHET *et al.* 1994. High level expression of the *B. microplus* Bm86 antigen in the yeast *P. pastoris* forming highly immunogenic particles for cattle. J. Biotechnol. **33:** 135–146.
35. GARCIA-GARCIA, J.C., C. MONTERO, M. RODRIGUEZ *et al.* 1998. Effect of particulation on the immunogenic and protective properties of the recombinant Bm86 antigen expressed in *Pichia pastoris*. Vaccine **16:** 374–380.
36. TRACEY-PATTE, P.D., D.H. KEMP & L.A.Y. JOHNSTON. 1987. *Boophilus microplus*: passage of bovine immunoglobulins and albumin across the gut of cattle ticks feeding on normal or vaccinated cattle. Res. Vet. Sci. **43:** 287–290.

# Immune Response to *Babesia bigemina* Infection in Pregnant Cows

T. D. GARCÍA, M. J. V. FIGUEROA, A. J. A. RAMOS, M. C. ROJAS, A. G. J. CANTÓ, N. A. FALCÓN, AND M. J. A. ÁLVAREZ

*CENID-PAVET, INIFAP, CIVAC, Morelos, C.P. 62500, Mexico*

**ABSTRACT: Babesiosis is a tick-borne disease of cattle caused by *Babesia bigemina* and *Babesia bovis* and is transmitted by the tick vector *Boophilus microplus*. In this study, we investigate *B. bigemina* infection regarding the clinical infection, T cell distribution, and cytokine profile during the acute phase of an experimental infection in pregnant cows.**

**KEYWORDS: *Babesia bigemina*; *Babesia bovis*; *Boophilus microplus*; babesiosis; cows; immune response; infection; pregnant; T cells**

## INTRODUCTION

Babesiosis is a tick-borne disease of cattle caused by *Babesia bigemina* and *Babesia bovis* and is transmitted by the tick vector *Boophilus microplus*. It is one of the economically most important infectious diseases in the tropical regions of the world.[1] In Mexico, it is a serious obstacle to improve the genetic background in tropical areas.[2] Different studies have been performed *ex vivo* and *in vitro* in order to understand the immune response. For example, *Babesia bovis*–infected erythrocytes and a membrane-enriched fraction of merozoites are able to stimulate inducible nitric oxide synthase (iNOs) transcription and NO production, but there is no report on the induction of inflammatory cytokines by *B. bigemina in vivo*.[3] In this study, we investigated *Babesia bigemina* infection regarding the clinical infection, T cell distribution, and cytokine profile during the acute phase of an experimental infection in cows in the second trimester of the pregnancy.

## MATERIALS AND METHODS

### *Experimental Animals*

Twelve 24-month-old Holstein cows were estrus synchronized and divided at random into four groups (I, II, III, IV). Groups I and II were selected based on the diagnostic of pregnancy; groups III and IV were nonpregnant. Groups I and III were infected with *Babesia bigemina*; groups II and IV were noninfected controls.

Address for correspondence: M. J. A. Álvarez, CENID-PAVET, INIFAP, Apartado Postal 206, CIVAC, Morelos, C.P. 62500, Mexico. Voice: +52(777)3260848; fax: +52(777)3204455. alvareza@pavet.inifap.conacyt.mx

**Ann. N.Y. Acad. Sci. 1026: 298–301 (2004). © 2004 New York Academy of Sciences. doi: 10.1196/annals.1307.046**

### *Parasites and Infection*

The *Babesia bigemina* strain was derived from a field outbreak and was kept frozen. The cows were injected with infected red blood cells ($1 \times 10^7$) IM, whereas the control groups received noninfected red blood cells.

### *Clinical Monitoring and Sampling*

The cows were observed daily from day 0 to day 11 postinfection (PI) in order to determine clinical findings, rectal temperature was registered, PCV was measured by the microhematocrit method, and the presence of the parasites was determined by microscopical examination of Giemsa-stained smears. Blood samples were taken daily by using vacuum tubes to separate sera and peripheral blood mononuclear cells (PBMCs).

### *Flow Cytometry Analysis*

PBMCs were isolated over Lymphoprep gradients (Nycomed Pharma As, Oslo, Norway). T cell distribution was determined by one-color staining analysis (fluorescein-isothiocyanate, FITC) as described previously.[4] Monoclonal antibodies used were anti-bovine $\gamma\delta$ (mouse anti-bovine ILA29, IgG1), anti-bovine $CD4^+$ (CACT138A, IgG1), and anti-bovine $CD8^+$ (BAQ111A1, IgM) (VMDR, Inc.). Secondary antibodies were rat anti-mouse IgG1-FITC and rat anti-mouse IgG2a-FITC (Pharmingen, Becton Dickinson, San Diego, CA). The analysis was performed on a FACS Vantage and data analyzed using Cell Quest Software.

### *RNA Extraction*

Total RNA was extracted from $1 \times 10^7$ PBMCs by using the RNeasy kit (QIAGEN Inc., CA), according to the manufacturer's recommendations.

### *RT-PCR*

Equal amounts of RNA were used from PBMCs from each animal for analysis. RT-PCR protocols were performed as previously described.[4] Both the cDNA synthesis and PCR were performed in a single tube by using the SuperScript One-Step RT-PCR with platinum *Taq* (Invitrogen, Carlsbad, CA). Forward and reverse primers (20 μM each) were added to each reaction in individual tubes, and the 2× reaction mix yielded the following: final concentration 1.2 μM $MgSO_4$, total RNA 0.15 μg/reaction, 200 μM each dNTP, RT/platinum *Taq* mix 1 μL, in a final volume of 50 μL. The cDNA synthesis was performed at 45°C for 15 min and 99°C for 5 min in a thermocycler model ICycler (BIO-RAD, Hercules, CA). The PCR amplification was performed under the following conditions: 93°C for 1 min, 55°C for 1 min, and extension at 72°C for 2 min for 35 cycles, with a final extension at 72°C for 5 min. The PCR products (25 μL) were electrophoresed on a 1.8% agarose gel containing ethidium bromide. Primers used were as follows: $G_3PDH$ forward, 5′-GGA GAA ACC TGC CAA GTA TGA T-3′, reverse 5′-TCG CTG TTG AAG TCG CAG GAG AC-3′ (120-bp product);[4] TNF-α forward, 5′-CCC AGA GGG AAG AGC AGT-3′, reverse 5′-CCC TGA AGA GGA CCT GTG-3′ (253-bp product);[5] IL-10 forward, 5′-TGT CTG ACA GCA GCT GTA TCC-3′, reverse 5′-CAC TCA TGG CTT TGT AGA

CAC-3′ (405-bp product);[5] IL-4 forward 5′-ACA TCC TCA CAA GCA GAA AG-3′, reverse 5′-GTC TTG GCT TCA TTC ACA GA-3′ (220-bp product);[5] iNOs forward 5′-TAG AGG AAC ATC TGG CCA GG-3′, reverse 5′-TGG CAG GGT CCC CTC TGA TG-3′ (372-bp product);[3] IL-12 forward 5′-TGG TAT CCT GAT GCT CCT GGA G-3′, reverse 5′-TGC TCC AAG CTG ACC TTC TCT G-3′ (444-bp product).[3]

### *Determination of IFN-γ in Plasma*

Plasma was harvested for IFN-γ analysis with a commercial ELISA kit (Bovigam) according to the manufacturer's directions.

## RESULTS AND DISCUSSION

Infected pregnant and nonpregnant animals were severely affected on days 5–7 PI (DPI), showed fever up to 41.5°C, and had decreased PCV up to 50%. The parasite was observed in groups I and II on day 6 PI (0.3%). Specific treatment was required in order to avoid death. By the flow cytometry analysis, the T cell subpopulations of $CD4^+$, $CD8^+$, and γδ T cells showed no significant changes in values at 6–9 DPI on peripheral blood when compared to the control groups (data not shown). The induction of inflammatory cytokines was demonstrated at two different periods of time; 0–3 DPI and 5–7 DPI. TNF-α was induced at 1–3 DPI (FIG. 1A), with a marked presence at 5–6 DPI (FIG. 1B). TNF-α mRNA expression has been induced by addition of *B. bovis* merozoites plus IFN-γ in monocyte-derived macrophages taken *ex vivo*.[6] When *B. bovis*–infected erythrocytes induce production of TNF-α, IFN-γ potentates the effect.[3] In this study, IFN-γ and IL-12 production were induced particularly during the clinical phase (5–7 DPI) (FIG. 1B); a similar condition was observed during an acute infection with WA1 *Babesia* in mice. These cytokines and induction of macrophage-derived effector molecules like NO are important elements of the response.[7,8] Another study showed that stimulation of NO was dependent on

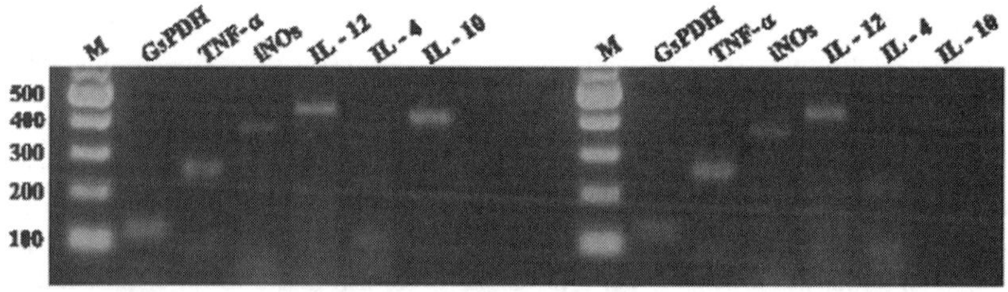


**FIGURE 1.** *Babesia bigemina* virulent strain (1 × $10^7$ infected red blood cells) enhances transcription of cytokine RNA in bovine PBMCs. PBMCs were obtained from 0 to 11 days postinfection (DPI), and RNA was isolated, subjected to DNase treatment, and analyzed by RT-PCR: **(A)** marker, analysis of TNF-α (253 bp), iNOs (372 bp), IL-12 (466 bp), IL-4 (220 bp), and IL-10 (405 bp) at 1 DPI in one pregnant cow infected with *B. bigemina*; **(B)** marker, analysis of TNF-α, iNOs, IL-12, IL-4, and IL-10 at 5 DPI in one pregnant cow infected with *B. bigemina*.

IFN-γ in mononuclear phagocytes exposed to *B. bigemina* merozoites.[6] iNOs showed a weak presence at 1 DPI (FIG. 1A) and 5 DPI (FIG. 1B). Some T cell lines have been shown *in vitro* to produce IFN-γ in response to antigen (*B. bovis* membrane) and antigen presenting cells, which activate macrophages to produce NO.[9,10] IL-12 production was better observed on 5–7 DPI. The *Plasmodium chabaudi* infection involves production of IL-12, IFN-γ, and TNF-α associated with an NO-dependent mechanism as protective immunity.[7,11] IL-4 was not detected (FIGS. 1A and 1B), whereas IL-10 was detected at 1 DPI. The pregnancy condition had no effect on the susceptibility to the infection nor on the immune response. This study highlights the importance of the innate immunity against *B. bigemina* and suggests that Th1 response is induced as a protective condition during the acute phase of the disease *in vivo*. T cell isolation and/or studies *in vivo* by using new techniques such as real-time PCR and microarrays are required for better understanding of the different elements involved in the immune response.

## ACKNOWLEDGMENTS

This research was supported in part by CONACyT Mexico, Project No. 34477-B.

## REFERENCES

1. RISTIC, M. 1981. Babesiosis. *In* Diseases of Cattle in the Tropics. Vol. 6, pp. 443–468. Nijhoff. The Hague.
2. FIGUEROA, J.V., *et al.* 1993. Use of a multiplex polymerase chain reaction–based assay to conduct epidemiological studies on bovine hemoparasites in Mexico. Rev. Elev. Med. Vet. Pays Trop. **46**(1–2)**:** 71–75.
3. SHODA, L.K.M., *et al.* 2000. *Babesia bovis*–stimulated macrophages express interleukin-1b, interleukin-12, tumor necrosis factor alpha, and nitric oxide and inhibit parasite replication *in vitro*. Infect. Immun. **68**(9)**:** 5139–5145.
4. SMITH, R., *et al.* 1999. Role of $CD8^+$ and $WC1^+$ γδ T cells in resistance to *Mycobacterium bovis* infection in the SCID-bo mouse. J. Leukocyte Biol. **65:** 28–34.
5. MWANGI, D.M., *et al.* 1998. Immunization of cattle by infection with *Cowdria ruminatum* elicits T lymphocytes that recognize autologous, infected endothelial cells and monocytes. Infect. Immun. **66:** 1855–1860.
6. GOFF, W.L., *et al.* 2002. IL-4 and IL-10 inhibition of IFN-γ and TNF-α dependent nitric oxide production from bovine mononuclear phagocytes exposed to *Babesia bovis* merozoites. Vet. Immunol. Immunopathol. **84:** 237–251.
7. AGUILAR-DELFÍN, I., *et al.* 2002. Resistance to acute babesiosis is associated with interleukin-12 and gamma interferon–mediated responses and requires macrophages and natural killer cells. Infect. Immun. **71**(4)**:** 2002–2008.
8. BROWN, W.C. & G.H. PALMER. 1999. Designing blood-stage vaccines against *Babesia bovis* and *B. bigemina*. Parasitol. Today **15:** 275–280.
9. BROWN, W.C., *et al.* 1998. Immunodominant T-cell antigens and epitopes of *Babesia bovis* and *Babesia bigemina*. Ann. Trop. Med. Parasitol. **92**(4)**:** 473–482.
10. STICH, R.W., *et al.* 1998. Stimulation of nitric oxide production in macrophages by *Babesia bovis*. Infect. Immun. **66:** 4130–4136.
11. NAHREVANIAN, H. & M.J. DASCOMBE. 2001. Nitric oxide and reactive nitrogen intermediates during lethal and nonlethal strains of murine malaria. Parasite Immunol. **23**(9)**:** 491–501.

# The Caribbean *Amblyomma* Program

## Some Ecologic Factors Affecting Its Success

RUPERT PEGRAM,[a] LISA INDAR,[a] CARLOS EDDI,[b] AND JOHN GEORGE[c]

[a]*FAO, Caribbean Amblyomma Program, Bridgetown, Barbados*

[b]*FAO, 00100 Rome, Italy*

[c]*USDA-ARS Tick Research Unit, Kerrville, Texas 78028, USA*

**ABSTRACT: The Caribbean *Amblyomma* Program has been operational for 8 years. However, owing to funding availability, some islands did not commence eradication activities until late 1997. During the past 2 years, 6 of the 9 islands (St. Kitts, St. Lucia, Anguilla, Montserrat, Barbados, and Dominica) under the program have attained the status of provisional freedom from the tropical bont tick (TBT). There are several administrative and technical reasons why the attainment of the program goals took longer than originally anticipated. This paper examines some of the ecologic factors that necessitated the prolongation of the treatment period and the recrudescence of TBT infestation in some islands. The introduction and subsequent spread of the cattle egret, *Bulbucus ibis*, in the 1960s and 1970s was most likely closely associated with the dissemination of the TBT in the region. At the national or island level, variations in land use are believed to have had a major impact on the eradication efforts in the different islands. Two islands, Antigua and Nevis, both opted out of sugar production several decades ago for economic reasons. Unfortunately, however, land from former sugar estates was not developed for other agricultural purposes and it became "unimproved free-grazing" areas for livestock. Thus, in both Antigua and Nevis, large numbers of livestock tend to become feral or free-ranging, making compliance with the mandatory treatment schedules impossible. In contrast, St. Lucia has large tracts of land allocated to banana plantations and St. Kitts to sugar plantations. Thus, feral or free-ranging livestock were rarely a problem in these islands. These differences in land use management are compared and discussed in relation to their perceived profound impact on TBT eradication efforts in the region.**

**KEYWORDS: Caribbean *Amblyomma* Program (CAP); tropical bont tick (TBT); ecologic factor; Antigua; Nevis; St. Kitts; St. Lucia**

## INTRODUCTION

Historical aspects of the introduction and spread of *Amblyomma variegatum* in the Caribbean have been documented previously.[1–3]

Address for correspondence: Rupert Pegram, FAO, Caribbean *Amblyomma* Program, P. O. Box 631c, Bridgetown, Barbados.
rpegram@cgnet.com

**Ann. N.Y. Acad. Sci. 1026: 302–311 (2004). © 2004 New York Academy of Sciences.**
**doi: 10.1196/annals.1307.047**

Anon (1894)[4] reviewed anecdotal information on the introduction of a large colored tick called the "Antigua Gold tick" into the Caribbean and noted "there appears to be a well-founded belief that the Gold tick was introduced into Antigua from the west coast of Africa about 30 years ago". Cattle from Senegal first disembarked in St. Kitts on route to Antigua, but there is little doubt that this tick was *Amblyomma variegatum.* The dates thus suggest its introduction was in about 1864 and not 1895 as widely reported previously. There is also a more recent report that the tick was introduced into Guadeloupe some 100 years earlier than reported previously.[5]

Progress in implementation of the Caribbean *Amblyomma* Program (CAP) has been reported regularly in the STVM forum.[6–10] St. Clair Phillip[10] described the magnitude of the impact of TBT and dermatophilosis in St. Kitts, where livestock numbers were decimated between 1984 and 1990 from 22,000 head down to 2200 head. Within 2.5 years of compulsory treatment against the TBT, however, livestock numbers had recovered 6-fold up to 13,500 head. Intensive, quantitative surveillance demonstrated that most of the island was free from TBT infestation, and the mandatory island-wide treatment of all livestock with Bayticol Pour-on was terminated after 33 months. Despite the dramatic positive impact of the campaign, the final stages of eradication were elusive, and elimination of the residual infestations in two or three hot-spot areas took a further 30 months. In contrast, the sister island of Nevis continues to have persistent residual infestations, albeit at a low level, more or less island-wide.

During the past 2 years, 6 of the 9 islands under the CAP have been certified as provisionally free from the TBT. These islands are St. Kitts and St. Lucia (November 2001), Anguilla and Montserrat (February 2002), and Barbados and Dominica (February 2003). St. Vincent[d] also qualifies for certification as they have not seen TBT during the past 18 months of intense quantitative surveillance.

Antigua, Nevis, and St. Maarten/St. Martin remain tick-infested. Nevis has been under blanket treatment for 7 years and is of major concern. Island-wide treatment on Antigua was suspended in 2000 after 2.5 years of inadequate treatment coverage with only 50–60% compliance. Thus, additional funding will be required to continue the efforts on these last 3 TBT-infested islands.

Initial estimates for the cost of the TBT eradication programs were about U.S.$17.0 million for Guadeloupe and Martinique, and U.S.$10.0 million for the CARICOM islands. However, it is believed that these estimates were grossly underestimated for several reasons. Notably, in 1995, the CAP recalculated costs for a conventional approach using public service delivery for treatment of livestock in the CARICOM English-speaking islands. It was estimated at that time that at least U.S.$5.0 million was required for Antigua alone based on the numbers of livestock reported. Subsequently, in 1998, after the Antigua eradication program was launched, it was realized that the livestock census data were clearly underestimated: cattle by one-third and small ruminants 4-fold. At that time, however, it was not known to the authors that the proposed budget for the USAID/USDA pilot project for Antigua[11] in the late 1980s was U.S.$4.8 million.

[d]St. Vincent was never classified as TBT-infested. Two male ticks were reported in 1988, but thereafter regarded as free until 2000 when an isolated outbreak was reported.

In this paper, we will examine and compare the main ecologic factors that are believed to impact significantly on successes and failures in the eradication of TBT. Particular attention will be given to St. Kitts and St. Lucia as examples of successful programs and to Antigua and Nevis as programs that have been less successful to date.

## MATERIALS AND METHODS

### *Ecologic Favorability for the TBT*

High-quality GIS maps incorporating climatic and ecologic data of Caribbean islands were not available either on the Web site or in the FAO GIS library and other internationally recognized sources. Thus, attempts were made to simulate "tick prediction maps" using the following critical factors for assessing the ecologic suitability for survival and development of *Amblyomma variegatum*:[12]

- climate,
- altitude,
- vegetation,
- land use.

Thus, we developed a rudimentary computerized system to simulate TBT favorability for each island, using the software, ArcView 3.0. This system allowed graphic data display of different layers of specific information, relating to altitude, climate, land use, and vegetation. These layers were then compiled into maps and combined with the biological and ecologic data for the TBT in Africa and in the Caribbean and with field observations in the specific countries.

The following ecologic parameters were considered:

- Climate parameters:
  - *(i)* temperature between 20°C and 30°C;
  - *(ii)* minimum temperature in the coldest months < 20°C;
  - *(iii)* maximum temperature in the warmer months > 30°C;
  - *(iv)* humidity over 75%, but over 85% may be detrimental;
  - *(v)* minimum rainfall: 19.5 inches (500 mm);
  - *(vi)* maximum rainfall: 110 inches (2800 mm).
- Altitude: sea level to 2600 meters.
- Vegetation: TBT-suitable habitats are varied, but are mainly composed of steppe and grassland in semiarid to humid zones. The TBT avoids equatorial forests, but occupies the clearings and the borders of the forest.
- Land use: crops such as sugarcane and banana limit the presence/movement of livestock in cultivated areas. Increasing areas of grazing land are now being used for housing development and tourism activities.

The resultant maps delineate three areas of suitability for the TBT:

- Green: Ecologically unsuitable TBT areas (rainfall > 110 inches; cloudy and moist forest);
- Yellow: Possible TBT-infested areas (used mainly for annual crops);
- Red: Ecologically suitable TBT areas.

### *Surveillance for the TBT*

Quantitative surveillance for the TBT was carried out according to the CAP protocol.[13] The surveillance system is based on the double binomial nested probability function. Surveillance rounds were carried out each quarter, and the data recorded and analyzed in the new, customized CAP database, "Tick*INFO*".

## RESULTS

A comparison of land mass and livestock numbers for the 4 islands is summarized in TABLE 1. Regarding the overall suitability for the TBT in each of the 4 representative islands, Antigua, Nevis, St. Kitts, and St. Lucia, the most dominant feature influencing habitat favorability is land use. Notably, a large amount of land in St. Kitts is used for sugar plantation, whereas in St. Lucia a large mass of the land in the center of the island is used for banana production. A brief description of each island follows.

### *Antigua*

Antigua was one of the first islands infested with *A. variegatum* during the 19th century (about 1865). The tick occurs throughout the island on cattle and other domestic livestock. Acute dermatophilosis, a skin infection caused by the bacterium, *Dermatophilus congolensis*, and cases of heartwater are associated with the widespread occurrence of theTBT.

The climatic conditions, geography, land use, and vegetation cover are particularly suitable for the development of the *A. variegatum* in Antigua. The only area that seems to be less favorable for TBT development is the southwest part of the island where higher rainfall patterns and a moist forest cover are reported. Notably, Antigua abandoned sugar production many years ago and now much of the grazing area is unimproved *Acacia* shrub land. In prolonged dry periods, animals are often released to wander unattended throughout the island. The inability to control loose livestock was one of the main reasons that the program was suspended in January 2000.

**TABLE 1. Livestock populations (2000)**

| Island | Size ($km^2$) | Human population | Cattle | Sheep | Goats |
|---|---|---|---|---|---|
| Antigua | 442 | 64,000 | 16,000 | 20,000 | 39,000 |
| St. Kitts | 168 | 34,000 | 3500 | 6500 | 5000 |
| St. Lucia | 610 | 154,000 | 7000 | 12,500 | 10,000 |
| Nevis | 93 | 9000 | 1000 | 10,000 | 18,000 |

### *Nevis*

Throughout the 1800s and early 1900s, Nevis was considered one of the richest islands in the Caribbean owing to its highly lucrative sugar industry. However, like Antigua, Nevis opted out of sugar production several decades ago and land was never developed for alternative agricultural development. It became a playground for speculative acquisition of land for residential and hotel development. Marginal land remained for opportunistic grazing areas. Lack of land-use planning and ineffective legislation led to many livestock owners turning out their animals to range freely over the island in search of grazing. Overstocking and periodic drought exacerbate the problem.

*Amblyomma variegatum* has been present on Nevis since 1977. It quickly became widespread and was associated with acute dermatophilosis, a severe skin disease, which caused a 10-fold reduction in cattle numbers from over 5000 to fewer than 500. The TBT prediction map shows a central area, around the top of the volcano, which is unsuitable for TBT development. Most of that area is covered by cloud and moist forest. From the lower part of the hillside around the volcano, gradually moving towards the coastline, the vegetation is characterized by steppe with shrub and low grass cover as a result of decreasing rainfall patterns. These areas were once cultivated with sugarcane and cotton, but are now available for grazing, and they are highly suitable for TBT development.

### *St. Kitts*

Specimens of *A. variegatum* were reported to have been collected from St. Kitts in about 1909, but apparently local farmers and the Chief Veterinary Officer did not recognize the tick until 1978 when it became established on the island. It then occurred in all cattle-rearing areas, with acute dermatophilosis affecting the cattle population almost to the same magnitude as on Nevis.

The TBT prediction map shows that the central mountainous area, occupied by cloud and moist forest, is unsuitable for TBT development. In addition, the widespread cultivation of sugarcane (most of the yellow areas) limits the livestock grazing areas. Livestock grazing areas are mainly steppe and grassland that coincide with favorable TBT habitats (shown in red). It is likely that large areas of land, now being made available from the declining sugarcane cultivation, will be used for livestock production under the agricultural diversification program. Although this may be seen as a positive aspect of agricultural diversification, it will increase the area suitable for TBT development.

### *St. Lucia*

The first report of *A. variegatum* in St. Lucia was in 1972, and the infestation spread to engulf the northern part of the island. In 1984, the TBT was then found in the south of the island and a control program was implemented. The TBT prediction map for St. Lucia shows a large central area that seems to be unsuitable to TBT development. This area is characterized by high rainfall patterns and by the presence of cloudy and moist forest. Importantly, because of banana production, the presence and the movement of livestock in the central regions are very limited. In contrast, the

**TABLE 2. Status of program (2002)**

| Island | Implementation date | Duration of mandatory treatment | Duration of hot-spot treatment | TBT status |
|---|---|---|---|---|
| Antigua | Aug. 1997 | 24 months[a] | | Infested |
| St. Kitts | Oct. 1995 | 33 months | 36 months | Provisionally free |
| St. Lucia | Nov. 1996 | 36 months[b] | 15 months | Provisionally free |
| Nevis | Oct. 1995 | 7.5 years | | Infested |

[a]Suspended because of lack of progress.
[b]St. Lucia had a different approach, progressively "pushing" the TBT northward and southward towards the sea.

drier conditions and the vegetation covering of the north and the south of St. Lucia seem to be highly favorable for the development of the TBT.

### *Progress in Eradication*

A summary of progress in the eradication of TBT is given in TABLE 2. The final stages of eradication in both St. Kitts and St. Lucia were elusive, and elimination of the residual infestations in remaining hot-spot areas took a further 15–30 months. Chronological, quantitative surveillance data for the TBT from 1998 to 2002 is summarized in TABLE 3.

The eradication campaign in St. Kitts was implemented in October 1995, and the mandatory treatment phase was extended until July 1998. St. Kitts stopped "blanket" treatment as of 31 July 1998, once they found a progressively decreasing number of ticks. Areas where TBT still occurred were declared "tick-infested areas" and were designated as quarantine areas in which government staff treated the animals. Mandatory treatment of all animals outside the quarantine areas ceased. Billboards and signs demarcated tick-infested areas; animal movement into and out of the infested area was restricted by law. The four locations remained under treatment, but St. Kitts was deemed to be provisionally free from the TBT by November 2001. However, after 1 year during which no female ticks were seen, St. Kitts reported a recrudescence of an active and persistent TBT infestation in the third quarter of 2002.

St. Lucia officially launched the TBT eradication campaign in late 1996, some 15 months after St. Kitts, although they had maintained an efficient control program for several years prior to that. The approach to eradication was slightly different in that St. Lucia gradually pushed the tick progressively further north and south, releasing areas from mandatory treatment as they were demonstrated to be free of the tick, but maintaining a high level of surveillance. After 3 years of acaricide treatment, St. Lucia appeared to be almost TBT free, and no reproductive infestations were seen throughout 2001. In the second quarter of 2002, however, a new active focus appeared in the southern tip of the island that justified an emergency slaughter policy to eliminate it.

The TBT eradication program in Nevis was launched in October 1995. Surveillance for TBT had to be suspended in 1999–2000 because of staff deficiencies. Although the overall prevalence is low, infestations persist throughout the island,

**TABLE 3. Epidemiology of TBT surveillance data 1998–2002**

| | 1998 | | | | 1999 | | | | 2000 | | | | 2001 | | | | 2002 | | | |
|---|---|---|---|---|---|---|---|---|---|---|---|---|---|---|---|---|---|---|---|---|
| | 1 | 2 | 3 | 4 | 1 | 2 | 3 | 4 | 1 | 2 | 3 | 4 | 1 | 2 | 3 | 4 | 1 | 2 | 3 | 4 |
| *Nevis* | | | | | | | | | | | | | | | | | | | | |
| Properties exam. | | | | 699 | 234 | 231 | | | | | | | 429 | 356 | 356 | 306 | 348 | 304 | 285 | 308 |
| Hosts exam. | | | | 14052 | 6938 | 7016 | | | | | | | 12478 | 8417 | 7173 | 5963 | 5359 | 5992 | 3426 | 3228 |
| No. TBT-positive | | | | 74 | 40 | 17 | | | | | | | 10 | 16 | 22 | 1 | 9 | 9 | 10 | 2 |
| Prevalence (%) | | | | 1.80 | 0.58 | 0.24 | | | | | | | 0.08 | 0.19 | 0.31 | 0.02 | 0.17 | 0.15 | 0.29 | 0.06 |
| Male TBT | | | | – | 55 | 20 | | | | | | | 24 | 23 | 60 | 1 | 18 | 26 | 103 | 9 |
| Female TBT | | | | – | 15 | 3 | | | | | | | 14 | 1 | 34 | 0 | 0 | 7 | 11 | 1 |
| *St. Kitts* | | | | | | | | | | | | | | | | | | | | |
| Properties exam. | 80 | 80 | 9 | 34 | 88 | 83 | 174 | 197 | 207 | 178 | 194 | 147 | 180 | 148 | 147 | 109 | 98 | 93 | 48 | 125 |
| Hosts exam. | 969 | 916 | 70 | 427 | 984 | 1015 | 761 | 412 | 434 | 390 | 602 | 613 | 657 | 528 | 422 | 334 | 321 | 293 | 164 | 796 |
| No. TBT-positive | 3 | 1 | 0 | 2 | 2 | 4 | 7 | 0 | 2 | 0 | 0 | 0 | 0 | 1 | 0 | 0 | 0 | 1 | 1 | 1 |
| Prevalence (%) | 0.31 | 0.11 | 0 | 0.47 | 0.20 | 0.39 | 0.92 | 0 | 0.46 | 0 | 0 | 0 | 0 | 0.19 | 0 | 0 | 0 | 0.34 | 0.61 | 0.13 |
| Male TBT | 5 | 2 | 0 | 4 | 1 | 4 | 11 | 0 | 2 | 0 | 0 | 0 | 0 | 1 | 0 | 0 | 0 | 1 | 11 | 1 |
| Female TBT | 3 | 0 | 0 | 0 | 1 | 3 | 12 | 0 | 0 | 0 | 0 | 0 | 0 | 0 | 0 | 0 | 0 | 0 | 0 | 0 |

**TABLE 3.** (*continued*) **Epidemiology of TBT surveillance data 1998–2002**

| | 1998 | | | | 1999 | | | | 2000 | | | | 2001 | | | | 2002 | | | |
|---|---|---|---|---|---|---|---|---|---|---|---|---|---|---|---|---|---|---|---|---|
| | 1 | 2 | 3 | 4 | 1 | 2 | 3 | 4 | 1 | 2 | 3 | 4 | 1 | 2 | 3 | 4 | 1 | 2 | 3 | 4 |
| *St. Lucia* | | | | | | | | | | | | | | | | | | | | |
| Properties exam. | 170 | 204 | 111 | 196 | 189 | 158 | 150 | 68 | 110 | 221 | 138 | 135 | 345 | 60 | 183 | 88 | 95 | 150 | 102 | 41 |
| Hosts exam. | 1101 | 1936 | 1068 | 1583 | 2416 | 2488 | 4068 | 1739 | 2157 | 2027 | 1801 | 2427 | 2753 | 1222 | 1122 | 1450 | 1551 | 2001 | 1205 | 311 |
| No. TBT-positive | 0 | 0 | 5 | 2 | 2 | 0 | 4 | 2 | 0 | 1 | 1 | 3 | 0 | 0 | 1 | 0 | 0 | 5 | 0 | 0 |
| Prevalence (%) | 0 | 0 | 0.47 | 0.13 | 0.08 | 0 | 0.10 | 0.11 | 0 | 0.05 | 0.06 | 0.12 | 0 | 0 | 0.09 | 0 | 0 | 0.25 | 0 | 0 |
| Male TBT | 0 | 0 | 10 | 2 | 3 | 0 | 10 | 3 | 0 | 0 | 0 | 5 | 0 | 0 | 4 | 0 | 0 | 10 | 0 | 0 |
| Female TBT | 0 | 0 | 5 | 3 | 6 | 0 | 7 | 3 | 0 | 1 | 1 | 0 | 0 | 0 | 0 | 0 | 0 | 3 | 0 | 0 |
| *Antigua* | | | | | | | | | | | | | | | | | | | | |
| Properties exam. | | | | | – | – | – | – | 103 | 34 | 73 | 200 | 388 | | | | | | 113 | 121 |
| Hosts exam. | | | | | 195 | 280 | 1037 | 1578 | 2378 | 579 | 801 | 1821 | 4716 | | | | | | 929 | 937 |
| No. TBT-positive | | | | | 1 | 7 | 3 | 1 | 2 | 0 | 12 | 8 | 4 | | | | | | 3 | 2 |
| Prevalence (%) | | | | | 0.51 | 2.50 | 0.29 | 0.06 | 0.08 | 0 | 1.50 | 0.43 | 0.08 | | | | | | 0.32 | 0.21 |
| Male TBT | | | | | 1 | 12 | 2 | 2 | 3 | 0 | 28 | 11 | 10 | | | | | | 2 | 2 |
| Female TBT | | | | | 0 | 4 | 6 | 0 | 1 | 0 | 20 | 3 | 2 | | | | | | 1 | 0 |

with a marked seasonal rise in the third quarter. Free-ranging and feral small ruminants, particularly goats, continue to be a problem and contribute to inadequate treatment compliance in some areas.

## DISCUSSION

Several factors unquestionably contributed to the varying levels of success in the attempts to eradicate the TBT. On the one hand, in St. Kitts and St. Lucia, both islands generally had dedicated leadership and consistent field teams, at least until 2000. On the other hand, Antigua and Nevis both suffered from serious droughts that exacerbated the "loose livestock" problems as owners turned their animals out indiscriminately in search of grazing. However, for the purpose of this discussion, we will focus on the impact of land use and management on the progress in the Caribbean *Amblyomma* Program.

The most comprehensive overview of the biology and ecology of *A. variegatum* in the Caribbean is presented by Barre and Garris.[14] Relevant ecologic factors (climate, topography) influencing survival and abundance of *A. variegatum*, together with land use and vegetation data, were used to compile the TBT prediction maps.

Previously, Barre and Garris had used their biological and ecologic data to estimate the duration of treatment programs for the elimination of *A. variegatum* in an area. Two scenarios were proposed: one of 46 months based on the accumulated maximum survival periods for eggs, larvae, nymphs, and adults; the second scenario, based on the maximum survival period of adults only, had a duration of only 20 months. In the second model, it was assumed that all tick stages would be concentrated within the grazing area or range of the hosts, and all livestock would be treated every 2 weeks.

The latter strategy of intensive treatment for 2 years was successful in Puerto Rico. It was perhaps erroneously assumed that it would work throughout the Caribbean. There are at least two reasons why this may not be so.

*Amblyomma variegatum* was not well established in Puerto Rico by the time intensive treatment programs were implemented. Thus, it was very unlikely that *A. variegatum* immature stages had adapted to feral hosts. In contrast, in most other Caribbean islands, the tick had been present at least 10–20 years prior to the outset of the intensive treatment programs.

Another important factor influencing widespread disbursement of the tick is related to livestock management systems. On enclosed or fenced commercial farms, it is very likely that all stages of ticks will be "exposed" simultaneously to acaricide-treated livestock hosts. In Puerto Rico, there is a much higher proportion of livestock kept commercially than in other Caribbean islands. In these latter islands, there are very few livestock managed commercially, except on government farms. Thus, most livestock are free-ranging and in some situations, for example, Antigua and Nevis, there is a high proportion of feral livestock.

In these situations, there could well be small pockets of residual infestations of *A. variegatum* persisting due to larvae and nymphs feeding on domestic small ruminant hosts (and possibly nondomestic hosts) for up to 2 years. Depending on drought and other factors influencing the host range grazing areas of domestic animals, especially goats, those animals may pick up unfed, mature adults some 24–48 months after the deposition of eggs. Notably, immature stages of *A. variegatum* are known to stay alive longer in sheltered bush scrublands than on well-managed or overgrazed pastures.

These factors, together with those associated with land-use management, that is, those related to sugar and banana plantations, are believed to have been associated with the persistence of infestations of TBT in the Caribbean islands. In the case of this study, land use in St. Kitts is based primarily on the sugar estates and in St. Lucia for bananas. Livestock are not allowed to roam freely over most of these islands. In Antigua and Nevis, however, there is little cultivation and a much higher proportion of undeveloped bush scrub, which is more conducive to tick survival.

## REFERENCES

1. BARRE, N. *et al.* 1996. Tropical bont tick eradication campaign in the French Antilles. Ann. N.Y. Acad. Sci. **791:** 64–76.
2. PEGRAM, R.G. *et al.* 1996. Eradicating the tropical bont tick from the Caribbean. FAO World Anim. Rev. **87:** 56–65.
3. PEGRAM, R.G., J.J. DE CASTRO & D.D. WILSON. 1998. The Caribbean *Amblyomma* Program. Public Health (Bayer) **14:** 60–67.
4. ANON. 1894. Agric. J. Leeward Islands, pp. 4–9.
5. MAILLARD, J.C. & N. MAILLARD. 1998. Cited in the ICCTD Newsletter **9:** 3.
6. PEGRAM, R.G., J.J. DE CASTRO & D.D. WILSON. 1998. The CARICOM/FAO/IICA Caribbean *Amblyomma* Programme. Ann. N.Y. Acad. Sci. **849:** 343–348.
7. PEGRAM, R.G., J.W. HANSEN & D.D. WILSON. 2000. Eradication and surveillance of the tropical bont tick in the Caribbean: an international approach. Ann. N.Y. Acad. Sci. **916:** 179–185.
8. PEGRAM, R.G., E.F. GERSABECK, D. WILSON & J.W. HANSEN. 2002. Eradication of the tropical bont tick in the Caribbean: is the Caribbean *Amblyomma* Program in a crisis? Ann. N.Y. Acad. Sci. **969:** 297–305.
9. ROSE-ROSETTE, F., N. BARRE & P. FOURGEAUD. 1998. Successes and failures in the tropical bont tick eradication campaigns in the French Antilles. Ann. N.Y. Acad. Sci. **849:** 349–354.
10. ST. CLAIR PHILLIP, K. 2000. Tropical bont tick (*Amblyomma variegatum*) eradication in the Caribbean. Ann. N.Y. Acad. Sci. **916:** 320–325.
11. ALEXANDER, F.C. 1990. Cowdriosis and dermatophilosis of livestock in the Caribbean region. CTA Seminar Proceedings, CARDI/CTA, Antigua.
12. CARIBBEAN *AMBLYOMMA* PROGRAM. 2000. Country profiles [www.caribvet.net].
13. CARIBBEAN *AMBLYOMMA* PROGRAM. 1999 (July). Monitoring and Surveillance for the TBT in the Caribbean. CAP Second Edition.
14. BARRE, N. & G. GARRIS. 1989. Biology and ecology of *Amblyomma variegatum* (Acari; Ixodoidea) in the Caribbean: implications for a regional eradication program. J. Agric. Entomol. **7:** 1–9.

# Reduced Incidence of *Babesia bigemina* Infection in Cattle Immunized against the Cattle Tick, *Boophilus microplus*

SATHAPORN JITTAPALAPONG,[a] WEERAPHOL JANSAWAN,[a]
OMAR O. BARRIGA,[b] AND ROGER W. STICH[c]

[a]*Department of Parasitology, Kasetsart University, Bangkok, Thailand*

[b]*Instituto de Ciencias Biomedicas, Facultad de Medicina, Universidad de Chile, Santiago, Chile*

[c]*Department of Veterinary Preventive Medicine, Ohio State University, Columbus, Ohio, USA*

**ABSTRACT: *Boophilus microplus* is an important vector of bovine disease agents having a major economic impact on cattle production in many tropical and subtropical countries. Components of tick saliva that enable ticks to feed may also facilitate establishment of tick-borne pathogens in the vertebrate host. It has been suggested that acquired resistance against molecules in tick saliva could inhibit parasite transmission, and there is increasing evidence to support this hypothesis. The effect of immune resistance to *B. microplus* on the incidence of tick-transmitted pathogens was the focus of this experiment. Groups of four dairy cows were injected with antigen extracts of tick salivary glands, midgut, adjuvant only, or PBS, prior to a grazing period in a pasture in Thailand where ticks are abundant and babesiosis is enzootic. These animals were then observed for evidence of babesiosis throughout the rainy season. A reduction in the incidence of clinical babesiosis was observed among cattle immunized with salivary gland preparations compared to nonimmunized controls ($P < 0.05$). Immunization with midgut or adjuvant only both resulted in a slight reduction in observed disease compared to the same negative control group. *B. bigemina* was detected in fewer ticks (24.43%) collected from salivary gland–immunized cattle than those collected from the remaining groups (≥44.57%). These results indicated that immunization with salivary gland antigens could affect pathogen transmission and appears promising for control of tick-borne diseases of cattle.**

**KEYWORDS: *Babesia bigemina*; *Boophilus microplus*; babesiosis; cattle; ticks**

## INTRODUCTION

Ticks and tick-borne diseases are important to livestock production in tropical Asia.[1] Control measures for these diseases have been based primarily on intensive

Address for correspondence: Sathaporn Jittapalapong, Department of Parasitology, Faculty of Veterinary Medicine, Kasetsart University, Bangkok 10903, Thailand.
fvetspj@ku.ac.th

**Ann. N.Y. Acad. Sci. 1026: 312–318 (2004). © 2004 New York Academy of Sciences.
doi: 10.1196/annals.1307.048**

tick control using acaricides. However, recent studies have shown that this approach may not be cost-effective in terms of increasing the productivity of cattle.[2]

The concept of using concealed antigens to immunize against ticks is traceable to the finding that bloodsucking arthropods could be controlled by raising host antibodies to parasite molecules such as hormones.[3] The use of concealed tick antigens for vaccination of cattle has served as the basis of a commercial vaccine,[4] but there are still potential drawbacks that should be addressed.[5] One issue is that immunity to nonsalivary gland antigens may not prevent tick feeding because salivary gland components are believed to be involved in establishment and regulation of the tick feeding sites as well as pathogen transmission.[6,7] Thus, current concealed antigen vaccines may not prevent damage to bovine hides or the transmission of tick-borne infections.

Vector-blocking immunity has been reported to protect vertebrate hosts from tick transmission of *Francisella tularensis*, *Borrelia burgdorferi*, tick-borne encephalitis virus, THO virus, and *Babesia* species.[8–14] *Boophilus microplus* species are considered important because of their roles as vectors as well as pests; thus, preventing pathogen transmission is as important as tick mortality. A vaccine based on tick molecules required for pathogen transmission would be highly desirable. The purpose of this experiment was to determine the effects of immunization with two complex sources of tick antigens on the incidence of naturally transmitted *B. bigemina* in an enzootic area of Thailand.

## MATERIALS AND METHODS

### *Antigen Preparation*

Adult female *B. microplus* that had fed for 3–7 days were collected from uninfected laboratory cattle used to rear and maintain our clean *B. microplus* colony. Ticks were placed into 0.15 M phosphate buffer saline (PBS, pH 7.4) and opened along their dorsal surface. Salivary glands (SG) were removed, dissected free of other tissues, placed into PBS at 4°C, and then suspended in 1% SDS and 5% 2-mercaptoethanol prior to incubation in a water bath at 56°C overnight followed by boiling for 5 min. The solution was cooled to room temperature, transferred into a 12,000–14,000 molecular weight cutoff dialysis tube (Spectra/Por7, Denver, CO), immersed into 1 L of PBS, and left at 4°C on a magnetic stirrer overnight (PBS was changed every 4–6 h). The protein concentrations of the extracts were estimated with the method of Bradford. These mixtures (0.5 mL, 2 mg protein/mL) were filtered and mixed with 0.5 mL of complete or incomplete Freund's adjuvant H37Ra (Difco Laboratories, Detroit, MI) immediately prior to immunization of the hosts.

Tick midgut (MG) samples were also removed and placed in PBS at 4°C. These organs were disrupted for 30 s in PBS at 4°C with a tissue homogenizer (KIKA T8.01, KIKA LABORTECHNIK, Germany) followed by sonication for 15 s (Branson Sonifier 450, USA), set between 30% and 50% duty cycle for 50–100 output power, for a total of 10 times. The homogenates were dialyzed in PBS at 4°C overnight and centrifuged at 16,000×*g* for 30 min at 4°C. MG preparations were sterilized with a 0.45-μm filter (Millipore, Bedford, MA), and 0.5 mL (2 mg of protein/mL) mixed thoroughly by sonication with 0.5 mL of complete or incomplete Freund's adjuvant prior to immunization.

### *Animals*

Sixteen crossbred dairy cows (82% Holstein-Friesian × 18% Thai native), 6 to 13 years old, with some previous exposure to *B. microplus*, were used in this study. These cows tested negative for *Anaplasma marginale*, *Babesia bovis*, *B. bigemina*, and *Theileria*; were randomly divided into immunized and control groups of 4 cows each; and were injected with SG, MG, adjuvant alone (ADJ), or PBS as untreated negative controls (CTR). Cattle were housed in pens and fed hay, concentrates, and water *ad libitum* to the conclusion of immunogen injections, after which they were turned out to the pasture and examined daily for evidence of host resistance. After 6 months, the animals were subjected to a booster immunization with the same antigen and/or adjuvant, and again observed for febrile illness and other clinical changes. Blood samples were collected weekly for hematological studies and the sera were stored at –20°C until ready for use.

### *Immunization*

The three immunized groups were intradermally injected, with 1 mg (total protein) of SG or MG, or an equivalent amount of ADJ or PBS (CTR), in 10 locations every 2 weeks for a total of 3 immunizations prior to the grazing period. After 6 months, each group was reimmunized with the same antigen and adjuvant, and observed for evidence of *B. bigemina* infection.

### *Natural Exposure to Ticks*

Cows were successively naturally infested with *B. microplus* larvae, nymphs, and adults during the grazing period. Engorged female ticks were collected during milking times at 6:00 A.M. and 4:00 P.M. Collected female ticks were weighed and maintained in a 12-h/12-h (light/dark) photoperiod at 30 ± 5°C, 70–80% relative humidity. Twenty percent of the female ticks were bisected and tested for *B. bigemina* with PCR.

### *PCR Assay*

PCR was used to assay for the tick-borne parasites, *B. bovis*, *B. bigemina*, *A. marginale*, and *Theileria* sp.,[15–18] in bovine blood as previously described.[16] The primer sets and references for these procedures are shown in TABLE 1. PCR was also used to detect *B. bigemina* DNA in tick samples. Adult ticks were cut in half and crushed with a plastic grinder in extraction buffer [10 mM Tris-HCl (pH 8.0), 100 mM EDTA, 0.5% SDS, 0.1 mg/mL proteinase K, 20 μg/mL RNase] and then incubated at 50°C for 3 h. Subsequently, the sample of DNA was isolated by phenol extraction and ethanol precipitation. Each DNA sample was suspended in 5 μL of TE [10 mM Tris-HCl (pH 8.0), 1 mM EDTA] and stored at –20°C until PCR analysis.

**TABLE 1. Primers used to detect tick-borne parasites with PCR**

| Parasite | Primer sequences (5′ to 3′) | Reference |
|---|---|---|
| *A. marginale* | TTGAAGGTTGAAGTGCAGGT<br>CCATATCGAATGCACCAAAC | 15 |
| *B. bigemina* | CGCAAGCCCAGCACGCCCCGGTGC<br>CCTCGGCTTCAACTCTGAGTCCAAAG | 16 |
| *B. bovis* | CACGAGGAAGGAACTACCGATGTTGA<br>CCAAGGAGCTTCAACGTACGAGGTCA | 17 |
| *Theileria* sp. | AGTTTCTGACCTATCAG<br>TTGCCTTAAACTTCCTTG | 18 |

### *Diagnosis of Babesiosis*

Evidence of babesiosis was based on clinical changes characteristic of the disease, including the presence of piroplasms in blood smears by light microscopy or PCR, and the demonstration of high fever (>104°C), anemia, and hemoglobinuria.

### *Statistical Analysis*

The effects of treatment (SG, MG, ADJ versus CTR) on measurements of tick fecundity and feeding parameters were determined by analysis of variance (ANOVA) for significant differences between immunized group means for each of the variables. Pairwise comparisons of least-squares (LS) means were accomplished with the Tukey-Kramer method and were carried out using the SAS program. We considered differences with $P < 0.05$ to be significant.

## RESULTS

The detection rate of severe clinical babesiosis among the SG-immunized cows was lower than that of MG, ADJ, and CTR groups (TABLE 2). Cows in all groups manifested febrile illness, and blood smears from affected animals were positive for *B. bigemina* at the peak of the fever. These results were confirmed with PCR specific for *B. bigemina*. Clinically, 2 of the 4 cows in the SG group showed a mild anorexia and recovered rapidly. Most of the cows in the MG and ADJ groups, and all cows in the CTR group, developed a severe form of the babesiosis that included hemoglobinuria, anemia, and high fever. These animals were treated with imidocarb dipropionate (3 mg/kg, sc) and supportive therapy until they recovered and resumed feeding.

While most cows were infected by *B. bigemina* via tick infestation, the recovery rates were noticeably different. The SG group had the lowest incidence of clinical babesiosis, which was statistically different from that of the CTR group, in which all of the animals developed clinical babesiosis ($P < 0.05$).

*B. bigemina* was also detected in ticks recovered from cows grazing in the contaminated pasture. Fewer engorged females and subsequent $F_1$ larvae were recovered from the SG group, and a statistically lower percentage of both tick stages collected from SG-immunized cattle tested PCR-positive for *B. bigemina* (TABLE 2).

**TABLE 2. Detection of *B. bigemina* in naturally exposed immune cattle and ticks**

| | Cattle | | *B. microplus* | |
|---|---|---|---|---|
| Group ($n = 4$) | PCR-positive for *B. bigemina* | Clinical babesiosis | % Adult ticks that are PCR-positive for *B. bigemina* (no. tested) | % Egg masses with $F_1$ larvae PCR-positive for *B. bigemina* (no. tested) |
| SG | 2 | $1^a$ | $24.43^a$ (60) | $16.41^a$ (50) |
| MG | 3 | $2^b$ | $44.57^b$ (102) | $38.82^b$ (50) |
| ADJ | 3 | $3^b$ | $46.93^b$ (132) | $40.19^b$ (120) |
| CTR | 4 | $4^c$ | $50.48^b$ (144) | $45.64^b$ (130) |

$^{a,b,c}$For each column, values with different superscripts differ ($P < 0.05$).

## DISCUSSION

*Boophilus microplus* is an important ectoparasite of cattle in Asia, especially in a tropical country like Thailand. Several groups have shown that host resistance to *B. microplus* can provide protection from tick burdens.[19–21] The common method of determining the resistance status of cattle against tropical cattle ticks is to determine direct effects on the ticks rather than measuring the effects of immunization with tick tissues on the transmission of tick-borne parasites.[22–24] This research was focused on the influence of tick immunity on pathogen transmission and, although immunized cattle in this study remained partially susceptible to transmission of *B. bigemina*, a clear reduction in the severity of clinical babesiosis in the SG group emerged. In addition, fewer *B. bigemina* PCR-positive ticks were collected from the SG group, suggesting a potential synergistic effect between immunity to tick SG and babesial infection that could be detrimental to the infected vectors.

These results raise interesting possibilities regarding the control of tick-borne diseases such as babesiosis, which is considered an important tick-borne disease in Thailand. The majority of dairy cows in this country are European breeds (82% Holstein-Friesian × 18% Thai native) that are very susceptible to tick infestations and tick-borne diseases. The role of host resistance to the vector of *B. bigemina*, *B. microplus*, could facilitate blocking transmission of the pathogen by interference with vector salivary gland function. Reducing the complexity of the SG antigen preparation to include only those antigen targets associated with reduced pathogen transmission could increase the efficacy of transmission-blocking immunity provided by this approach. Therefore, further work is required to identify protective antigens associated with vector transmission of *B. bigemina* and to determine what vector functions are targeted by such transmission-blocking immunity.

## ACKNOWLEDGMENTS

We thank Kasetsart University Research Development Institute (KURDI-KIP.90.42 for financial support of this research.

## REFERENCES


1. McLeod, R.S. 1995. Cost of major parasites to the Australian livestock industries. Int. J. Parasitol. **25:** 1363–1367.
2. Norval, R.A.I., P.L. Donachie, M.I. Meltzer & S.L. Deem. 1995. The relationship between tick (*Amblyomma hebraeum*) infestation and immunity to heartwater (*Cowdria ruminatium*) infection in calves in Zimbabwe. Vet. Parasitol. **58:** 335–352.
3. Galun, R. 1975. Research into alternative arthropod control measures against livestock pests (part 1). *In* Workshop on the Ecology and Control of External Parasites of Economic Importance on Bovines in Latin America, pp. 155–161. Centro Internacional de Agricultura Tropical (CIAT). Columbia.
4. Willadsen, P., P. Bird, G.S. Cobon & J. Hungerford. 1995. Commercialisation of a recombinant vaccine against *Boophilus microplus*. Parasitology **110:** 543–550.
5. Sahibi, H., A. Rhalem & O.O. Barriga. 1997. Comparative immunizing power of infections, salivary extracts, and intestinal extracts of *Hyalomma marginatum marginatum* in cattle. Vet. Parasitol. **68:** 359–366.
6. Ribeiro, J.M.C. 1989. Vector saliva and its role in parasite transmission. Exp. Parasitol. **69:** 104–106.
7. Mulenga, A., C. Sugimoto, K. Ohashi & M. Onuma. 2000. Characterization of an 84 kDa protein inducing an immediate hypersensitivity reaction in rabbits sensitized to *Haemaphysalis longicornis* ticks. Biochim. Biophys. Acta **1501:** 219–226.
8. Bell, J.F., S.J. Stewart & S. K. Wikel. 1979. Resistance to tick-borne *Francisella tularensis* by tick-sensitized rabbits: allergic klendusity. Am. J. Trop. Med. Hyg. **28:** 876–880.
9. Nazario, S., S. Das, A.M. de Silva *et al.* 1998. Prevention of *Borrelia burgdorferi* transmission in guinea pigs by tick immunity. Am. J. Trop. Med. Hyg. **58:** 780–785.
10. Wikel, S.K., R.N. Ramachandra, D.K. Bergman *et al.* 1997. Infestation with pathogen-free nymphs of the tick *Ixodes scapularis* induces host resistance to transmission of *Borrelia burgdorferi* by ticks. Infect. Immun. **65:** 335–338.
11. Votyakov, V.I. & N.P. Mishaeva. 1980. Investigation of a possibility of protecting vertebrates against transmissive infection with tick-borne encephalitis virus. Vop. Virusol. **2:** 170–172.
12. Jones, L.D. & P.A. Nuttall. 1989. The effect of virus immune hosts on Thogoto virus infection of the tick, *Rhipicephalus appendiculatus*. Virus Res. **14:** 129–139.
13. Francis, J. & D.G. Little. 1964. Resistance of droughtmaster cattle to tick infestation and babesiosis. Aust. Vet. J. **40:** 247–253.
14. Vanegas, L.F., S.A. Parra, C.G. Vanegas & J. de la Fuente. 1995. Commercialization of the recombinant vaccine Gavac against *Boophilus microplus* in Columbia. *In* Recombinant Vaccines for the Control of Cattle Tick, pp. 196–199. Elfos Scientiae. La Habana, Cuba.
15. Figueroa, J.V., L.P. Chieves, G.S. Johnson & G.M. Buening. 1993. Multiplex polymerase chain reaction based assay for the detection of *Babesia bigemina*, *Babesia bovis*, and *Anaplasma marginale* DNA in bovine blood. Vet. Parasitol. **50:** 69–81.
16. Inokuma, H. & D.H. Kemp. 1998. Establishment of *Boophilus microplus* infected with *Babesia bigemina* by using *in vitro* tube feeding technique. J. Vet. Med. Sci. **60:** 509–512.
17. Suarez, C.E., C.H. Palmer, D.P. Jasmer *et al.* 1991. Characterization of the gene encoding a 60-kilodalton *Babesia bovis* merozoite protein with conserved and surface exposed epitopes. Mol. Biochem. Parasitol. **46:** 45–52.
18. Tanayuthawongese, C., K. Saengsombul, W. Sukhumsirichat *et al.* 1999. Detection of bovine hemoparasite infection using multiplex polymerase chain reaction. Sci. Asia **25:** 85–90.
19. Opdebeeck, J.P., J.Y.M. Wong, L.A. Jackson & C. Dobson. 1988. Vaccines to protect Hereford cattle against the cattle tick, *Boophilus microplus*. Immunology **63:** 363–367.
20. Kemp, D.H., R.D. Pearson, J.M. Gough & P. Willadsen. 1989. Vaccination against *Boophilus microplus*: localization of antigens on tick gut cells and their interaction with the host immune system. Exp. Appl. Acarol. **7:** 43–58.

21. BARRIGA, O.O., S.S. DA SILVA & J.S.C. AZEVEDO. 1995. Relationships and influences between *Boophilus microplus* characteristics in tick-naïve or repeatedly infested cattle. Vet. Parasitol. **56:** 225–238.
22. KEMP, D.H., R.I.S. AGBEDE, L.A.Y. JOHNSTON & J.M. GOUGH. 1986. Immunization of cattle against *Boophilus microplus* using extracts derived from adult female ticks: feeding and survival of the parasite on vaccinated cattle. Int. J. Parasitol. **16:** 115–120.
23. JOHNSTON, L.A.Y., D.H. KEMP & R.D. PEARSON. 1986. Immunization of cattle against *Boophilus microplus* using extracts derived from adult female ticks: effects of induced immunity on tick populations. Int. J. Parasitol. **16:** 27–34.
24. JOHNSTON, T.H. & H.J. BRANCROFT. 1918. A tick-resistant condition in cattle. Proc. R. Soc. Queensl. **30:** 219–317.

# Laboratory Evaluation of the Compatibility and the Synergism between the Entomopathogenic Fungus *Beauveria bassiana* and Deltamethrin to Resistant Strains of *Boophilus microplus*

THIAGO C. BAHIENSE AND VÂNIA R. E. P. BITTENCOURT

*Departamento de Parasitologia Animal, Universidade Federal Rural do Rio de Janeiro, Seropédica, Rio de Janeiro, Brazil, 23890-000*

**ABSTRACT: *Beauveria bassiana* is one of the most promising agents for use as a bioacaricide to control the cattle tick *Boophilus microplus*, responsible for economic losses and transmission of infectious diseases. With the aim of optimizing the efficacy of chemical products, as well as the use of entomopathogens, the objective of the present study was to evaluate the compatibility and the synergism between this fungus and the drug deltamethrin on a strain of *B. microplus* that is resistant to this product.**

**KEYWORDS: *Beauveria bassiana*; *Boophilus microplus*; deltamethrin**

## INTRODUCTION

*Beauveria bassiana* (Ascomycota: Clavicipitaceae) is one of the most promising agents for use as a bioacaricide to control the cattle tick *Boophilus microplus*, responsible for economic losses and transmission of infectious diseases. Entomopathogens are considered important factors of insect population reduction. Thus, there is the necessity of entomopathogen conservation, if it occurs naturally, to be applied or introduced with the objective of controlling insects. However, for this conservation, it is important to know the pathogen compatibility with other control measure practices to avoid losses of control efficiency.[1]

The control of *B. microplus* is still based exclusively on the use of chemical acaricides, which have been used indiscriminately, causing damage to the ecosystem and resulting in the development of resistance in ticks. The treatment of resistant strains of *Boophilus* currently involve either the use of an increased dosage of a product to which ticks have become resistant or treatment with a different chemical formulation. The first alternative leads to increased environmental damage as well as

Address for correspondence: Vânia R. E. P. Bittencourt, Departamento de Parasitologia Animal, Universidade Federal Rural do Rio de Janeiro, BR 465, Km 7, Seropédica, Rio de Janeiro, Brazil, 23890-000. Voice/fax: +55 21 26821617.
vaniabit@ufrrj.br

**Ann. N.Y. Acad. Sci. 1026: 319–322 (2004). © 2004 New York Academy of Sciences.**
**doi: 10.1196/annals.1307.049**

health risks to the applicator and treated animals and their progeny, while the second measure leads to a situation that limits the use of different chemical groups.[2]

With the aim of optimizing the efficacy of chemical products, as well as the use of entomopathogens, the objective of the present study was to evaluate the compatibility and the synergism between this fungus and the drug deltamethrin on a strain of *B. microplus* that is resistant to this product.

## MATERIALS AND METHODS

*Boophilus microplus* engorged females, resistant to pyrethroid compounds, were collected from naturally infested animals at the Dairy Cattle Station PESAGRO, Rio de Janeiro State, Brazil. After collection, females were transported to the laboratory, where they were kept under controlled conditions of temperature and humidity (27 ± 1°C; RH > 80%) and negative photoperiod.

The *B. bassiana* isolate used was the 986 (ESALQ) and the conidia suspensions were prepared from fungi that had been cultivated on rice inside polypropylene bags. Conidia suspensions were prepared at concentrations of $10^8$, $10^7$, $10^6$, and $10^5$ conidia/mL, and solutions of synthetic deltamethrin pyrethroid were prepared at concentrations of 6.25, 3.12, 1.56, 0.78, and 0.39 ppm. Each treatment group was composed of 10 repetitions, with an aliquot of 50 mg of *B. microplus* eggs. These aliquots were placed into test tubes and were treated 15 days after eclosion of larvae.[3] The experiment comprised 30 treatments that included the suspensions of *B. bassiana* fungi, the synthetic pyrethroid solutions, and the associations between them. The control group did not receive any treatment. The results were evaluated considering the mortality percentage at 10 days after treatment. The Kruskal-Wallis test followed by Student's $t$ test ($P < 0.05$) and the mathematic calculation of $LC_{50}$ and $LC_{90}$ lethal concentrations[4] were used for statistic analysis. The resistance factor was calculated using the ratio between the $LC_{50}$ of the examined strain and the $LC_{50}$ of the reference strain Mozo. Synergism was determined by evaluating the capacity of the entomopathogen and the chemical agent to carry on its actions simultaneously, without suffering reciprocal interference.

## RESULTS AND DISCUSSION

The results obtained are shown in TABLES 1 and 2. The $LC_{50}$ for this *B. microplus* strain in 1988 was 1.6 ppm,[5] demonstrating that the continuous and indiscriminate use of this pyrethroid increased the resistance to it.

The associations between the *B. bassiana* fungi and the synthetic pyrethroid resulted in higher mortality rates than those observed for the respective nonassociated doses. Increase in mortality percentage was directly proportional to the concentration of the associations. Deltamethrin, as well as the fungi, showed higher efficacy when associated. The associations of *B. bassiana* fungi, at the different concentrations, presented good results and can be classified as synergic and compatible with all concentrations of deltamethrin.

The calculated resistance factors showed, consequently, the same results as their respective $LC_{50}$. It is interesting to highlight that the resistance factor found for the

**TABLE 1. Medium percentage of larvae mortality of *Boophilus microplus* obtained with treatments with *Beauveria bassiana*, deltamethrin, and their associations at different concentrations**

| Deltamethrin concentrations (ppm) | | | *Beauveria bassiana* concentrations (con/mL) | | | | | | | |
|---|---|---|---|---|---|---|---|---|---|---|
| | | | $2.4 \times 10^5$ | | $2.4 \times 10^6$ | | $2.4 \times 10^7$ | | $2.4 \times 10^8$ | |
| | | | 3 | A, a | 4.3 | A, bc | 10.5 | B, b | 61 | C, bc |
| 0.39 | 1.1 | A, a | 9.3 | B, a | 3.2 | A, b | 19 | C, c | 47 | D, b |
| 0.78 | 7.1 | A, b | 19 | B, b | 8.3 | A, c | 27 | B, c | 72 | C, c |
| 1.56 | 7.8 | A, b | 19 | BC, b | 17 | B, d | 26 | C, c | 77 | D, cd |
| 3.12 | 18 | A, c | 43.5 | BC, c | 22.5 | AC, d | 71.5 | D, d | 71 | D, c |
| 6.25 | 39 | A, d | 53.5 | A, d | 52.5 | A, e | 88.5 | B, d | 97.6 | B, d |
| Control | 2.6 | A, a | 2.6 | A, a | 2.6 | A, ab | 2.6 | A, a | 2.6 | A, a |

NOTE: Means followed by the same uppercase letter in the same row did not differ by Student's *t* test ($P < 0.05$). Means followed by the same lowercase letter in the same column did not differ by Student's *t* test ($P < 0.05$).

**TABLE 2. Values of $LC_{50}$ of deltamethrin (d) and the association with *Beauveria bassiana* (Bb) on *Boophilus microplus* larvae and the respective confidence limits, regression equation, and resistance ratio**

| Treatment | $LC_{50}$ | Confidence limits (95%) | Regression equation | Resistance ratio |
|---|---|---|---|---|
| d | 10.65 | 4.05–27.97 | $Y = 3.24 + 1.70X$ | 1238 |
| Bb $10^5$ + d | 5.26 | 2.99–9.24 | $Y = 4.14 + 1.19X$ | 611 |
| Bb $10^6$ + d | 6.87 | 4.06–11.61 | $Y = 3.71 + 1.53X$ | 798 |
| Bb $10^7$ + d | 1.78 | 1.37–2.30 | $Y = 4.55 + 1.76X$ | 206 |
| Bb $10^8$ + d | 0.37 | 0.19–0.74 | $Y = 5.46 + 1.09X$ | 43 |

association of deltamethrin and the *B. bassiana* fungi at a concentration of $10^8$ con/mL was lower than that found in 1988,[5] suggesting that this association may restore the sensibility of a resistant strain.

At the end of the experiment, all treatments were inoculated into potato-dextrose-agar culture medium for identification.[6]

Based on the results obtained in the present experiment, we can state that the association of *B. bassiana* fungi and deltamethrin resulted in synergism since deltamethrin and the fungi acted on tick larvae. They acted in different sites, promoting the death in a more efficient manner than when they were used separately. These results confirm that the entomopathogenic fungi *B. bassiana* can be used in association with deltamethrin, at the tested doses, even for tick strains that are resistant to this chemical group. This practice may allow restoring susceptibility of a tick strain to a determined chemical agent, making it possible to control infestations.

## ACKNOWLEDGMENTS

This work was supported by CNPq and CAPES/Brazil.

## REFERENCES

1. HIROSE, E., P.M.O. NEVES & J.A.C. ZEQUI. 2001. Effect of biofertilizers and neem oil on the entomopathogenic fungi *Beauveria bassiana* and *Metarhizium anisopliae*. Braz. Arch. Biol. Technol. **44**(no. 4)**:** 419–423.
2. FURLONG, J. & J.R.S. MARTINS. 2000. Resistência dos carrapatos aos carrapaticidas. Circ. Téc. n. 59. Embrapa Gado de Leite. Brazil.
3. GRILLO TORRADO, J.M.G. & R.O. GUTIERREZ. 1969. Método para medir la actividad de los acaricidas sobre larvas de garrapata. Rev. Investig. Agropecu. **6:** 135–158.
4. FINNEY, D.S. 1971. Probit Analysis. Third edition. Cambridge University Press. London/New York.
5. LEITE, R.C. 1988. *Boophilus microplus*: susceptibilidade, uso atual e retrospectivo de carrapaticidas em propriedades das regiões fisiogeográficas da Baixada do Grande Rio e Rio de Janeiro. Uma abordagem epidemiológica. Ph.D. thesis. UFRRJ, Brazil.
6. PETCH, T. 1935. Notes on entomogenous fungi. Trans. Br. Mycol. Soc. **19:** 55–75.

# Dedication: Conrad Yunker

KATHERINE KOCAN AND EDMOUR F. BLOUIN

*Department of Veterinary Pathobiology, College of Veterinary Medicine, Oklahoma State University, Stillwater, Oklahoma 74078, USA*

**ABSTRACT: This volume of the proceedings of STVM-03 is dedicated to Dr. Connie Yunker for his many contributions to tropical veterinary medicine and for being a good colleague and friend.**

**KEYWORDS: Conrad Yunker; dedication; tropical veterinary medicine; research**

Dr. Conrad Yunker, or Connie as most of us know him, received his M.S. and Ph.D. from the University of Maryland, College Park, in 1954 and 1958, respectively. His Master's research thesis involved a survey of acarine parasites of bats in Maryland and surrounding states, and his Ph.D. thesis research involved studying the ecology of parasitic mites and ticks of burrowing rodents in the Egyptian desert. He did his doctoral research at the U.S. Naval Medical Research Unit, Cairo, Egypt, in the laboratory of Dr. Harry Hoogstraal. Dr. Yunker's distinguished career has afforded him the opportunity to work in several countries and settings. Connie has published over 150 papers and has made many original contributions to the fields of acarology, tropical veterinary medicine, and invertebrate cell culture.

While working in Egypt, Connie also did research in Kenya, Tanzania, and Uganda, where he described new acarina taxa. After returning from Egypt, he taught acarology courses with Drs. Wharton, Baker, and others, and conducted field studies with Dr. Robert Heubner, Chief of Infectious Diseases, NIAID, NIH, on polyoma virus in rodents and their mite parasites in Harlem, New York City. He then identified and described mites of economic importance for Canada Agriculture, Ottawa. Through a commission in the U.S. Public Health Service at Rocky Mountain Laboratory (RML), Hamilton, Montana, Connie was assigned with Dr. James Brennan to the Middle American Research Unit, Balboa, Panama, where he collected, identified, and tested parasites of acari for viral and rickettsial agents and described new forms of parasitic mites. After returning to RML, he began a long-term project on development and application of vector tissue cultures for the isolation and characterization of viral and rickettsial organisms. He was assigned to the Bolivian Haemorrhagic Fever Project (BHF) in San Joaquin, Bolivia, where he collected, identified, and tested various species of biting and sucking arthropods for infection with a "mystery agent" that was later named Machupo virus. He then did field

Address for correspondence: Katherine Kocan, Department of Veterinary Pathobiology, College of Veterinary Medicine, Oklahoma State University, 250 McElroy Hall, Stillwater, OK 74078-2007. Voice: 405-744-7271; fax: 405-744-5275.
kmk285@cvm.okstate.edu

**Ann. N.Y. Acad. Sci. 1026: 323–324 (2004). © 2004 New York Academy of Sciences.**
**doi: 10.1196/annals.1307.050**

studies of seabird-parasitizing ticks on offshore islands of the northwestern United States in collaboration with Drs. Carleton Clifford and James E. Kerians. He tested argasid and ixodid ticks for evidence of viral infection, resulting in the isolation, characterization, and description of numerous arboviruses.

In 1985, Connie was recruited by Dr. Andy Norval, Veterinary Research Laboratory, Harare, Zimbabwe, to initiate a new lab for the study of heartwater disease. He became Chief-of-Party and served in this position until 1991. During this time, Connie led the team in cultivating the heartwater (HW) organisms in endothelial cells, provided DNA for research at the University of Florida, developed DNA probes for HW, and (with Dr. Suman Mahan) used these probes to detect HW in the bont tick. In collaboration with Andy Norval, he developed and tested pheromone-baited traps for collection of ticks. In 1992, Connie accepted a post at the Onderstepoort Veterinary Institute, S.A., with Dr. Durr Bezuidenhout, where he expanded existing tissue culture facilities for large-scale production of HW organisms for use in molecular studies. He worked with Dr. Jean du Plessis and others to establish a permanent cell line from the bovine saphenous vein and demonstrated the use of the cell line for cultivation of HW. In collaboration with Dr. Nigel Bryson, Connie applied the pheromone-baited trap technology for collection of wild bont ticks.

After a long and distinguished career, Connie retired in 1995 to Seattle, Washington, where he is enjoying spending time with his wife (Samira), children, and four grandchildren, and sailing in the Puget Sound. We dedicate STVM-03 to Dr. Connie Yunker for his many contributions to tropical veterinary medicine and for being a good colleague and friend.

# Index of Contributors